Lecture Notes in Computer Science 763

Edited by G. Goos and J. Hartmanis

Advisory Board: W. Brauer D. Gries J. Stoer

B. Richter R. Moreno-Díaz (Eds.)

Computer Aided Systems Theory – EUROCAST '93

A Selection of Papers from the
Third International Workshop on
Computer Aided Systems Theory
Las Palmas, Spain, February 22–26, 1993
Proceedings

Springer-Verlag
Berlin Heidelberg New York
London Paris Tokyo
Hong Kong Barcelona
Budapest

F. Pichler R. Moreno Díaz (Eds.)

Computer Aided Systems Theory – EUROCAST '93

A Selection of Papers from the
Third International Workshop on
Computer Aided Systems Theory
Las Palmas, Spain, February 22-26, 1993
Proceedings

Springer-Verlag

Berlin Heidelberg New York
London Paris Tokyo
Hong Kong Barcelona
Budapest

Series Editors

Gerhard Goos
Universität Karlsruhe
Postfach 69 80
Vincenz-Priessnitz-Straße 1
D-76131 Karlsruhe, Germany

Juris Hartmanis
Cornell University
Department of Computer Science
4130 Upson Hall
Ithaca, NY 14853, USA

Volume Editors

Franz Pichler
Institute of Systems Science, Johannes Kepler University
Altenbergerstraße 69, A-4040 Linz, Austria

Roberto Moreno Díaz
Dept. of Computer Science and Systems, Univ. of Las Palmas de Gran Canaria
P. O. Box 550, 35080 Las Palmas, Spain

CR Subject Classification (1991): H.1, J.6, I.6, I.2, J.7, J.3

ISBN 3-540-57601-0 Springer-Verlag Berlin Heidelberg New York
ISBN 0-387-57601-0 Springer-Verlag New York Berlin Heidelberg

Typesetting: Camera-ready by author
45/3140-543210 - Printed on acid-free paper

Preface

This volume contains a selection of papers presented at the third European CAST workshop, EUROCAST'93, which was held at the Universidad de Las Palmas de Gran Canaria, Spain, in February 1993.

Following the tradition of the former workshops, EUROCAST'93 again emphasized interdisciplinarity with the specific goal of creating a synergy between fields such as systems theory, computer science, systems engineering and related areas. One aim of the workshop was to enable specialists to meet others in related fields and to learn from each other by communicating on a "systems level".

The papers in this volume contain workshop contributions which are strongly related to current problems in CAST research. Concerning systems theory, they certainly emphasize an engineering point of view. Since the computer is the essential instrument in CAST research, close relations to specific topics in computer science naturally exist. This should be a legitimation for the publication of this volume in the Lecture Notes in Computer Science.

EUROCAST'93 was organized by the Facultad de Informatica of the Universidad de Las Palmas de Gran Canaria, Canary Islands, Spain. The organizers are grateful for the cooperation with the International Federation of Systems Research (IFSR) and the Instituto Technológico de Canarias, S.A. (ITC).

The editors of this volume would like to thank Professor Gerhard Goos for his constructive criticisms and suggestions for more closely correlating CAST research with related research topics in computer science, especially in the field of formal methods in programming and software engineering. This would enable an even greater degree of cooperation between systems research and computer science. A final word of thanks goes to the staff of the Springer-Verlag in Heidelberg for their help in publishing this volume.

November, 1993

Franz Pichler
Roberto Moreno-Díaz

Contents

Contents

1 Systems Theory and Systems Technology

2 Specific Methods

3 Applications

1 SYSTEMS THEORY AND SYSTEMS TECHNOLOGY

Systems Theory and Engineering

Franz Pichler

Johannes Kepler University Linz
Austria

Abstract. This paper addresses the question concerning the direction in which traditional Systems Theory should be extended in order to meet the requirements of new technologies which have been developed for the engineering professions. It is of primary importance to include the available computer technology. Although computer science and computer engineering offer a highly developed theoretical framework, it seems that new approaches, which more fully take the systems aspects into account, have to be developed. In the near future an extension of the traditional Systems Theory is required to enable it to provide theoretical support to the application of "Systems Technology" for complex computer-based engineering systems. Different activities such as the IEEE task force on CBSE or the ESPRIT project "ATMOSPHERE" show that there is common awareness of this problem area. Of specific interest would be such an extension to newly established fields such as automation engineering and mechatronics.

1 Introduction

Systems Theory, as it evolved along with the scientific treatment of problems in communications engineering and control engineering, is still a relatively young discipline. Because of its specific nature, however, it is permanently in danger of being isolated from the different engineering disciplines.

In this paper we try to show that the existence of Systems Theory as an independent scientific discipline should be encouraged. It is even necessary that new research activities be started in Systems Theory in order to meet the requirements for modeling complex computerized systems, which is a typical endeavor of contemporary engineering.

For our purposes it seems appropriate to first discuss the historical evolution of Systems Theory, both from the research as well as from the educational viewpoint. This is followed by a short discussion of the present state of the art of Systems Theory. The formulation of some proposals for future research involving important problems in the field of engineering of complex computerized systems will conclude the paper.

2 Historical Development

In communication engineering Karl Küpfmüller's book "Die Systemtheorie der Elektrischen Nachrichtenübertragung", published in 1949, is one of the first (and perhaps even the very first) which defines the topic of systems theory for this field of engineering.

Küpfmüller [1] writes:

> In systems theory transmission lines are only characterized by a small number of variables and their values, regardless of the possibility of their final realization. Therefore all fundamental questions remain independent of problems caused by specific selected arrangements of circuits; problem solving will be facilitated, since only the essential features are considered. In consequence, real systems can be constructed by the approximation of the assumed properties. The conclusion is that Systems Theory enables the study of the most fundamental properties of the process of information transmission.

Basically, this definition of Küpfmüller's is still valid today. It stresses the fact that Systems Theory provides means for the functional specification of information transmission systems and also methods for architectural refinement until a level is attained at which established technologies exist for the realization with engineering building blocks and components.

Systems Theory in the sense of Küpfmüller did not provid a completely new approach in communication engineering. The theory of electrical filters and networks and especially the "Vierpoltheorie" (4-port theory) which was introduced in 1921 by Breisig and worked out in detail by Richard Feldtkeller had similar goals in mind. According to Feldtkeller [2],

> a "Vierpol" (4-port circuit) consists of an electrical circuit whose inner details are not known. It is completely characterized as a black box by measurements performed on the four ports.

From the mathematical point of view a "Vierpol" can be specified by a complex 2x2 matrix. However, Feldtkeller [2] pointed out, that

> the "Vierpoltheorie" can become merely a useless game with complex numbers unless it is related to the goals and the experience of engineering design praxis.

That means that the interpretation of the mathematical formalism has to be an essential part of the theory.

In analogy, Systems Theory not only has to deal with the mathematics of formal models but also with the engineering problems which are related to the design process. Systems Theory has to be closely correlated with practical requirements. This applies not only to functional requirements as defined by the specifications provided by a client but also to design constraints which have to be met in the final implementation. Therefore, to be an expert in the application of Systems Theory one must not only be familiar with the logical/mathematical methods but one must also have practical insight. This is not a contradiction but certainly a requirement which is not always easy to fulfill.

Michael Arbib, well known for his theoretical contributions to computer science and biology, wrote in one of his numerous books [3] concerning the field of cybernetics:

> *Cybernetics is interdisciplinary. Thus, at its best, it allows a scientist the pleasure of working simultaneously in several fields - such as engineering, psychology, mathematics, and physiology. At its worst, it allows a not-very- good engineer to find refuge from his own problems by doing incompetent theoretical biology - his pride being saved because he knows too little biology to realize what a fool he is making of himself. If cybernetics is to shed all taint of pseudoscience, then people who work in it must be more than competent in at least one of the present-day divisions of science and really conversant with one or more other relevant ones. Teams of narrow specialists, too ignorant of one another's specialities to communicate effectively, cannot meet the need.*

For Systems Theory, which has much in common with cybernetics, a similar statement could be made.

But why was Systems Theory originally developed in the field of electrical engineering (and there mainly for applications in communication engineering and control engineering)? To what extent is it known in other engineering disciplines, especially in mechanical engineering?

To answer these questions we must briefly consider the historical development of engineering. When it began - at the time of the building of the gothic cathedrals, the invention of mining machinery, water mills etc. - engineering was considered as a kind of "art", its knowledge being confined to a few people and transferred from one generation to the next. This kind of engineering knowledge was not easy to communicate and to preserve. Albrecht Dürer's [4] warning words "Denn gar leichtiglich verlieren sich die Künst und aber schwerlich und durch lange Zeit sind sie wieder erfunden" (english translation: The knowledge of the arts (= engineering and technology) is easily lost, however it is very difficult and it takes a long time to reinvent it), were in these times certainly true.

With the development of mathematics and its applications to astronomical and physical problems by scientists such as Kepler, Newton, Euler, Lagrange and others, engineering - especially the field of mechanics - aquired a scientific basis. A breakthrough was the founding of the Ecole Politechnique in Paris in 1794, where famous scientists such as Lagrange, Monge, Poncelet and Cauchy were engaged as researchers and teachers. The practical methods of mechanical engineering thereby received a theoretical scientific foundation. Based on this, it was possible to develop a theory of machinery. A fundamental contribution was that of Ferdinand Redtenbacher from the Politechnical School of Karlsruhe and his pupil Franz Reuleaux from the "Berliner Gewerbeakademie" (Berlin Academy of Commerce), the founding institution of the "Technische Universität Berlin".

Redtenbacher (born in Steyr, Austria, traditionally one of the main centers for iron manufacturing in central Europe), stressed the importance of engineering drawings in machine construction. He writes [5] *not only for design, ... also for implementation, drawings are very essential. They determine all measures and forms precisely, such that the manufacturing process merely consists of following the guidance provided by the drawings.*

Reuleaux [6] writes in his important book on "Theoretische Kinematik" of the year 1875:

Today, the number of different mechanical mechanisms has immeasurably, likewise the number of methods for their application. It seems to be impossible to structure them to find appropriate pathways in the current labyrinth. For the development of engineering science it is therefore important to reduce research and academic teaching concerning too specific matters, but instead to stress the discovery of fundamental laws by which it is possible to integrate the specific ones. We have to combat the tendency to emphasize the description of every individual detail by numbers and formulas. Instead, we should look for general laws which allow the derivation of such specific results. The goal is to build a science of machinery which is based on deduction.

It seems that both Ferdinand Redtenbacher and Franz Reuleaux can be considered as founders of the Systems Theory (altough they did not use this term) for mechanical engineering. Today, due to the existence of modern CAD tools for machine specification and design, close relations to systems engineering as well as to Systems Theory can be seen. In addition, in the field of "stability and control", which was responsible for the development of control engineering, a further connection of mechanical engineering to Systems Theory can be found.

System-theoretical problems of communications engineering deal mainly with steady state response phenomena. Control engineering, on the other hand, very often depends on transient state response phenomena. In Systems Theory this led, as we know, to the development of the Laplace transform as an extension of the Fourier transform and the operational calculus of Oliver Heavyside.

At this point, we should remember Charles Proteus Steinmetz, the "inventor" of the use of the theory of complex numbers to deal with alternating current phenomena in electrical engineering. Born in Breslau (today Wroclaw) as Karl Rudolf Steinmetz, he studied mathematics and physics at the local university. His primary interest, however, was in electrical engineering, which was just in its beginnings. In 1889 he emigrated to the United States of America, where he, as a lifelong employe of General Electric, became one of the most famous engineers.

His lecture at the International Electrical Exhibition in Chicago in 1893 on "Application of complex numbers in Electrical Engineering" - published later in the Elektrotechnische Zeitschrift in German - showed that by means of the theory of complex numbers it is possible to analyze alternating current phenomena by the same fundamental laws (such as Ohm's law or the Wheatstone-Kirchhoff rules) that apply to direct current phenomena. On the basis of this work Steinmetz can be considered as one of the founders of Systems Theory.

Modeling of control processes by systems of differential equations led to the development of the "state space methods" in Systems Theory. The concept of a "dynamical system with input and output" was thereby brought to the attention of system theorists. Until today this concept has turned out to be one of the most fruitful ones in systems theory. Depending on the generation principle - differential equations, difference equations, abstract machines (e.g. finite state machines or cellular automata in the sense of Neumann-Codd), different classes of related dynamical systems are derived. The related theory constitutes an important part of Systems Theory.

A central question in the theory of dynamical systems concerns the availability of effective computational methods for decomposition - taking a dynamical system apart to form a network of components. Of special interest here are methods which lead to hierarchical decompositions in the form of multi-level systems. Further important problem areas in dynamical systems concern their stability (in the sense of Ljapunov), their controllability and their observability. A further modern problem area deals with the phenomena of "chaos" and "irreversibility" as observed in connection with dynamical systems. For both areas it seems that we will still have to wait for a proper embedding of the theory into the framework of Systems Theory.

Systems Theory, as defined for communication engineering and control engineering, is today a well-established and fully accepted discipline. The classical frequency methods based on the Fourier- and Laplace transformations together with the state space methods, which deal with systems behavior in the time-domain, constitute an integral part of engineering knowledge and are to a great extent already integrated into practical tools. This reflects an ideal situation: Systems Theory served to enhance the development of a powerful systems technology for the design of systems in communication- and control engineering. Such a technology allows engineering designs on a systems level, being to a large extent independent of the kind of basis technology which is used to finally implement the systems.

To sum up: In classical engineering disciplines such as mechanical engineering and electrical engineering, Systems Theory and the related philosophy of problem solving has found acceptance by practitioners. It has been widely integrated into practical tools.

3   Systems Theory and Computer Engineering

The invention of the computer caused, as we know, a revolution in many engineering domains. Today the components of engineering systems are often realized by computers. This puts new requirements on a related Systems Theory. We want to discuss this in the following paragraphs.

First, let us briefly look at the history of computers. A starting point was certainly the design of the "analytical engine" by Charles Babbage, about which he reported in 1840 in Turin, Italy. However, as we know, because of the lack of an appropriate basis technology, the "analytical engine" could not be implemented.

Konrad Zuse finished his work on the Z3, the first computer realized by relays as used in telephone-engineering, in Berlin in 1941. Mauchly and Eckert built the first electronic computer using 18000 vacuum tubes at the Moore School of the University of Pennsylvania in Philadelphia, USA. It became operational in 1945 and stayed in service until 1964.

With the invention of the transistor and the development of highly integrated electronic circuits in semiconductor technology, computer engineering progressed rapidly. This involved not only "computer-hardware", but also - and maybe even to a larger extent - computer programming, that means "computer-software".

How has Systems Theory been influenced by the appearence of computers in engineering systems? There are at least two aspects to consider. One of them concerns the use of computers to improve the theoretical "tool box" of an engineer. The other concerns the appearance of new classes of engineering systems and the development of an appropriate Systems Theory - insofar as this is possible - for them.

The availability of computers for the realization of tools has led, as we know, to CAD tools for different branches of engineering. Some of the most advanced CAD tools can be found in the field of VLSI-design in computer engineering, digital signal processing and control engineering.

Although most of the methods supported by such CAD tools relate to design steps which concerns typical engineering work (e.g. wiring of prefabricated components, simulation to study the validity of the design or to study the fulfillment of nonfunctional requirements) there are also a number of methods which have their origin in Systems Theory. One of the goals of the recently initiated CAST research [7]-[10] (CAST = Computer Aided Systems Theory) is to emphasize the integration of systems-theoretical methods into existing and future CAD tools. For the future it seems to be of special interest to concentrate upon the field of microsystem-design. Besides microelectronics as the basis-technology, other "micro-technologies" such as micromechanics or microoptics are applied to this discipline [11], [12]. For the integration of the already existing methods a systems-theoretical approach seems to be strongly needed [13]. Therefore, CAST research in microsystem-engineering is today an important field of Systems Theory.

In the field of engineering, as we mentioned above, computers are, not only technological means to improve the tool-boxes for engineering design, but they in many cases realize important parts of engineering systems. By their nature they are, of course, always part of the "nervous system" of a machine or a technological process; that means that they are responsible for information-processing. We find computers in the administration of any business firm, computers control the working of steel-mill's and computers supervise the control of electrical power-stations and aeroplanes. Computers form today the integral part of a telephone switching exchanges and they are indispensable for the realization of any modern intelligent machine, such as a robot in a automated factory. The conclusion is, that the development of computers creates a new class of machines and engineering systems. For the design (synthesis) and the maintenance (analysis) of this new type of system a theory which deals with systems aspects is very much needed. The situation can be compared with that of mechanical engineering in the last century, when Ferdinand Redtenbacher and Franz Reuleaux started to develop a theoretical framework for mechanical engineering or it can be compared with the situation of electrical engineering and control engineering about fifty years ago, when engineers and scientists, like Karl Küpfmüller, Karl Willi Wagner or Norbert Wiener brought Systems Theory to the attention of the engineering community. As we have experienced, it took about fifty years of research work until Systems Theory (as defined for communication engineering and control engineering) reached the current stage. Many engineers and applied mathematicians such as Lotfi Zadeh, Gerhard Wunsch, Mihajlo Mesarovich, Rudolf Kalman, George Klir, Yasuhiko Takahara and others, have made important contributions to it. It seems to be reasonable to think that the development of a Systems Theory for this new class of computer-based engineering systems will need some time and the intense effort of engineers and applied mathematicians.

Within the engineering community there is certainly an awareness of the importance of the development of CAD tools and a related conceptual framework for complex computer-based engineering systems. As an example we can refer to the IEEE task force on CBSE (CBSE = Computer Based Systems Engineering) [14].

In the CBSE workshops, international experts from industry and from universities discuss possible conceptual frameworks, related methods and tools for the design, implementation and maintenance of complex computer-based engineering systems. As an European initiative the ESPRIT project "ATMOSPHERE", in which eight prominent European companies participated, should also be mentioned. ATMOSPHERE had a research goal, similar to that of the CBSE IEEE-task force and the results have been published in three books [15], [16], [17]. They show, however, that for the development of a proper Systems Theory for complex computer-based engineering systems intensive research efforts are still needed.

4 Systems Theory and Mechatronics

Today, the traditional engineering fields, such as mechanical engineering, electrical engineering and control engineering (all disciplines, which, as we have pointed out, support a systems-theoretical viewpoint) are strongly influenced by the development of electronics and computers. In education, different universities take this into account and offer engineering curricula in which the related technologies are integrated. New curricula on "automation engineering" and on "mechatronics" deserve in this respect special attention. Mechatronics is a new engineering discipline with the goal of integrating mechanical engineering and electronics (including computer engineering). In addition to the basis technologies which are used in the implementation of "mechatronic machines" (a modern copying machine is a typical example) a proper "systems technology" for the upper levels of design will certainly be needed. In consequence, the existence of a proper Systems Theory for mechatronics is essential. Such a Systems Theory will certainly have to include not only the traditional concepts (as available for electrical- and control-engineering) but also the concepts necessary for the application of realization technologies of mechatronic machines, such as mechanical technologies or computer technology (both in hardware and in software).

Systems Theory for mechatronics certainly has to in many respects be similar to Systems Theory for CBSE. However, due to the narrower definition of a "mechatronic machine", it seems to be easier feasible. Future activities, such as the planned conference on "Computer Aided Systems Technology" [18], will have to explore the development of such an extension of traditional Systems Theory.

5 Conclusion

This paper addresses the question concerning the direction in which traditional Systems Theory should be extended in order to meet the requirements of new technologies which have been developed for the engineering professions. It is of primary importance to include the available computer technology. Although computer science and computer engineering offer a highly developed theoretical framework, it seems that new approaches, which more fully take the systems aspects into account, have to be developed. In the near future an extension of the traditional Systems Theory is required to enable it to provide theoretical support to the application of "Systems Technology" for complex computer-based engineering systems. Different activities such as the IEEE task force on CBSE or the ESPRIT project "ATMOSPHERE" show that there is common awareness of this problem area. Of specific interest would be such an extension to newly established fields such as automation engineering and mechatronics.

References

1. Küpfmüller, K.: Die Systemtheorie der Elektrischen Nachrichtenübertragung, S. Hirzel Verlag, Stuttgart (1949)
2. Feldtkeller, R.: Einführung in die Vierpoltheorie der elektrischen Nachrichtentechnik, Verlag von S. Hirzel in Leipzig (1937)
3. Arbib, M.: Brains, Machines and Mathematics, McGraw Hill, New York (1964)
4. Dürer, A.:Underweysung der messung mit dem zirckel un richtscheyt 1525, Verlag Walter Uhl, Unterschneidheim (1972)
5. Spur, G.: Vom Wandel der industriellen Welt durch Werkzeugmaschinen, Carl Hanser Verlag München Wien (1991)
6. Reuleaux, F.: Theoretische Kinematik, Vieweg und Sohn Braunschweig (1875)
7. Pichler, F., Schwärtzel, H.: CAST - Computerunterstützte Systemtheorie, Springer Verlag Berlin (1990)
8. Pichler, F., Moreno-Diaz, R.(eds): Computer Aided Systems Theory: EURO-CAST'89, Springer Verlag Berlin (1990)
9. Pichler, F., Moreno-Diaz, R.(eds): Computer Aided Systems Theory: EURO-CAST'91, Springer Verlag Berlin (1992)

10. Pichler, F., Schwärtzel, H. (eds): CAST Methods in Modelling, Springer Verlag Berlin (1992)
11. Reichl, H.(ed): MICRO SYSTEM - Technologies 90, Springer Verlag Berlin (1990)
12. Reichl, H. (ed): MICRO SYSTEM - Technologies 92, vde-verlag gmbh, Berlin-Offenbach (1992)
13. De Man, H.: Microsystems: A Challenge for CAD Development, in: MICRO SYSTEM - Technologies 90, Springer Verlag Berlin (1990) 3-8
14. Thom, B. (ed): Task Force on Computer Based Systems Engineering-Newsletter, IEEE Computer Society/TF CBSE, 1/1 (1993)
15. Thom, B. (ed): Systems Engineering: Principles and Practice of Computer based Systems Engineering, John Wiley & Sons, Chichester (1993)
16. Kronlöf, K. (ed): Method Integration: Concepts and Case Studies, John Wiley & Sons, Chichester (1993)
17. Schefström, D., van den Broek, G. (eds): Tool Integration: Environments and Frameworks, John Wiley & Sons, Chichester (1993)
18. CAST'94: Fourth International Workshop on Computer Aided Systems Technology, Ottawa (1994)

Computer-Aided Systems Technology: Its Role in Advanced Computerization

Tuncer I. Ören

Computer Science Department, University of Ottawa
Ottawa, Ontario Canada K1N 6N5
oren@csi.uottawa.ca

Abstract. Next generation intelligent machines which will work autonomously or semi-autonomously as well as computer-aided problem solving environments including computer-aided design and computer-aided software system design environments will have sophisticated computerization requirements. It appears that Computer-Aided Systems Technology (CAST) may provide the necessary powerful scientific framework to handle the inherent complexity associated with such requirements. A system theoretic concept, coupling, has been shown to provide a reach paradigm for this purpose. The following types of couplings are discussed: Data coupling, stamp coupling, control coupling, external coupling, common coupling, and content coupling; 1-system coupling, n-system coupling; external and internal couplings; nested, feedback, cascade, and conjunctive couplings; time-varying coupling and multimodels; coupling applied to events, processes, and multimodel process models; time-varying coupling and multifacetted models; hard and soft couplings; and coupling and object-oriented formalism.

1 Introduction

The aim of computerization is not necessarily to develop software but to solve problems with the assistance of computers. The importance of the assistance of computers in problem solving is also emphasized by Schwärtzel and Mizin: "During the last few years, computers have evolved from pure number crunching machines to 'intelligent' problem solving tools." [27, p. v]. This shift of paradigm is essential to attack the computerization problem properly.

It is already a well established fact that most of the application code should not be written by humans. Code generators should replace the labor-intensive and error-prone manual code generation process [2, 13]. Such an attitude necessitates problem specification environments where systems can be specified (i.e., systems can be designed or models can be formed) and problems (i.e., what to do with the system specification) can be formulated, consistency checks can be done by knowledge-based tools, and computer code can be generated by an appropriate code generator. In this approach, a user can concentrate on the specification of the system and the problem to be solved, rather than instructing a computer how to do what needs to be done.

For advanced computerization, what is important is the realization of the necessary shifts of paradigms in software engineering as well as in systems theories to assure their synergies in order to develop model-based problem solving environments [16] and to establish computer-aided problem solving as a reliable, easy, efficient, productive, and profitable process.

Computer-aided problem solving consists of design, analysis, or control problems:

A *design problem* has two aspects: (1) development of a set of requirements for a system; (2) development of a model (or design) based on which the real system that will satisfy the requirements can be built or developed .

An *analysis problem* has three aspects: (1) specification of the goal of the analysis problem; (2) development of a model of the given system; (3) validation of the model to ensure that it can be used to represent the system under investigation for the intended purpose. The model developed during the analysis process can be used for understanding, evaluating, modifying, managing, or controlling the system to satisfy previously established or updated goals.

A *control problem* necessitates finding the operational conditions for a given system to satisfy a set of requirements.

Design, analysis, or control problems are all model-based activities [31]. Therefore, computer-aided problem solving is a model-based activity. As it was discussed elsewhere, specification and processing of models are primordial in model-based activities and system theories provide powerful modelling and model processing knowledge necessary for the computer-aided environments to support them [16].

In this article, the aim is to highlight some of the possible contributions of Computer-Aided Systems Technology (CAST) to computer-aided problem solving in general and to computer-aided design (CAD) and computer-aided software engineering (CASE) in particular.

2 Advanced Computerization

Computerization has two aspects: The first aspect involves the use of computers directly as knowledge processing machines.

The second aspect refers to any system (or tool, or machine) which has knowledge processing ability in addition to the physical abilities necessary to carry out the tasks for which it is designed. Such systems are also called computer-embedded systems or machines. (When the emphasis is on the computer system embedded in the total system, the term "embedded system" is used). The primary task of such systems is not knowledge processing. But with their knowledge processing abilities they can perform their task better. Examples include automatic cameras, self-repairing copying machines, and industrial robots. Autonomous or semi-autonomous systems used in domestic, industrial, or military applications as well as near-and deep-space applications also belong this category of machines [17]. Consideration of the second aspect of computerization is becoming more and more important with the advances in the computer-embedded systems which also include intelligent machines and systems [22] as well as intelligent autonomous systems and robots [35].

Some of the computerization requirements of next-generation intelligent systems, regardless whether they are autonomous or semi-autonomous, are as follows:

(1) Automatic downloading of software (or system and problem specifications) from a knowledge base. This requires the ability to identify the task to be performed and to match the task to the corresponding modules in the knowledge base.

(2) Automatic tailoring of the software (or the specifications) during the down loading.

(3) Automatic synthesizing of software (or the specifications) from library modules with or without tailoring.

(4) Automatic synthesizing of software (or the specifications) from methodological knowledge, modelling formalisms, and other knowledge from observation and knowledge bases.

(5) Ability to learn and to adapt; that is, ability to monitor its own performance (self-monitoring) or the performance of another computer-embedded system to modify its own or of the observed system's software (or the specifications).

(6) Ability to set hypotheses and to test them.

(7) Ability to perform experimentations on real systems or with models to solve analysis, design, and control problems. This ability requires abilities in design of experiments, instrumentation, modelling, model validation, goal setting and goal-directed knowledge processing.

Software development and maintenance environments exist in three levels: they are manual, CASE, and intelligent software environments. Manual software environments support the minimal functionalities to develop and maintain software manually. Computer-Aided Software Engineering is the basis of software tools and environments. Intelligent software tools and environments are realized through Knowledge-Based Software Engineering (KBSE) (KBSE-7 1992, Lowry 1991) or through Artificial Intelligence in Software Engineering (AISE) (Ören 1990).

3 Computer-Aided Systems Technology (CAST)

As part of applied mathematics, deductive, inductive, hierarchical, and fuzzy systems theories have been maturing for some time [11, 20, 28]. Furthermore system theories provide solid mathematical bases for several types of important goal-oriented systems [29] such as anticipatory systems [24], autopoietic systems , evolving systems, self-organizing systems [6, 26], and self-reproducing systems. Theoretical foundations of systems engineering were developed by Wymore [30, 31]. Computer-Aided Systems Engineering exists for some time [5]. The developments in the Computer-Aided Systems Theory have also been reported in the literature [21, 22].

Computer-Aided Systems Technology aims to solve practical yet complex problems by applying systems theoretic concepts with the assistance of computers. Since most engineering and scientific problems deal with complex systems, CAST applications would be very beneficial in such areas. Some important applications of CAST are in computer-aided design in particular and in computer-aided problem solving in general. Some specific application areas are software systems engineering, telecommunications, autonomous intelligent systems and robotics, very large scale integration, computer integrated manufacturing, information technology in business, government, and health care, and in ecological systems.

4 CAST for Advanced Computerization

The aim of Computer-Aided Systems Technology for the first aspect of computerization is to help human problem solvers directly. The aim of Computer-Aided Systems Technology for the second aspect of computerization is to help computer-embedded systems to solve computerization problems autonomously or semi-autonomously.

Some important categories of possible contributions of CAST to advanced computerization (or the implications of CAST-based computerization) are as follows:

(1) To provide system specification (i.e., design or modelling) paradigms to represent complex systems and to facilitate specification and processing of problems related with such system specifications.

(2) To provide the much needed theoretical and powerful formalisms for relevant computer-aided environments. Since system specifications would be based on system theoretic formalisms, specification environments would be based on solid theoretical bases instead of programming paradigms.

(3) To provide scientific and coherent knowledge for the symbolic processing of the models developed based on different modelling formalisms.

(4) To provide the possibility to maintain (i.e., to update) system designs and problem specifications directly, rather than obliging the user (or a software maintenance system, in automatic programming) to do the modifications on the computer code. This characteristic is very desirable in systems design as well as in software engineering.

(5) To provide the possibility of system models to be processed symbolically for analysis such as qualitative simulation and for transformation purposes such as for model simplification [9].

(6) To ease understanding of system and problem specifications for better use and updating. In program understanding, the understanding process becomes much easier if the level of the specification is higher. For example,Gest, being a modelling and simulation language based on a system theory [30], can be read directly to get information about several aspects of a simulation problem [15].

Having a strong and conceptually rich theoretical foundation is imperative for the two aspects of advanced computerization. Some of the important concepts of software engineering such as modularity, information hiding [19], and coupling [23] have their powerful counterparts in systems theories [11, 30]. These systems theoretic concepts are very promising in providing the long-needed solid background for software engineering, computer-aided design, and computer-aided problem solving.

The developments in systems theory-based modelling and simulation [15,32, 33] and associated environments [1] and generalization of the model-based simulation concept to model-based problem solution concept [16] may be useful to emulate in developing CAST-based software and more importantly, CAST-based design and CAST-based problem solving environments.

In developing CAST-based software, CAST-based design, and CAST-based problem solving environments , the synergy of related fields such as systems theories, simulation methodologies, software engineering, and artificial intelligence would be most beneficial [7]. Part of this synergy may also lead to ICAST (Intelligent Computer-Aided Systems Technology).

5 An Example: Coupling

Coupling specifies input/output relationship of component model. By allowing modular representations, coupling renders the complexity of systems more easily manageable. However, in software engineering and in systems theories therefore in CAST, the emphases are different.

5.1 Coupling in Software Engineering

Coupling is an important concept in software engineering. Several types of couplings are defined between software modules. These are: data, stamp, control, external, common, and content couplings.

Data coupling occurs when data are passed from one module to another. *Stamp coupling* occurs when a portion of data structure is passed to another module. *Control coupling* exists between two modules if a control data is passed from one module to the other. *External coupling* occurs when software is connected to environment external to it. *Common coupling* occurs when two modules use data or control information maintained in a global area. *Content coupling* occurs when a module uses data or control information maintained within another module or in a global area.

However, the coupling concept is used only partially —mostly as a metric—in software engineering. "Coupling is a measure of interconnection among modules in a software structure. ... Coupling depends on the interface complexity between modules ..." [23, p. 336)]

5.2 Coupling Concepts for CAST-Based Design and Software Engineering

A CAST-based paradigm for computer-aided design, including computer-aided software design, where input/output relationships of the components, including software modules, can be specified as their coupling, might prove to be very useful. In such a paradigm, one can have several types of coupling specifications of the components and the associated computer assistance.

As a design concept, coupling promotes modularity. Since it obliges modules to communicate through their input/output variables only, it is also a good mechanism to ensure encapsulation. A system theoretic definition of coupling is given by Wymore [30, 31]. The concept of coupling has been applied by Ören [15] to the specification of simulation models; however, the concept is general enough for application to any type of model specification.

In a coupled model (or a resultant model), there is at least one and normally several component models. Accordingly, one can define *1-system coupling* and *n-system coupling*.

External and Internal Couplings. The inputs and outputs of a coupled model can be called its external inputs and external outputs, respectively. The inputs and outputs of the component models can be distinguished as internal inputs and internal outputs. In a coupling, an internal input variable can receive information from at most one external input variable or from at most one internal output variable. An internal output variable can be connected to one or more internal input variables and at most to one external output variable. For a coupled model, two types of couplings have to be specified; these are external and internal couplings and can be specified in any order.

External coupling specifies input/output relationship of a coupled model with its environment. It can be specified in two steps: (1) Each external input has to be connected to one or more internal inputs of one or more component models. (2) Each external output has to be connected to one internal output of a component model.

To specify the *internal coupling*, one can consider each component model taken in any order, as follows: For a component model, consider all non-committed inputs, that is, those which are not used in the specification of the external coupling. Each non-committed input of each component model has to be associated with an output variable of the same or another component model to receive information.

If the nature, range of acceptable values, physical units, and other characteristics are associated with input and output variables, some checks can be performed before finalizing the coupling specification. These checks can be useful especially for graphical specification of the coupling. If all the component models have already been specified, coupling allows a bottom-up approach for the specification of the resultant model. A top-down specification is also possible; in this case, for every component models which are not fully specified, one has to provide input/output models where component models would be described only by they input/output variables; additional specifications including their internal dynamics, given typically by state transition functions and output functions, would be completed later.

Nested, Feedback, Cascade, and Conjunctive Couplings. The coupling definition allows system specification based on component models and comprises several types of system coupling such as feedback coupling, cascade coupling, and conjunctive coupling.

In *nested couplings*, component models can be coupled models. A modelling formalism which allows nested coupling is said to be closed under coupling. Nested coupling is very convenient way to handle complexity and to represent hierarchical systems.

Feedback coupling can be defined on one component model (1-system coupling) or on a set of several component models. Feedback coupling on a single component model implies that at least one input variable receives its values from an output variable of the same component model. 1-system coupling can only be defined for component models having at least one state variable. *Memoryless models*, that is, models without state variables cannot be used in 1-system couplings. Feedback coupling in an n-system coupling occurs when an internal input of a component model M_S receives information from a component model M_T which in turn receives information (through one of its input variables) directly or indirectly from M_S. In a feedback coupling, one can have memoryless models; however, in the loop, one needs to have at least one *memory model*, i.e, a component model with state variable(s).

In a *cascade coupling*, there is an ordering of the component models and the i th component model can provide input to j > i th component model.

In a *conjunctive coupling*, there is no internal coupling; i.e., there is no input/output relationship between the component models. In this case, all internal inputs and outputs have to be connected to external inputs and outputs, respectively.

Time-Varying Coupling, Model Update and Multimodels. A generalization of the concept of coupling, as formulated by Ören, is *time-varying coupling* where input-output relationships of component models or the component models may vary in time [14].

Starting with the concept of discontinuity in systems described by a set of ordinary differential equations, model update and multimodel formalisms have been developed by Ören [18]. In the *model update* formalism, state variables and/or parameter values can be re-initialized under given conditions. A *multimodel* formalism consists of a set of submodels; only one of which is active at a given time. Transitions between subsystems are defined as the transitions in a finite state automaton where each state represents a subsystem and each transition condition corresponds to an input to a state. Transition conditions can depend on time directly (i.e., to time events in the terminology of combined continuous/discrete system simulation) or indirectly (state events).

The model update and multimodel formalisms provide powerful concepts to define time-varying coupling where component models can change during the life-span of a coupled model. Therefore, time-varying coupling with updatable models and/or with multimodels can be used to describe coupled models where some component models may change in time. In polymorphic models, evolutionary system models, adaptive system models, and learning system models, the time-varying coupling with updatable models and/or multimodels may be very useful.

Coupling Applied to Events and Processes. "Event" was formalized by Zeigler [33]. An interesting and important generalization of the concept of coupling to multimodels which include discrete event modules has been done by Fishwick and

Zeigler [8]. Since, process-oriented paradigm in discrete simulation is basically a set of interrelated discrete event modules where interrelations depend on time or state conditions, the generalization of process-oriented modelling—as multimodels— is also possible. In the generalized process-oriented paradigm, that is, in a *multimodel process model*, the transitions between event routines can be more general than the original case; i.e., one does not need to scan the event routines in a sequential way. Instead, the transition can be defined from any event model (of the process model), to any other event model, based on the occurrence of different time or state conditions.

Two other implications of event are their use in interrupt mechanisms and in demons. An interrupt in a system is basically corresponds to a message sending (soft coupling) to a special interrupt handling module. A demon can monitor some component models and is responsible for the initiation of a special event under a pre-specified condition.

Time-varying Coupling and Multifacetted Models. Time-varying coupling with multimodels can also be a powerful way of implementing the pruning mechanism as developed by Zeigler for multifacetted models [33]. In this case, given an aspect of the model (or a goal to select an aspect), only the selected aspects can be made active.

Hard and Soft Couplings. Based on the nature of knowledge channels, two categories of component model communications are possible: hard couplings and soft couplings. Couplings as discussed until now are hard-wired couplings, or hard couplings, in short. A *hard coupling* correspond to fixed input/output communication channels between communicating modules. Hard couplings can be permanent (time-invariant) or time-varying couplings, as discussed above.

A *soft coupling* corresponds to a flexible input/output communication channel and can exist between a sender entity and a set not empty of destination entities. It allows a packet of knowledge (a message) to be passed under certain conditions. Message passing can be one-to-one, one-to-some, and one-to many, corresponding to knowledge transmission, selective knowledge dissemination, and knowledge broadcasting, respectively. Knowledge broadcasting can be to a common area or to an individual area. These correspond to knowledge posting (blackboard or bulletin board model) and knowledge distribution (via journals or lecture notes). Since messages may initiate corresponding events, well known discrete event formalisms including DEVS formalism developed by Zeigler [33] can be also used to represent and process message mechanisms.

Coupling and Object-Oriented Formalism. Benefiting from the synergy of object-oriented simulation formalism [34], as well as CAST and CAD, one can generalize the concept of permanent and temporary entities as defined and used in Simscript II.5 [25]. One can therefore define two types of entities, such as, permanent and temporary entities (or objects, modules, or component models). A desired number of permanent entities can be generated automatically, each or groups of them having different values for their attributes (such as parameter values and initial values of their state variables). The temporary entities can be generated or destroyed according to the class concept which is well known since the time of the Simula language [4] and currently with the advent of contemporary object-oriented languages such as Smalltalk [12] and C++ [3].

A direct application of the coupling concept to object-oriented formalism would be the addition of hard coupling to permanent objects. Furthermore, time-varying coupling can be applied to multimodel and multifacetted models.

6 Conclusion

As it is seen even in one concept, such as coupling, systems theories provide rich paradigms to represent and process complexity. Computer assistance based on such system theoretic concepts would definitely provide powerful tools and environments to tackle complexity inherent in design problems.

I am sure system theories will also be extended and computerized tools and environments will be developed for CAST-based design, CAST-based software engineering, and CAST-based problem solving.

References:

1. K.Z. Aytaç, T.I. Ören: MAGEST: A model-based advisor and certifier for GEST programs. In: M.S. Elzas, T.I. Ören, B.P. Zeigler (eds.): Modelling and simulation methodology in the artificial intelligence era. Amsterdam: North-Holland, pp. 299-307 (1986)
2. Bell Canada: Trillium - Telecom software product development capability assessment model (Draft 2.2). Montréal: Bell Canada (1992)
3. Borland: Turbo C++ programmer's guide. Scotts Valley, CA.: Borland (1993)
4. O.J. Dahl, K. Nygaard: Simula - An Algol-based simulation language, CACM 9:9, 671-678 (1966)
5. H. Eisner: Computer-aided systems engineering. Englewood Cliffs, NJ.: Prentice-Hall (1988)
6. S.J. Farlow (ed.): Self-organizing methods in modelling, GMDH type algorithms. New York: Marcel Dekker (1982)
7. P.A. Fishwick: An integrated approach to system modeling using a synthesis of artificial intelligence, software engineering, and simulation methodologies. ACM Transactions on Modeling and Computer Simulation, 2:4 (Oct.), 307-330 (1992)
8. P.A. Fishwick, B.P. Zeigler: A multimodel methodology for qualitative model engineering. ACM Trans. on Modeling and Computer Simulation, 2:1 (Jan.), 52-81. (1992)
9. H.J. Greenberger, J.S. Maybee: Computer-assisted analysis and model simplification. New York: Academic Press (1981)
10. KBSE-92 (1992). Knowledge-based software engineering. Proc. of the 7th Conf., Sept. 20-23, 1992, McLean, VA. Los Alomitos, CA.:IEEE CS Press,
11. G.J. Klir: Facets of systems science. New York: Plenum Press (1991)
12. W.R. LaLonde, J.R. Pugh: Inside Smalltalk. Englewood Cliffs, NJ.: Prentice-Hall, (1990)
13. M.L. Lowry: Software engineering in the twenty-first century. In: M.L. Lowry, R.D. McCartney (eds.): Automating software design. Menlo Park / Cambridge: AAAI Press / The MIT Press, pp. 627-654 (1991)
14. T.I. Ören: Simulation of time-varying systems. In: J. Rose (ed.): Advances in cybernetics and systems. England: Gordon & Breach Science Publishers, pp. 1229-1238 (1975)

15. T.I. Ören: GEST - A modelling and simulation language based on system theoretic concepts. In: T.I. Ören, B.P. Zeigler, M.S. Elzas (eds.): Simulation and model-based methodologies: An integrative view. Berlin: Springer, pp. 281-335 (1984)

16. T.I. Ören: Model-based activities: A paradigm shift. In: T.I. Ören, B.P. Zeigler, M.S. Elzas (eds.): Simulation and model-based methodologies: An integrative view. Berlin: Springer, pp. 3-40 (1984)

17. T.I. Ören: A paradigm for artificial intelligence in software engineering. In: T.I. Ören (ed.): Advances in artificial intelligence in software engineering - Vol. 1. Greenwich, Connecticut: JAI Press, pp. 1-55 (1990)

18. T.I. Ören: Dynamic templates and semantic rules for simulation advisors and certifiers. In: P.A. Fishwick and R.B. Modjeski (eds): Knowledge-based simulation: Methodology and application. Berlin: Springer, 53-76 (1991)

19. D.L. Parnas: On criteria to be used in decomposing systems into modules, CACM, 14:1 (April), 221-227 (1972)

20. F. Pichler: Mathematische Systemtheorie - Dynamische Konstruktionen. Berlin: Walter de Gruyter (1975)

21. F. Pichler, R. Moreno-Díaz (eds.): Computer aided systems theory: EUROCAST'91. Berlin: Springer (1992)

22. F. Pichler, H. Schwärtzel (eds.): CAST methods in modelling: Computer aided systems theory for the design of intelligent machines. Berlin: Springer (1992)

23. R.S. Pressman: Software engineering. New York: McGraw-Hill (1992)

24. R. Rosen: Anticipatory systems. Oxford: Pergamon Press (1984)

25. E.C. Russell: Building simulation models with Simscript II.5. La Jolla, CA.: CACI (1983)

26. W.C. Schieve, P.M. Allen (eds.): Self-Organization and dissipative structures: Applications in the physical and social sciences. Texas Univ. Press. (1982)

27. H. Schwärtzel, I. Mizin (eds.): Advanced information processing. Berlin: Springer (1990).

28. J.P. van Gigch: Applied general systems theory. New York:Harper & Row (1974)

29. M. Weir: Goal-directed behavior. New York: Gordon and Breach (1984)

30. A.W. Wymore: A mathematical theory of systems engineering—The elements, New York: Wiley (2nd Edition, Kriger, 1977) (1967)

31. A.W. Wymore: Model-based systems engineering: An introduction to the mathematical theory of discrete systems and to the tricotyledon theory of systems design. Boca Raton: CRC (1993).

32. B.P. Zeigler: Theory of modelling and simulation. Wiley (1976)

33. B.P. Zeigler: Multi-facetted modelling and discrete event simulation. Academic-Press. (1984)

34. B.P. Zeigler: Object oriented simulation with hierarchical, modular models: Intelligent agents and endomorphic systems. Academic Press (1990)

35. B.P. Zeigler, J. Rozenblit: AI, simulation and planning in high autonomy systems. Los Alamitos, CA.: IEEE Computer Society Press (1990)

Computer Aided Nonlinear System Design Based on Algebraic System Representation and on Nonlinear Bundle Graphs

H. J. Sommer ,   H. Hahn

Control Engineering and Systems Theory Group, Department of Mechanical Engineering (FB 15), University of Kassel (GhK), 34125 Kassel,Mönchebergstraße 7, Germany. Phone: 05618043261, Fax: 05618042330, E-mail: hahn@hrz.uni.kassel.de.

Abstract: A Bundle Graph representation of nonlinear systems is defined and is used to solve problems in control theory with algorithmic methods.

Key Words: Nonlinear systems,Bundle graphs,Decoupling,Stability.

Introduction

System theory usually is based on system description in terms of differential equations or difference equations. In state space representation a nonlinear system is modelled by equations of the form:

$$\dot{x}(t) = f(x(t), u(t)), \quad y(t) = h(x(t), u(t)) \quad t \in I\!\!R, x(t) \in R^n, u(t) \in I\!\!R^p, y(t) \in R^q$$

or

$$x(i+1) = f(x(i), u(i)), \quad y(i) = h(x(i), u(i)) \quad, i \in Z\!\!\!Z \quad x(i) \in I\!\!R^n, u(i) \in I\!\!R^p, y(i) \in I\!\!R^q \quad.$$

Many results related to controllability, observabillity, decoupling and linearization of these systems have been obtained using differential geometry and manifold theory ([1],[2],[4],[6],[7],[10],[15],[22]). But very often a direct computer realisation of these theoretical results may lead to principal problems:

(A) A Theorem may contain conditions which can not be decided by the computer ([26]).

(B) Classical nonlinear system theory is basically a local theory. Numerical realisations of global results often lead to NP-hard problems ([31]).

(C) A theorem has to be formulated compactly in one or several short sentences. Various useful and approved solution strategies cannot be formulated adequately in terms of theorems. (Search algorithms ,genetic algorithms, neural networks, "often solving" algorithms etc.)

Computer aided systems theory is a new field of research that combines system theory with computer based methods and with artificial intelligence. The main challenge of Computer aided system theory has been summarized by Pichler ([24]): " Global systems tend to become complex in the sense that "problem solving" becomes very difficult or even impossible. New system methods have to be developed for modelling. General system theory has the task of contributing to that problem area by developing a proper conceptual framework and related methods. Computer aided system theory supports general system theory by providing the tools. Computer aided system theory tools should allow for the effective application of system theoretical methods in complex problem solving. Computer aided system theory research has the important task of providing a common framework to be able to have an integrated view for all existing modelling tools."

Based on these ideas, a computer adequate solution strategy in nonlinear system theory is presented in this paper (Figure 1).

Nonlinear control problems are in general formulated in differential geometrical language.

In a first step the problem is mapped here into a simpler frame. This problem frame will be formulated in a pure algebraic language and is adapted to a given problem.

In a second step computer algorithms are directly applied to provide numerical results.

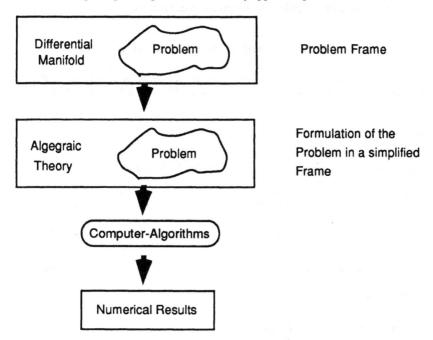

Figure 1: Solution Path in a Computer Aided System Theory

These two solution steps are the topic of this paper which is organised as follows:

- Taking pattern from J.C.Willems [32] and M.Fliess [4,7,12], a general algebraic system definition is introduced in Definitions 1 and 2 of the first Section , illustrated by various examples of different nonlinear systems and characterized by additional definitions.

- In the second Section, the definition of system structure due to Fliess ([4,7,12]) is discussed and interpreted in the language of Definitions 1 and 2. This algebraic definition has turned out to be a strong tool for abstract investigations of nonlinear dynamical systems. Extensions of basic theorems from linear to nonlinear systems have been done in the framework of this theory. Four of those are cited in Section 2, two of which have been proved by the authors ([29]). Numerical algorithms for a concrete nonlinear system analysis and design based on these methods are not yet available. It appears very difficult to construct numerical algorithms directly from the proceeding approach ([4,7,12]), (compare Section 2, Note 3).

- As a consequence another approach, called bundle graph techniques, has been introduced in the third Section according to step 2, to derive numerical algorithms.

This bundle graph representation will be defined for the system concept introduced in Definition 2. Using this bundle graph representation, system theoretical questions (as e.g. i/o-decoupling and controllability) can be attacked by numerical algorithms in combination with interactive computer aided solution methods. Thus, numerical methods from the theory of algorithms and artificial intelligence (as e.g. search algorithms and adaptive learning), can be applied in system theory. This will be demonstrated by the i-o-decoupling of a nonlinear system example and by introducing a stability test for polynomial systems.

1. General Concept of a Dynamical System

According to J.C. Willems ([32]), a system is a set of functions, called universum, restricted by relations, called behavioural equations. Here the following system definition is used.

Definition 1: (Class of Dynamical Systems)

A class of dynamical systems is defined as a triple $\left(F,R,\delta\right)$ where

-(i) F is a linear subspace of functions

$$F \subseteq \left\{w:IR^1 \to IR^1\right\} \tag{1a}$$

or

$$F \subseteq \left\{w:\mathbb{Z}^1 \to IR^1\right\}. \tag{1b}$$

-(ii) R is a linear subspace of mappings

$$R \subseteq \left\{r \mid r:F^j \to F, j\in IN\right\} \tag{2}$$

which is closed under composition: $\tag{3}$

$$r:F^j \to F, \quad r_i:F^{j_i} \to F, \quad rr_1,\ldots,r_j \in R \quad implies \quad r\left(r_1,\ldots,r_j\right):F^{\sum\limits_{i=1}^{j} j_i} \to F, \quad r\left(r_1,\ldots,r_j\right)\in R$$

-(iii) $\delta :F \to F$ is an operator, where for each $r_j \in R$ there exists an element

$$r_{2j}' \in R : \delta\left(r_j\left(w_1,\ldots,w_n\right)\right)=r_{2j}'\left(w_1,\ldots,w_n,\delta\, w_1,\ldots,\delta\, w_n\right). \tag{4}$$

Example 1a: (Class of Linear Time Continuous Systems):

$$F:=C^\infty\left(IR^1\right) \qquad \text{(smooth functions)} \tag{5a}$$

$$\text{(linear mappings)} \tag{5b}$$

$$R := \left\{ r_j : I\!R^{\,j} \to I\!R^{\,1} \,\middle|\, r_j \begin{pmatrix} x_1 \\ \vdots \\ x_j \end{pmatrix} := \sum_{i=1}^{j} r_{j,i} \cdot x_i \quad where \quad \begin{pmatrix} r_{j,1} \\ \vdots \\ r_{j,j} \end{pmatrix} \in I\!R^{\,j} \right\}$$

$$\delta\left(w(t)\right) := \frac{d}{dt} w(t) \ . \quad \text{(time differential operator)} \tag{5c}$$

Example 1b: (Class of Linear Time Discrete Systems):

$$F := \{ w : Z\!\!\!Z \to I\!R \} \qquad \text{(time discrete functions)} \tag{6a}$$

$$\text{(linear mappings)} \tag{6b}$$

$$R := \left\{ r_j : I\!R^{\,j} \to I\!R^{\,1} \,\middle|\, r_j \begin{pmatrix} x_1 \\ \vdots \\ x_j \end{pmatrix} := \sum_{i=1}^{j} r_{j,i} \cdot x_i \quad where \quad \begin{pmatrix} r_{j,1} \\ \vdots \\ r_{j,j} \end{pmatrix} \in I\!R^{\,j} \right\}$$

$$\delta\left(w(t)\right) := w(t+1) \ . \qquad \text{(time shift operator)} \tag{6c}$$

Example 2: (Class of Time Continuous Polynomial Systems):

$$F := C^{\infty}\!\left(I\!R^{\,1}\right) \qquad\qquad \text{(smooth functions)} \tag{7a}$$

$$\text{(polynomial mappings)} \tag{7b}$$

$$R := \left\{ r_j : I\!R^{\,j} \to I\!R^{\,1} \,\middle|\, r_j \begin{pmatrix} x_1 \\ \vdots \\ x_j \end{pmatrix} := \sum_{\{i_1,\dots i_p\} \subseteq \{1,\dots j\}} r_{j,i_1\dots i_p} \cdot x_{i_1} \cdot \dots x_{i_p} \quad wherer \quad r_{j,i_1\dots i_p} \in I\!R^{\,1} \right\}$$

$$\delta\left(w(t)\right) := \frac{d}{dt} w(t) \ . \quad \text{(time differential operator)} \tag{7c}$$

Example 3: (Class of Fuzzy Systems):

In fuzzy systems, the values of the system functions $w = \left(w_1, \dots, w_v \right)$ only describe tendencies t: great, small, normal, etc., called fuzzy terms. A fuzzy term t is defined by its membership function $\psi_t : I\!R \to [0,1]$ which gives a measure of affiliation of $s \in I\!R^1$ to t. For example, if t_o is the term "old", $\psi_{t_o}(70) \approx 1$

means that normally a man with 70 years is called old. Let $Fu(I\!R)$ denominate the

set of fuzzy terms over $I\!R^1$. For fuzzy terms $t_1, t_2 \in Fu(I\!R)$ the following

relations are defined:

Addition: $\qquad \psi_{t_1 + t_2}(s) := \displaystyle\int_{s = s_1 + s_2} \psi_{t_1}(s_1) \cdot \psi_{t_2}(s_2) \, ds_1 \, ds_2 \qquad$ (8a)

Multiplication: $\psi_{t_1 \cdot t_2}(s) := \displaystyle\int_{s = s_1 \cdot s_2} \psi_{t_1}(s_1) \cdot \psi_{t_2}(s_2) \, ds_1 \, ds_2 \qquad$ (8b)

And: $\qquad \psi_{t_1 \wedge t_2}(s) := min\left\{ \psi_{t_1}(s), \psi_{t_2}(s) \right\} \qquad$ (8c)

Or: $\qquad \psi_{t_1 \vee t_2}(s) := max\left\{ \psi_{t_1}(s), \psi_{t_2}(s) \right\} \qquad$ (8d)

Discrete (or time continuous) fuzzy systems are defined as triples (F, R, δ) :

$$F = \{ w : \mathbb{Z} \to Fu(I\!R) \} \qquad (or \quad F = C^\infty (Fu(I\!R))) \qquad (9a)$$
$$R = \{ r : F \to F \mid r \quad generated \quad by \quad +, \quad \cdot \quad , \quad \wedge \quad and \quad \vee \;(9b)$$
$$with \quad coefficients \quad from \quad Fu(I\!R) \}$$
$$\delta (w(t)) := w(t+1) \qquad (or \quad \delta (w(t)) := \frac{d}{dt} w(t) \;). \qquad (9c)$$

Definition 2: (Dynamical Systems) ([32])

Given a class of dynamical systems (F, R, δ). A dynamical system is a tupel

(U, B), where $U := F^\nu$ $(\nu \in N)$ is called universum of the system and

$$B := \left\{ w = (w_1, ..., w_\nu) \in U \;\middle|\; r_i(w, \delta \, w_i) = 0, i = 1, ... n \right\} \qquad (10)$$

is called behaviour of the system, defined by n fixed elements $r_i \in R$.

Example 4a: (Time Discrete Polynomial System in Implicit Form)

Let $U=\left\{w=\left(w_1,w_2,w_3,w_4,w_5,w_6\right)\ \left|w_i\ :\mathbb{Z}\rightarrow I\!R\right.\right\}$

with behavioural equations:

(11a)

$$w_3(k+1)-w_3(k)-w_4(k)-w_3(k)^2\cdot w_4(k)^2-w_1(k)-w_2(k)\cdot w_3(k)=0$$

$$w_4(k+1)-w_4(k)-w_4(k)^2-w_3(k)\cdot w_4(k)-w_2(k)-w_1(k)\cdot w_4(k)=0$$

$$w_5(k)\ -w_3(k)-w_3(k)^3-w_1(k)-w_2(k)\cdot w_3(k)=0$$

$$w_6(k)\ -w_4(k)-w_3(k)\cdot w_4(k)-w_2(k)-w_1(k)\cdot w_4(k)=0$$

Notes:

1. Definitions 1 and 2 introduce a **pure algebraic** system concept without using analytical tools. They cover most systems from the areas of engineering and natural science. All subsequent results are independent of concrete interpretations of F, R and δ .

2. If the mappings r_i don't depend explicitly on time, the associated systems are **time invariant**.

Definition 3: (Causal Systems)

A system is called **causal** , iff the relations (12)

$$r_i\left(w,\delta\ w_i\right)=0\ ,\quad i=1,...,n \tag{12}$$

have unique solutions $\delta\ w_i\in F$ for all $w\in F^V$. In case, the solution can be written as

$$\delta\ w_i=r_{i0}(w)\ ,\quad i=1,...,n\quad , \tag{13}$$

where $r_{i0}\in R$, the system representation is called **explicit** .

Example 4b: (Explicit Representation of a Time Discrete Polynomial System)

Substituting $u=\left(u_1,u_2\right)=\left(w_1,w_2\right)$, $x=\left(x_1,x_2\right)=\left(w_3,w_4\right)$ and

$y=\left(y_1,y_2\right)=\left(w_5,w_6\right)$

into (11a), the explicit system representation (11b) is obtained:

$$x_1(k+1)=x_1(k)+x_2(k)+x_1(k)^2\cdot x_2(k)^2+u_1(k)+u_2(k)\cdot x_1(k) \qquad \text{(11b)}$$

$$x_2(k+1)=x_2(k)+x_2(k)^2+x_1(k)\cdot x_2(k)+u_2(k)+u_1(k)\cdot x_2(k)$$

$$y_1(k) =x_1(k)+x_1(k)^3+u_1(k)+u_2(k)\cdot x_1(k)$$

$$y_2(k) =x_2(k)+x_1(k)\cdot x_2(k)+u_2(k)+u_1(k)\cdot x_2(k)$$

Definition 4:   (Generator System)

- Let $u_1,...u_p$ components of $w=\left(w_1,...,w_v\right)$. A component w_i of w is

  called dependent on $u=\left(u_1,...u_p\right)$ iff there exist $r\in R$, $\mu,\sigma\in N$ with

  $$r\left(w_i,\delta\ w_i,...,\delta^{\ \sigma}w_i,u,\delta\ u,...,\delta^{\ \mu}u\right)\equiv 0 \qquad \text{(14)}$$

- The set of components $u=\left(u_1,...u_p\right)$ is called free iff no component u_i is

  dependent on $\left(u_1,...,u_{i-1},u_{i+1},...,u_p\right)$, $i=1,...,p$.

- A finite subset of components $\left(u_1,...u_p\right)$ of $w=\left(w_1,...,w_v\right)$ is called

  generator system of W iff all other components of W are dependent on those.

Note 3: (Input and Output Functions)

If it is possible to find p components $u=\left(u_1,...u_p\right)$ of $w=\left(w_1,...,w_v\right)$ which
are free and simultaneously a **generator system** of W , they are called system
input. The existence of such components is discussed in Theorem 2.

Output functions are arbitrarily selected components $y=\left(y_1,...,y_m\right)$ of
$w=\left(w_1,...,w_v\right)$.

Definition 5 : (Controllability   ([32]))

Let the universum U=(T,W) be defined by functions $w:T\rightarrow W$ with the set of
time values $T\subseteq Z$ or $T\subseteq IR^1$ and $W\subseteq IR^v$ as signal space. Let $w|_{T'}$ be the

restriction of w to T' for $T' \subseteq T$.

Then the **concatenation of two functions** $w_{[t_1,t_2]}$ and $w_{[t_3,t_4]}$

$(t_1 < t_2 \leq t_3 < t_4)$ is defined as (compare [32]):

$$w_{[t_1,t_2]} \wedge w_{[t_3,t_4]} := \begin{cases} w_{[t_1,t_2]}(t) & for \quad t \in [t_1,t_2] \\ w_{[t_3,t_4]}(t) & for \quad t \in [t_3,t_4] \end{cases} \qquad (15)$$

The System $\Sigma = (U,B) = (T,W,B)$ is called **controllable** $:\Leftrightarrow$

$$\left\{ w_{[t_1,t_2]}, w_{[t_3,t_4]} \in B \text{ with } t_1 < t_2 \leq t_3 < t_4 \right. \qquad (16)$$

$$\Rightarrow \exists w_{[t_1,t_4]} \in B \quad where \quad w_{[t_1,t_4]}\big|_{[t_1,t_2]} = w_{[t_1,t_2]} \quad and \quad w_{[t_1,t_4]}\big|_{[t_3,t_4]} = w_{[t_3,t_4]} \right\}$$

2. Structural System Description

The "system structure" introduced by Fliess [4,7,12] will be presented in the form given by Conte, Perdon and Moog [5] and will be interpreted in the language of Definitions 1 and 2. This algebraic definition has turned out to be a strong tool for abstract investigations of dynamical systems. Examples of general theorems which can be easily proved in the framework of this theory are extended from linear to nonlinear systems in this paper.

Definition 6: (System Structure of Fliess [4,7,12])

Given a dynamical system (U,B) , with $U := F^V$, $v \in I\!N$ and

$$B := \left\{ w = (w_1,...,w_v) \in U \mid r_i(w,\delta w_i) = 0, i = 1,...n \right\} \qquad (17)$$

Let

$$R_0 := \left\{ \psi \in R \mid \psi(0,...0) = 0 \right\} ,$$

$$J_0 := span_{I\!R} \left\{ \delta^n r_i \mid i = 1,...,p, n \in I\!N \cup \{0\} \right\} \qquad (18a)$$

($span_{I\!R} \{v_i\}$ denotes the vector space spanned by v_i over $I\!R$) and

$$\qquad\qquad\qquad\qquad\qquad\qquad\qquad\qquad\qquad (18b)$$

$$J_{k+1} := span_R \left\{ \psi(j_1,...,j_\mu,...,j_\rho) \mid \psi \in R_0 \quad , \quad j_\mu \in J_k, \quad j_1,...j_\rho \in F, \right.$$

The **ideal** generated by $\{r_1,...r_p\}$ in R is the set of functions:

$$\left(r_i(w,\,\delta w_i)\ \mid i=1,...,p\right):= span_R \bigcup_{k=0}^{\infty} J_k \quad . \tag{19}$$

The **structure** for the system (U,B) is defined as:

$$K:= {\left(F^v\right)^N} \Big/ {\left(r_i \mid i=1,...,p\right)\left(\left(F^v\right)^N\right)} \tag{20}$$

where K is the function set $\left(F^v\right)^{IN}$ modulo the equivalence relation "$\cong$".

Two elements $f,f' \in \left(F^v\right)^{IN}$ are equivalent ($f \cong f'$) iff:

$$\forall r \in R:\ r(f)+J=r(f')+J \tag{21}$$

$$with\ \ J:=\left(r_i,i=1,...,p\right)\left((F)^N\right):=\left\{ r(f)\Big| r\in\left(r_i\mid i=1,...,p\right),f\in(F)^{IN}\right\}\ .$$

Note 4 : (Comparison of System Definitions 2 and 6)

Definition (2) declares a system by means of the universum U and by the constraints generated by the behavioural equations $r_i(w,\delta\,w_i)=0$.

The system structure of Definition 6 represents the free behaviour of the system through the information contained in the universum of the system, disturbed by the mappings $r_i\ (i=1...p)$. These two definitions are equivalent. They provide complementary views of the same idea.

Using the representatives $k \in \overline{k}=k+J \in K$, the operations of $R \cup \{\delta\}$ can be defined over K .

The system structure has the advantage that it can be used without asking for the solvability of the behavioural equations. It was the idea of Fliess, that starting system theory with this system structure many results are easily obtained by means of the strong tools of algebra. On the other hand it appears very difficult to derive algorithms operating with equivalence classes defined in (21) .

The following Theorems 1 and 2 are generalizations of the results of Nieuwenhuis and Willems ([20]) and of Fliess ([11]) from linear systems to nonlinear systems. These two theorems are proved in ([29]).

Theorem 1: (Approximation Theorem)

Assume that the mappings $r \in R$ are continuous over F .

Let $r_{\varepsilon,i}$ $(i=1,...p)$ define the behaviour B_ε over the universum U, let $\bar{r} \in R$ be a fixed element and assume:

$$r_{\varepsilon,i} \rightarrow r_{0,i} \text{ for } \varepsilon \rightarrow 0 \text{ in } R \quad (i=1...p)$$

$$w_\varepsilon \in B_\varepsilon \ (\varepsilon>0) \ , \ w_\varepsilon \rightarrow w_0 \text{ for } \varepsilon \rightarrow 0 \text{ in } F \quad \text{and} \quad \bar{r}(w_\varepsilon)=0 \ .$$

Then

$$w_0 \in B_0 \text{ satisfies } r(w_0)=0 \ .$$

Theorem 2: (Controllability Criterion)

Let R' denote the set of uniquely solvable relations r of R (according to (13)), and assume that the following property (22) holds :

$$\forall r_1, r_2 \in R', t_1 < t_2 < t_3 < t_4 \in I\!R \ \exists r \in R' : \tag{22}$$

$$r\big|_{[t_1, t_2]} = r_1\big|_{[t_1, t_2]} \quad \text{and} \quad r\big|_{[t_3, t_4]} = r_2\big|_{[t_3, t_4]}$$

Then the following statements are equivalent:

(a) the system (U,B) has a finite free set of generators $u = \left(u_1,...,u_p \right) \subseteq w$.

(b) the system (U,B) is controllable.

The general result of the following Theorem 3 is due to Fliess and Delaleau [4]. A constructive algorithm based on bundle graph techniques for decoupling of nonlinear systems will be presented in Example 5 of Section 3.

Theorem 3 : (Decoupling Criterion)

A right invertible system is decouplable by state feedback using the input values

$$u, \delta\, u,...,\delta^{\ \mu} u \ .$$

Theorem 4 provides another very useful result of Fliess [9] based on Definition 6.

Theorem 4: (Normal Form Representation)

The system of Definition 6 can be represented in normal form:

$$x_1 = x_2$$
$$x_2 = x_3$$
$$\vdots \quad \vdots$$

$$x_{n-1} = x_n$$
$$r\left(\delta x_n, x_n, \ldots x_2, x_1, u, \delta u, \ldots \delta^\mu u\right) = 0 \quad .$$

3 . Nonlinear Bundle Graphs

Based on the system concept introduced in Definitions 1 and 2 a network representation of nonlinear systems is introduced called dotted bundle graph or bundle graph . This special nonlinear graph representation provides guidelines for generating algorithms for a concrete computer aided system analysis and for design steps of those systems.

Definition 7 : (Bundle Graph Representation of Nonlinear Systems)

Let R of Definition 1 be a linear space spanned by $\left\{\tau_i \mid i \in N \right\}$ .Then every $r \in R$ can be written as

$$r = \sum_{k=1}^{K} \alpha_k \tau_k \quad \text{with} \quad \alpha_k \in IR \quad .$$

A **directed bundle graph** is a set of **nodes** N (represented by circles) connected by a set of **directed edges** E (represented by lines marked with dots), where the edges are combined by **bundles**.

The set of **nodes** N represents the set of functions $w = \left(w_1, \ldots, w_\nu\right) \in F^\nu$ of Definition 2.

The set of **bundle edges** stands for the set of mappings $r_{i0} \in R$ of Definition 2, which represent the behavioural equations, written in explicit form as

$$\delta w_i = r_{i0}(w) \quad i = 1, \ldots n .$$

The set of **directed edge bundles** E includes two subsets E_δ and E_u, where

$$E = E_\delta \cup E_u \quad . \tag{23}$$

The subset E_δ includes all edge bundles, defined by the behavioural equations of the form

$$\delta w_i = \sum_{k=1}^{K} \alpha_{i,k} \cdot \tau_k(w) \quad with \quad \alpha_{i,k} \in IR \quad , \tag{24}$$

where each τ_k is represented by a **dotted edge bundle**. The $\delta - $ dependency is marked by a **dot** in the edge bundle output. The behavioural equations

$$w_i = \sum_{k=1}^{K} \alpha_{i\,k} \cdot \tau_k(w) \qquad with \quad \alpha_{i\,k} \in I\!R \tag{25}$$

are represented by **undotted edge bundles** .

The **length of a path in the graph** is defined by the number of dots contained in the path.

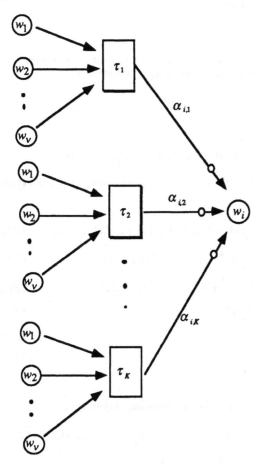

Figure 2:   **Dotted bundle graph of Relation (24).**

Decoupling Results

Our main result obtained by bundle graph techniques is stated in Theorem 5. It provides a constructive algorithm related to Theorem 3 .

Theorem 5 : (Decoupling of Nonlinear Systems)

The nonlinear input- output decoupling problem can be solved algorithmically for systems represented by bundle graphs.

The **decoupling algorithm** is based on the rules:

(1) **Find a shortest coupling path.**
(If there are various those path, use one of them.)

(2) **Compensate this path.**

A repeated application of these rules provides a system decoupling by feedback or shows that system decoupling is impossible (using this approach).

This will be applied to the Example 5.

Example 5: (Decoupling by Bundle Graph Algorithms)

Given the discrete time polynomial system (26), represented by the bundle graph of Figure 3.

$$x_1(i+1) = x_1(i) + x_1(i)^2 \cdot x_2(i)^2 + u_1(i) + u_2(i) \cdot x_2(i) \qquad (26a)$$

$$x_2(i+1) = x_2(i) + x_2(i)^2 + x_1(i) \cdot x_2(i) + u_2(i) \qquad (26b)$$

$$x_3(i+1) = x_1(i) + x_1(i)^3 + u_1(i) + u_2(i) \cdot x_2(i) \qquad (26c)$$

$$x_4(i+1) = x_2(i) + u_2(i) \qquad (26d)$$

$$y_1(i) = x_3(i) \qquad (26e)$$

$$y_2(i) = x_4(i) \qquad (26f)$$

Application of the decoupling algorithm

Following rule 1, the shortest coupling path from u_i to y_j , $(i=1,2\,;j=1,2\,;i \neq j)$

has to be selected. This is the path (u_2, x_3, y_1) with length 1. It includes the term $u_2(i) \cdot x_2(i)$ in equation (26c).

The compensation of this path is realized by a feedback term $-u_2(i) \cdot x_2(i)$, as shown in Figure 4a, where only the significant edges used for compensation are drawn.

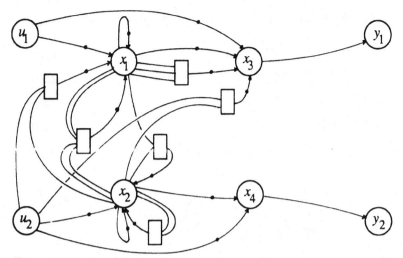

(The multiplication of the inputs to a node is marked by a square.)
Figure 3:   Bundle graph of system (26)

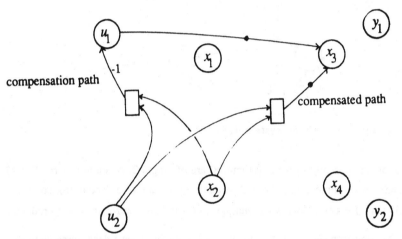

Figure 4a: Compensation of the coupling term $u_2(i) \cdot x_2(i)$

Introducing ,

$$u_1(i) = u_1'(i) - u_2(i) \cdot x_2(i) \tag{27a}$$

$$u_2(i) = u_2'(i) \quad , \tag{27b}$$

inserting (27a) into (26) and renaming $u_1'(i)$ into $u_1(i)$ yields the equations:

$$x_1(i+1) = x_1(i) + x_1(i)^2 \cdot x_2(i)^2 + u_1(i) \tag{28a}$$

$$x_2(i+1)=x_2(i)+x_2(i)^2+x_1(i)\cdot x_2(i)+u_2(i) \tag{28b}$$

$$x_3(i+1)=x_1(i)+x_1(i)^3+u_1(i) \tag{28c}$$

$$x_4(i+1)=x_2(i)+u_2(i) \tag{28d}$$

$$y_1(i)=x_3(i) \tag{28e}$$

$$y_2(i)=x_4(i) \tag{28f}$$

The graph of equation (28) is drawn in Figure 4b.

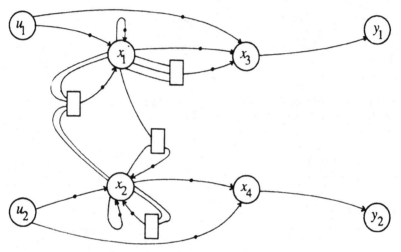

Figure 4b: **Graph of system (28)**

The shortest coupling paths of system (28) are of length 3. A coupling term $x_1^2\cdot x_2^2$ of (28a) for example cannot be compensated by means of a decoupling controller $u_1(i)=u_1'(i)-x_1^2(i)\cdot x_2^2(i)$ in analogy to (27). This controller would produce a new coupling from u_2 via x_3 to y_1. As a consequence a decoupling controller must be found that decouples x_3 from u_2. As x_4 and y_2 are not strictly influenced by x_3, a decoupling controller of the form (29a) is suitable. The decoupling controller from u_1 to x_4 of (29b) is designed in analogy to (29a).

$$u_1(i)=u_1'(i)+F\big(x_1,x_2,u_1',u_2'\big) \tag{29a}$$

$$u_2(i)=u_2'(i)+G\big(x_1,x_2,u_1',u_2'\big) \tag{29b}$$

The controllers $G\big(x_1,x_2,u_1',u_2'\big)$ and $F\big(x_1,x_2,u_1',u_2'\big)$ are determined subsequently, where the time arguments of the included variables will be determined, too. Inserting (29) into (28) yields:

$$x_1(i+1) = x_1(i) + x_1(i)^2 \cdot x_2(i)^2 + u_1'(i) + F\left(x_1, x_2, u_1', u_2'\right) \tag{30a}$$

$$x_2(i+1) = x_2(i) + x_2(i)^2 + x_1(i) \cdot x_2(i) + u_2'(i) + G\left(x_1, x_2, u_1', u_2'\right) \tag{30b}$$

$$x_3(i+1) = x_1(i) + x_1(i)^3 + u_1'(i) + F\left(x_1, x_2, u_1', u_2'\right) \tag{30c}$$

$$x_4(i+1) = x_2(i) + u_2'(i) + G\left(x_1, x_2, u_1', u_2'\right) . \tag{30d}$$

The graph of equations (30) is drawn in Figure 4c:

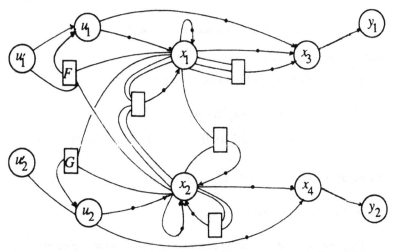

Figure 4c: **Graph of system (28) with the feedback law (29).**

Inserting (30a) into the time shifted equation (30c) yields:

$$x_3(i+2) = \left(x_1(i) + x_1(i)^2 \cdot x_2(i)^2 + u_1'(i) + F\left(x_1, x_2, u_1', u_2', i\right)\right)^3 + \tag{31a}$$

$$+\left(x_1(i) + x_1(i)^2 \cdot x_2(i)^2 + u_1'(i) + F\left(x_1, x_2, u_1', u_2', i\right)\right) + u_1'(i+1) + F\left(x_1, x_2, u_1', u_2', i+1\right)$$

The function $F\left(x_1, x_2, u_1', u_2', i+1\right)$,too, must compensate all terms of (31a)

depending on the unwanted coupling variable x_2 . As a consequence

$F\left(x_1, x_2, u_1', u_2', i+1\right)$ has to compensate the term:

$$\left(x_1(i) + x_1(i)^2 \cdot x_2(i)^2 + u_1'(i) + F\left(x_1, x_2, u_1', u_2', i\right)\right) +$$

$$+\left(x_1(i) + x_1(i)^2 \cdot x_2(i)^2 + u_1'(i) + F\left(x_1, x_2, u_1', u_2', i\right)\right)^3$$

This implies the definition

$$\overline{F}\left(x_1,x_2,u_1',u_2',i\right):=F\left(x_1,x_2,u_1',u_2',i+1\right)= \tag{32a}$$

$$=-\left(x_1(i)+x_1(i)^2\cdot x_2(i)^2+u_1'(i)+\overline{F}\left(x_1,x_2,u_1',u_2',i-1\right)\right)$$

$$-\left(x_1(i)+x_1(i)^2\cdot x_2(i)^2+u_1'(i)+\overline{F}\left(x_1,x_2,u_1',u_2',i-1\right)\right)^3$$

where $\overline{F}\left(x_1,x_2,u_1',u_2',i-1\right)$ at time step i-1 is equal to $F\left(x_1,x_2,u_1',u_2',i\right)$ at

time step i. $\overline{F}\left(x_1,x_2,u_1',u_2',i\right)$ is known in time step i and therefore

$\overline{F}\left(x_1,x_2,u_1',u_2',i+1\right)$,etc., can also be obtained from equation (32a).

This implies the decoupling of x_3 from x_2 .
G is determined in analogy to F . Equations (30b) and (30d) yield:

$$\tag{31b}$$

$$x_4(i+2) = x_2(i)^2+x_2(i)+x_1(i)\cdot x_2(i)+u_2'(i)+$$

$$+\overline{G}\left(x_1,x_2,u_1',u_2',i-1\right)+u_2'(i+1)+\overline{G}\left(x_1,x_2,u_1',u_2',i\right)$$

where

$$\overline{G}\left(x_1,x_2,u_1',u_2',i\right)=-x_1(i)\cdot x_2(i)-\overline{G}\left(x_1,x_2,u_1',u_2',i-1\right)\ . \tag{32b}$$

Then x_4 and y_2 are independent of x_1 .

System (28) together with the decoupling controller (32) is drawn in Figure 4d:

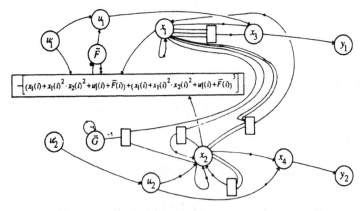

Figure 4d: **System (28) with decoupling controller (32).**

The combination of the two decoupling controllers (27) and (32) provides the following overall decoupling controllers of system (26):

(33a)

$$u_1(i) = u_1'(i) + \overline{F}\left(x_1, x_2, u_1', u_2', i\right) - \left(u_2'(i) + \overline{G}\left(x_1, x_2, u_1', u_2', i\right)\right) \cdot x_2(i)$$

$$u_2(i) = u_2'(i) + \overline{G}\left(x_1, x_2, u_1', u_2', i\right) \tag{33b}$$

where $\overline{F}$ and $\overline{G}$ are introduced in (32). (Note that the realisation of $\overline{F}$ and $\overline{G}$ contains cycles.)

Controllability Results

Pure graph theoretical investigations don't provide necessary and sufficient conditions for controllability and observability even in linear systems. To overcome this problem, the concept of structural controllability has been introduced by Lin [18] . This has not yet been successfully done for nonlinear systems . On the other hand we show in Definition 8 and Theorem 6 that the following weak controllability statements of linear graphs (compare [18] and[25]) can be easily extended to nonlinear bundle graph systems.

Definition 8 :   (Dilatation)

A bifurcation in the forward direction of the bundle graph is called a dilatation (cf. 5a).

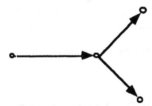

Figure 5a : Dilatation of a graph.

Theorem 6 : (Controllability Aspects)

(a) If the bundle graph has a dilatation, then the system is not controllable (cf. 5a) .
(b) If the bundle graph of a system can be decomposed by decoupling into subsystems, all of which are controllable, then the overall system is controllable (cf. 5b).

Proof: Theorem 6 is a direct extension of [18] .

Note 5: (Controllability and Decoupling)

The following Example 6 shows, that the assumed controllability of the decomposed nonlinear subsystems of Theorem 6b is a necessary condition.

Example 6: (Controllability and Decoupling)

The system $y(k+1) = y(k) \cdot u(k)^2$ $with$ $u, y \in IR$ is decoupled but not controllable (cf. 4c). A point on a trajectory $y(k) < 0$ cannot be connected to a point on a trajectory $y(k) > A > 0$.

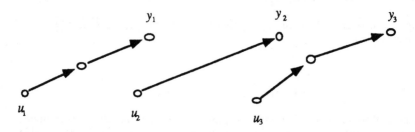

Figure 5b: Graph of a decoupled system.

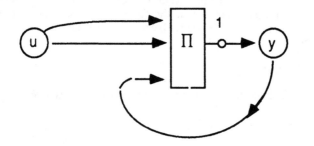

Figure 5c: Bundle graph of the system of Example 6.

Bounded input-bounded output stability test of polynomial systems

A simple BIBO-stability test for polynomial systems, including additional components with bounded output (for example Fuzzy controllers) is given by the rules:

Rules: (BIBO-Stability Test)

(1) Isolate all concentric cycles of the system graph.

(2) Eliminate cycles with bounded elements (as e.g. Fuzzy-controllers).

(3) Test cycle families without nonlinear couplings by the Hurwitz rule.

(4) Test the compensation of the nonlinear path. If a compensation is not possible in the system bundle graph, then a polynomial system is not BIBO-stable.

The **proof of these rules** is simple:

-(1) follows from the fact that instabilities can be only produced by cycles.
-(2) is proved by the following argument: for bounded inputs, a bounded element in a
     cycle reduces the values in all nodes of the cycle below an upper bound.
-(3) is an application of a result from linear theory.
-(4) is a consequence of the fact that due to the Hurwitz criterion, a dominating cycle
     (this is a cycle with a very high weight relative to the weights of the other
     cycles) can not be compensated.

The application of the proceeding BIBO-stability test rules to system (26) of Example
5 is shown in Figure 6 where the dominating cycle is marked by heavy lines. Due to
rule (4), system (26) is not BIBO-stable.

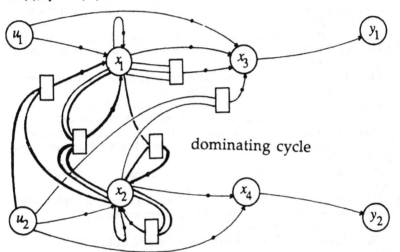

Figure 6: The dominating cycle in the bundle graph of system (26)

Conclusions:

The rules of Chapter 3 are an example of computer algorithms solving problems in
nonlinear system theory. Realisations of these algorithms lead directly to methods
well known in algorithm theory and artificial intelligence .

References:
[1] L.Berg, Allgemeine Operatorenrechnung, Überblicke Mathem. 6, BI 1973, 7-49.
[2] J.Birk,M.Zeitz, Nonlinear Control System Design IFAC Symp., Capri, 1989.
[3] J.Birk,M.Zeitz, Computer-Aided Analysis of Nonlinear Observation Problems,
    Proceedings of the NOLCOS 92, Bordeaux 251-256.
[4] E.Delaleau,M.Fliess, An Algebraic Interpretation of the Structure Algorithm with
    an Application to Feedback Decoupling,Proceedings of the NOLCOS 92
    Bordeaux, 489-494.
[5] G.Conte,A.M.Perdon,C.H.Moog, A Differential Algebraic Setting for Analytic
    Nonlinear Systems, Proceedings of the NOLCOS 92 Bordeaux, 203-208.

[6] H.Fehren, Multiparameteranalyse linearer zeitinvarianter Mehrgrößenregelkreise, Doctoral. Thesis, Department of Mech. Eng. (FB15) University of Kassel, 1993

[7] M.Fliess, A Note on the Invertibility of Nonlinear Input/output Systems, Systems Control Lett.8, 1986, 147-151.

[8] M.Fliess, Generalized linear systems with lumped or distributed parameters and differential vector spaces, Int.J.Control,Vol.49,No.6,1989,1989-1999

[9] M.Fliess, Generalized controller canonical forms for linear and nonlinear dynamics, IEEE Trans Autom.Control 35 1990 ,994-1001.

[10] M.Fliess, Some basic structural properties of generalized linear systems, Systems Control Lett.15, 1990, 391-396.

[11] M.Fliess, A remark on Willems´trajectory characterization of linear controllability, Systems Control Lett.19,1992,43-45

[12] M.Fliess, Some Remarks on a New Characterization of Linear Controllability, 2.IFAC Workshop on System Structure and Control 1992, 8-11

[13] C.Heij, Exact Modelling and Identifiability of Linear Systems, Automatica, Vol.28,No.2,1992,325-344

[14] A.Isidori , Nonlinear Control Systems, Second Edition, Springer-Verlag 1989.

[15] A.J.Krener, A Isidori, Linearization by output injection and nonlinear observers, Systems Control Lett. 3, 1983, 47-52.

[16] A.J.Krener, W.Respondek, Nonlinear observers with linearizable error dynamics, SIAM J.Control.Optim.23, 1985, 197-216.

[17] D.Laugwitz, Zahlen und Kontinuum, BI-Wissenschaftsverlag 1986

[18] C.T.Lin, Structural Controllability, IEEE Trans.AC-19,1974,201-208.

[19] R.Moreno-Diaz Jr, K.N.Leibovic,R.Moreno-Diaz, Systems Optimization in Retinal Research, EUROCAST`91, Springer Lecture Notes in Computer Science 585, 1991, 539-546

[20] J.W. Nieuwenhuis, J.C. Willems, Continuity of Dynamical Systems, Math. Control Signals Systems (1992) 5: 391-400

[21] H.Nijmeijer, A.J.v.d.Schaft, Nonlin. Dyn. Control Systems, Springer, 1990.

[22] U.Oberst,Multidimensional constant linear systems and duality, Preprint,University of Innsbruck,Austria,1988

[23] U.Oberst,Multidimensional Constant Linear Systems, EUROCAST`91, Springer Lecture Notes in Computer Science 585, 1991,66-72

[24] F.Pichler, General Systems Theoriy Requirements for the Engeneering of Complex Models, EUROCAST`91, Springer Lecture Notes in Computer Science 585, 1991, 132-141

[25] K.J.Reinschke, Multivariable Control, Springer, 1988.

[26] C.P.Schnor, Rekursive Funktionen und ihre Komplexität, Teubner, 1974

[27] H.Schwartz,Nichtlineare Regelungssysteme, Oldenburg-Verlag, 1991

[28] H.Sira-Ramirez, Dynamical sliding mode control strategies in the regulation of nonlinear chemical processes, Int.J.Control, Vol.56, No.1, 1992, 1-21

[29] H.J.Sommer, H.Hahn, Systemtheoretische Ansätze von Willems und Fliess, IMAT-Bericht, 1992

[30] H.J.Sommer, Einf. flexibler Begriffe mittels Fuzzy-Logik und eine Diskussion ihrer Anwendung, Vortrag 1.Workshop Fuzzy Control, Dortmund 1991

[31] C.P.Suarez Araujo, R.Moreno-Diaz, Neural Structures to Compute Homothetic Invariances for Artificial Perception Systems, EUROCAST`91, Springer Lecture Notes in Computer Science 585, 1991, 525-538.

[32] J.C.Willems, Paradigmas and Puzzles in the Theory of Dynamical Systems, IEEE Transactions on Automatic Control, 36 No 3,1991,259-294

[33] M.Zeitz, The extended Luenberger observer for nonlinear systems, Systems Control Lett, 9,1987,149-156,

A New Model-Based Approach to the Co-Design of Heterogeneous Systems

Prof. Dr. D. Monjau, St. Kahlert
TU Chemnitz, Germany

Dr. K. Buchenrieder, Ch. Veith
Siemens AG Munich, Germany

Abstract. This paper describes an approach to the design of heterogeneous Hardware-Software systems. It defines a strict sequence of transformations that begins with a system specification, and leads to an implementation of the system. The steps in the sequence are defined by system models on decreasing levels of abstraction, and every step in the sequence transforms an input model into an output model. The final output model is equivalent to the system implementation. The objective of our method is to add rigor to the prevailing inexactness in the development of heterogeneous systems by introducing well-defined formal synchronization points - the models - into the design process. The feasibility of our approach has been proven with two examples. Further we discuss the use of existing tools to support our method.

1   Introduction

Most of today's computer based systems are complex and have to operate in real-time. While engineers struggle with tight schedules and increasing complexity, the design of heterogeneous systems containing hardware (HW) and software (SW) components is not well understood.

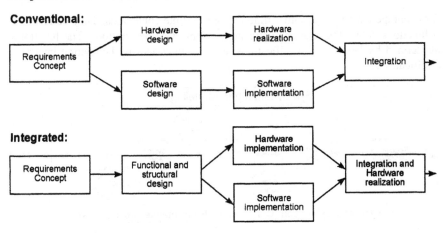

Figure 1. Ways to design a heterogeneous system

The motiviation for our method arises from the lack of structured approaches in the development of Hardware/Software systems, and the intent to narrow the gap between the designer's intent and the behavior of the resulting system implementation. By looking at numerous designs it became clear that the design of heterogeneous systems must be based on a formal approach. Our method comprises two aspects of formalization:

- we use models on every level of abstraction in the design process, and
- we introduce a formal approach to the design process itself.

In our view, Hardware/Software systems are usually complex, operate in real-time and consist of concurrent components that are related (see figure 2).

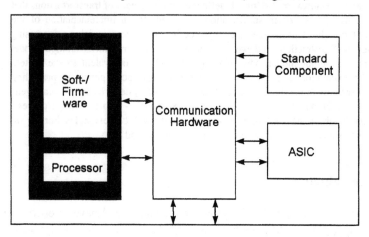

Figure 2. Hardware/Software-System

These components may be either time-discrete or time-continuous. They correspond to functions, while the relations correspond to interface descriptions. The functions are defined by behavioral specifications, and the interfaces by behavioral interface specifications.

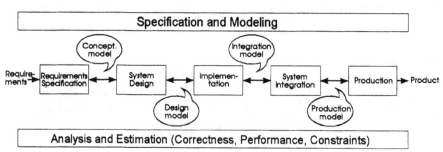

Figure 3. System Engineering Model

By suitable combination or partitioning of behavioral specifications, a structural description of the system can be achieved. Finally, the consideration of implementation constraints allows the transformation of the structural description to an implementation description, which is close to the final system implementation.

The entire process is guided by nonfunctional attributes, such as timing constraints, cost, power dissipation, and reliability issues. This view of Hardware/Software systems provides the basic framework of our approach. Our central model is the behavioral model, because it constitutes the first executable model of the system during design. Behavioral models can also be viewed upon as executable specifications of systems. They are also central in the sense that they are more conducive to analysis and estimation than, for instance, conceptual models, which lack the semantic completeness of behavioral models. The behavioral models are based on the concept of Parallel Random Access Machines (PRAMs), which will be described in section 2. In section 3, we describe our approach in detail.

In section 4, we describe two examples. In this section we also explain the use of existing tools to support our method. Section 5 contains our results and conclusions.

2   Parallel Random Access Machines

In this section, we introduce a new type of behavioral models for the design of Hardware/Software systems: parallel random access machines (PRAM) [Hopc79] [Gura89]. Random access machines (RAM) are abstract computer models which may simulate any existing computer.

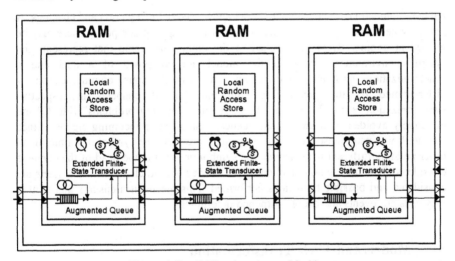

Figure 4. Parallel Random Access Machine

The RAM formalism is as powerful as the Turing formalism. A proof is given in Theorem 7.6, p.166-167 in [Hopc79]. RAM works somewhat like a Turing Machine, but its allowable operations resemble more closely than of an actual programming language. It consists of a Finite State Machine (FSM) and some finite random access store. Since RAMs are computationally equivalent to bounded-tape Turing machines, any existing computer or algorithm can be described using RAMs. This is especially advantageous because powerful theoretical methods exist to analyze Turing transducers [Hopc79].

As shown in Figure 4, the PRAMs input/output frame encapsulates several RAMs and thereby supports an explicit model of the environment in which the designed system is assumed to function. RAMs encapsulated in the I/O frame consist of an extended finite-state transducer (EFSM), local storage (STORE) and one or more input/output ports. Transducer extensions relate to the handling of external activities such as program start and termination (the start of programs while being in a state or during state-transition and signal exchange between transducer and program), access of the local store and state-history management, postponement of signals and events, and timing related behavior and timing-conditions.

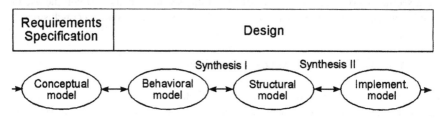

Figure 5. Model Based System Design

All communication activities in the model are carried out using ports and channels. A port may be connected via a channel to one or more appropriate ports. These may be located at another RAM or attached to the embracing frame that represents the system's environment. Ports in the model are composed of a data and a synchronization terminal. Synchronous and asynchronous communication is differentiated by the presence or the omission of connects at respective synchronization terminals. So called "short pulse catching problems" [Flet80] are effectively avoided with augmented queues, which buffer incoming signals. Since parallel finite-state transducers may consume more than one symbol at a time, in an order possibly different from arrival, the symbol or event at the head of the queue is considered first. In case no transition can be carried out, all elements in the queue are considered and possible transitions performed. This mechanism ensures that neither messages nor events are lost or postponed indefinitely.

3   Models and Their Transformation

Based on our view of Hardware/Software systems, we introduce conceptual models, behavioral models, structural models, and implementation models as synchronization points in the design process. As shown in figure 5 the design process is defined as a sequence of transformations between individual models, while every model in turn is subject to a stepwise refinement process. Note that the individual refinement steps are important to our approach. They considerably facilitate the transformation steps, because we do not require the transformation of an input model into a final output model, in which case the whole process would boil down to a nonfeasible sequence of automatic synthesis steps. Instead we perform - relatively easy - transformations of input models to initial output models. The conversion to the final output model is done by designers. The conceptual model corresponds to the initial specification of the system. It consists of a directed graph, where the nodes correspond to functions, and the arcs correspond to the hierarchical relations (figure 7) and the data dependencies (figure 8) between the functions. The nodes are hierarchical. The elements of the graph can be augmented with functional and non-functional textual attributes. Usually, functional requirements are related to single elements in the graph, while non-functional requirements are attached to sets of elements. Because of the clearness and comprehensibility of the description it is advantageous to introduce hierarchy into the model. So it is possible to divide functions into subfunctions and data dependencies into subdependencies.

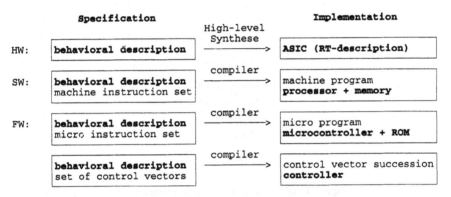

Figure 6. Ways of implementation

Once the conceptual model is finished, the transformation to a behavioral PRAM model is relatively easy: Each function is translated to a process represented by a PRAM, while each data dependency is translated to a message represented by a signal carried by a channel with terminals (ports) on each end. For buffering of incomming signals a queue is attached to every input port of a PRAM. So each function is characterized through an extended finite state machine in form of a PRAM and a set of input and output signals. The input signals are used to sychronize the work of the PRAMs and to transmit data. Because of the queues at the input ports of the PRAMs no signal can be lost. Subsequently, the PRAM model will be stepwise refined. At this moment it is not determined which part of the system will be implemented in hardware, software or firmware. Some additional requirements related to functions define characteristics of PRAMs, others, e.g. power dissipation, are just handed over to the behavioral model for later use. The behavioral model is the first executable model in the design process. So it is possible to verify the behavior of the system at a high level of abstraction through direct simulation or translation into C or VHDL.

The structural model provides the basis for the architecture of the system. It is essentially equivalent to the behavioral model. The structural model consists of implementation-independend design units and communication channels. The communication channels execute necessary datatype conversions.

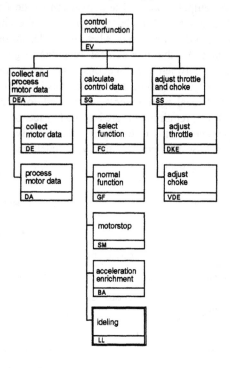

Figure 7. AKL functional tree of the motor control

The transformation of a behavioral model into a structural model (synthesis I) is done by a partitioning of the entire system through a sequence of combination steps that is guided by the still unresolved requirements. In this process behavioral elements are grouped together to form units and signals are combined to communication channels between this units. The structural model is the basis for the analysis and estimation of the model. It is also used to decide which part of the system should be realized as hardware and which part should be implemented as software or firmware.

The final model, the implementation model, is a collection of Hardware and/or Software modules. At this point, all requirements must be resolved, otherwise an error has occurred, and a trace-back will be necessary. The transformation of the structural model into an implementation model (synthesis II) consists of two steps:

- the decision whether a module will be implemented in Hardware or Software
- if possible, reuse Hardware and Software components from existing libraries.

As shown in figure 6 there are four different possibilities for the realization of components. The first way to implement the system is that all components are realized as digital hardware. As far as possible standard hardware components from existing libraries should be used. But normaly it is necessary to create additional application specific hardware components (ASICs). This is done by a synthesis step. The communication between the components is realized via special communication hardware synthesized from the communication channel descriptions. Secondly it is possible to implement all components exclusively as software. If this way is gone a standard or nearly standard hardware system with a known machine instruction set may be used as basis for the implementation. The communication between the software modules can be implemented in different ways but always as software. The third possibility for the realization of the system is the implementation as firmware. It is similar to the software implementation. The only difference is that there is a micro instruction set or a set of control vectors instead of a machine instruction set.

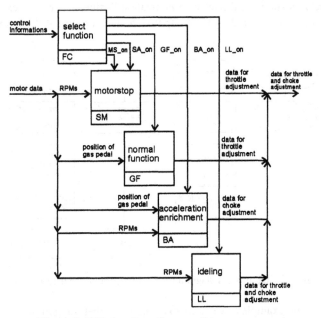

Figure 8. AKL data flow diagram of the motor control

The fourth and most common case is a heterogeneous implementation of the system. Hardware, firmware and software components itself are realized as shown above. The software components are based on a hardware consisting of a standard processor, memory and additional units e.g. I/O subsystems. The basis of the firmware components are controllers and ROMs. Because of this facts the communication between the hardware, firmware and software components of the system can be reduced to the communication between hardware components.

Without a detailed explanation should be pointed out that an analysis of the respective model has to be done in all phases of the design process (see figure 3). In the design steps linked with the structural and the implementation model functional and timing correctness as well as performance and the keeping of given boundary conditions and constraints is checked. Correctness and performance are determined through simulation.

4 Examples

To demonstrate the usefulness of our approach, we chose two examples. The first is an electronic motor control unit. We settled for this example, because it represents typical projects in the industrial world. The motor-control unit (called Ecotronic) is a closed-loop system that consists of a controlled engine, several sensors, and a micro-processor based brain-box.

The second example is a generic but simple telephone switch that is intended for the use in relay stations in the telecommunication business. For both examples, we developed the models according to our approach, and carried out the transformation steps.

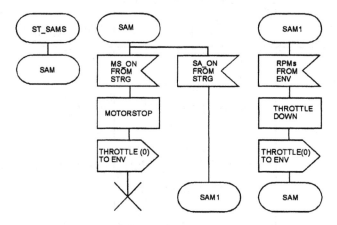

Figure 9. SDL process diagram for a subfunction of the motor control

To carry out our examples, we used existing tools that support our models with varying degree. For the capture of the conceptual model and its transformation into a PRAM model we used KLAR [Klar] and DAX [Dax], which were developed by Siemens. The behavioral models were developed with Statemate [Seif91], $Matrix_x$ [Mtx] and SDL [Sdl]. The transformation from the behavioral to the structural level was done manually. For the final transformation step we used the code generators of SDL and Statemate, a hardware synthesis system, and SIDECON, a system for the configuration of computers.

In this paper we only want to present a part of the example "motor-control unit". The complete description is published in [Kahl92]. Figure 7 shows the AKL functional tree diagram as result of the first step of the requirements specification. It defines the decomposition of the systems function into subfunctions. Each sub-function will be realized either as a hardware or a software component. The interfaces between the components are defined in an AKL data flow diagram (see figure 8). The next step is the transformation of the AKL model into a SDL description under use of the tool DAX (DAta eXchange). This transformation is based on the leaves of the functional tree and the related AKL flow diagrams. Every function is transformed automatically into a SDL task symbol, every incomming data flow into a SDL input symbol and every outgoing data flow into a SDL output symbol. The designer has to order the symbols and to add states manually. Figure 9 is the part "motorstop" of the resulting SDL description of the "motor-control" PRAM that describes the behavior of the motor-control unit. This SDL description can be translated into VHDL code for simulation with the SYNOSYS VHDL simulator.

5   Conclusions and Future Work

We have developed a new approach for the design of Hardware/Software systems. Its usefulness was shown with two example designs. By exercising the examples we noticed that we added structure to the design process at the cost of losing some flexibility. This is, however, not a major drawback, because the additional structure helped to manage the projects. Our method is quite well supported in the area of functional and behavioral models, but the handling of requirements needs improvement. The transformation of behavioral models to structural models is so far a manual step, although there exist some ideas to support this step with tools. This is an area of future work. Another area of future work is the general handling of requirements in the process.

References

[Buch92]   K. Buchenrieder, St. Kahlert, D.Monjau. Methodik und Werkzeuge für den Entwurf komplexer heterogener Systeme. In: Proceedings der ITG/GME/GI-Fachtagung Rechnergestützter Entwurf und Architektur mikroelektronischer Systeme. Darmstadt. November 1992. In ITG Fachbericht 122. vde-Verlag. Berlin (in German).

[CCITT92] CCITT Recommendation Z.100: Specification and Description Language (SDL). Geneva, 1992.

[Dax]   DAX - User Manual. Siemens AG. Munich, 1991.

[Flet80]   W. Fletcher. An Engineering Approach to Digital Design. Prentice-Hall. Englewood Cliffs, N.J., 1980.

[Gura89]   E. Gurari. An Introduction to the Theory of Computation. Computer Science Press. Rockville, 1989.

[Hare86]   D. Harel. On visual formalisms. Communications of the ACM 31(5):514-530, May 1986.

[Hoar85]   C.A. Hoare. Communicating Sequential Processes. Prentice Hall. Englewood Cliffs, N.J., 1985.

[Hofm91]   F. Hofmann. Betriebssysteme: Grundkonzepte und Modellvorstellungen. Leitfäden der Informatik. Teubner Verlag. Stuttgart, 1991 (in German).

[Hopc79]   J. Hopcroft and J. Ullmann. Introduction to Automata Theory. Languages and Computation. Addison-Wesley, 1979.

[Kahl92]   St. Kahlert. Systementwurf mit einheitlichen Beschreibungsmitteln für Hard- und Software. Diplomarbeit. Siemens AG, ZFE BT SE 52 / TU Chemnitz, Fachbereich Informatik, September 1992 (in German).

[Klar]   SIGRAPH-SET-KLAR - Ein Werkzeug zur Aufgabenklärung. Benutzeranleitung. Siemens AG. München, Juni 1991 (in German).

[Mtx] $Matrix_x$/SystemBuild - Version Description Document for Version 2.4. Integrated Systems Incorporation. Santa Clara, California, 1991.

[Sdl]   SIGRAPH-SET-SDL - Werkzeuge für Systementwurf und Implementierung. Benutzeranleitung. Siemens AG, Mai 1988 (in German).

[Seif91]   M. Seifert. Statemate: a new method for the design of complex systems. Proceedings of the Echtzeit '91 Conference on Real-Time Systems, June 1991 (in German).

Towards an "Erlangen Program" for General Linear Systems Theory

Part II: Space-time of $\mathcal{D}$-stationary systems

Reiner Creutzburg[1], Valerij G. Labunets and Ekaterina V. Labunets[2]

[1] University of Karlsruhe, Institute of Algorithms and Cognitive Systems
P. O. Box 6980, D–7500 Karlsruhe 1, Germany
phone: +49–721–608 4325, fax: +49–721–696 893)
e-mail: creutzbu@ira.uka.de
[2] Ural Polytechnical Institute, Faculty of Radioelectronics
Department of Automatics and Telemechanics
620002 Jekaterinburg, Russia
phone: +7–3432–449 779, fax: +7–3432–562 417

Abstract. This is the second part in the series of papers written on the algebraic foundation of an abstract harmonic analysis based on a generalized symmetry principle [LLC91a,LLC91b,LLC92]. The first part of this work [CLL92] was devoted to the theory of generalized shift operators (GSO), an algebraic classification of signals and systems, and a generalized harmonic analysis of signals and systems.
This second part deals with the space-time of $\mathcal{D}$-stationary systems.

1 Introduction – Space-time in the theory of signals and systems

A replacement in the process of analyzing $\mathcal{D}$-stationary systems (where $\mathcal{D} = D, R$ or L denotes a generalized shift operator (GSO)) leads to far-going consequences, since both the analysis and synthesis of linear systems are based on these decompositions and Fourier transformations: operational method, Wiener-Chinchin transformation the concept of impedance, transfer function, bandpasses, indeterministic functions etc. Therefore, each time we pass over to linear $\mathcal{D}$-system with new eigenfunctions we shall have to make up the mentioned software a new. To overcome this inconvenience in the present part of our work it is offered to reduce the $\mathcal{D}$-stationary linear dynamical system (LDS) to its stationary (in the common sense) equivalent LDS by changing the time scale and amplitude scale.

This allows, firstly, to apply all the well-known mathematical methods of classical Fourier transform to $\mathcal{D}$-stationary LDS analysis and, secondly, to prove the relativity of the notions "space" and "time" in the theory of non-stationary and non-homogenous spatio-temporal LDS.

The concept of systemic space and time (or eigenspace and eigentime) was evidently first introduced by *H. Bergson* [Ber29], and later it was developed into the theory of "biological space-time". The first scientist who formulated all basic principles of this theory was *V. I. Vernadsky* [Ver32]. He believed that one of *E. Cartan* geometries could be the geometry of that space. It is possible that there insight exist more than one "biological space-time". The physical calendar time is usually measured by steady even flow of events. From this standpoint it is possible to suppose that within a certain closed system (or organism) time is also measured by a number of monotonous events repeating one after another in one flow. The "inside" and "outside" observer's judgement of the uniformity of the events may be different. Only in this case the inside biological time can become non-homogenous; this non-homogeneity of inside biological time, as different from homogeneous physical time, can be interpreted as eigentime of the system. Within which the uneven (from the "outside" observer's point of view) distribution of uniform events exists. For instance, an existing hypothesis suggested that biological time t_δ is a logarithmic function of physical time $t : t_\delta = c \log(t)$, where c is constant.

The concept of "systems space-time" was first introduced into technical literature by *A. S. Vinitsky* [Vin69]. When he analyzed oscillation contours with periodically changing parameters he pointed out some of the features typical for such systems. Firstly, outside effects can be resonance for periodical contours only in case they reproduce witch constant factor accuracy the form of their own oscillation in the contour (the form of Hill's functions) without losses. Secondly, in steady state periodical contours extract from arbitrary outside influence Hill's functions (not harmonic constituents), bringing them into "independent existence". These two effects change the concept of sinusoid as universal elementary oscillation. That is why when analyzing periodical contours it is necessary to use periodical contour eigenfunction decomposition instead of harmonic decomposition. Thus we can introduce the concept of generalized resonant curve (transfer function), which differs from the usual resonant curve in that when it is taken harmonic signals are replaced by Hill's functions characteristic of the contour under analysis.

The response of any linear system to the elementary external influence takes the form of its free oscillation. The free oscillation period is a natural time scale, which can be used to measure the rhythm of all processes inside the contour; there is no other time "standard" except its free oscillation period. In harmonic contours reactive parameters L and C are constant, its free oscillation frequency being also constant. That is why they usually have regular intervals between counts of the processes inside the contour in the absolute time scale t the modulated contour the parameters L and C as well as the free oscillation period are constantly changing according to a complicated law. In the "coordinate system" of the contours its free oscillation period is perceived as constant scale to which the rhythm of all processes inside the contour is compared. Inside such contours only processes which "match" the rhythm of their free oscillation are perceived as regular.

In this connection it is useful to introduce the time scale graduated with the oscillation periods of the contours. Let us call it the eigentime scale of the contour. It is clear that the time scale which is reduced to the modulated contour will never match the absolute time scale. Alongside the eigentime scale it is commonly necessary to give the contour an unequally sampled amplitude eigenscale which is directly connected with the first one and in relation to which the oscillation amplitude of the contour is perceived as constant [Vin69]. For example, let the modulation laws of inductance and its capacitance be represented as

$$L(t) = L_0 + \Delta L(t) = L_0[1 + m_l l(t)],$$

$$C(t) = C_0 + \Delta C(t) = C_0[1 + m_c c(t)].$$

In this case the formulas that link absolute (calendar) time and eigentime (systemic time) also physical and amplitude eigenscales will take the following form

$$t_{sys} = \frac{1}{\omega_0} \int_0^t \omega(t) dt, \quad A_{sys} = A(t) \sqrt{\frac{\omega(t)}{\omega_0}},$$

where

$$\omega_0 = \sqrt{\frac{1}{LC}} \quad \text{and} \quad \omega(t) = \sqrt{\frac{1}{L(t)C(t)}}$$

are instantaneous and carrier oscillation frequencies of the contour. Substitution of t_{sys} and A_{sys} in the equations of Hill functions changes them into simple harmonic oscillations.

Thus, Hill functions in systemic space-time look like ordinary harmonic oscillations and periodical contour is reduced to harmonic one (the system with constant parameters).

In this paper we shall be demonstrating that *A. S. Vinitsky's* theory can be applied without great changes to linear dynamical systems, which are invariant with respect to some "exotic" (generalized) shift operators [LLC91a], [LLC91b].

The classical approach to the theory of stationary LDS describing the input/output behaviour of LDS in terms of the black-box model essentially bases on the following three isomorphism methods:

- Local methods on the basis of a differential equation with constant coefficient:

$$\sum_{k=0}^{n} a_k \frac{d^k}{dt^k} y(t) = \sum_{l=0}^{m} b_l \frac{d^l}{dt^l} x(t),$$

where $x(t)$ is input and $y(t)$ is output of a system.
- Global methods on the basis of an integral equation of the convolution type:

$$y(t) = \int_{-\infty}^{+\infty} h(t - \tau) x(\tau) d\tau,$$

where $h(t - \tau)$ is the impulse response of the stationary LDS.

– Frequency methods on the basis of frequency representation of input/output signals and impulse response:

$$Y(\omega) = H(\omega)X(\omega),$$

where $Y(\omega) = \mathcal{F}\{y(t)\}$, $X(\omega) = \mathcal{F}\{x(t)\}$, $H(\omega) = \mathcal{F}\{h(t)\}$.

Connections between these methods are illustrated in Fig. 1 and reflect the properties of the Fourier transform.

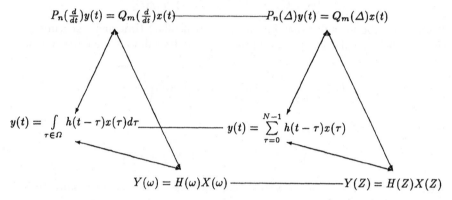

Fig. 1.: Connection between methods describing input/output behaviour of continuous and discrete stationary LDS.

From Fourier transform theory it is known that if a signal $x(t)$ has the spectrum $X(\omega)$, then its differentiation in the spectral domain is equivalent to a spectrum multiplication by $i\omega$:

$$\frac{d^n}{dt^n}x(t) = \mathcal{F}^{-1}\left[(i\omega)^n X(\omega)\right]. \tag{1}$$

For such differentiation operators the harmonic signals are eigenfunctions, corresponding to the eigenvalues $(i\omega)^n$:

$$\frac{d^n}{dt^n}\left[\exp(i\omega t)\right] = (i\omega)^n \exp(i\omega t). \tag{2}$$

This property lies on the basis of the algebraic approach to differential equations with constant coefficents, that are a well-known description of the behaviour of stationary LDS. In fact, if

$$\sum_{k=0}^{n} a_k \frac{d^k}{dt^k}y(t) = \sum_{l=0}^{m} b_l \frac{d^l}{dt^l}x(t) \tag{3}$$

or

$$P_n\left(\frac{d}{dt}\right)y(t) = Q_m\left(\frac{d}{dt}\right)x(t),$$

are differential equations, where

$$P_n\left(\frac{d}{dt}\right) := \sum_{k=0}^{n} a_k \frac{d^k}{dt^k}, \quad Q_m\left(\frac{d}{dt}\right) := \sum_{l=0}^{m} b_l \frac{d^l}{dt^l},$$

then in the spectral domain we obtain instead of these the algebraic equation

$$\sum_{k=0}^{n} a_k (i\omega)^k Y(\omega) = \sum_{l=0}^{m} b_l (i\omega)^l X(\omega)$$

or

$$P_n(i\omega)Y(\omega) = Q_m(i\omega)X(\omega). \tag{4}$$

On the other hand if

$$y(t) = \int_{-\infty}^{+\infty} h(t-\tau)x(t)d\tau,$$

then

$$Y(\omega) = H(\omega)X(\omega) \tag{5}$$

and it follows

$$\frac{Y(\omega)}{X(\omega)} = \frac{P_n(i\omega)}{Q_m(i\omega)} = H(\omega).$$

This expression is called the transition function of the stationary LDS.

Analogous connections exist between description methods of discrete stationary LDS. For $\mathcal{D}$-stationary LDS we work out two last methods. In this chapter we attend to the first method.

Let $\{\varphi_\alpha(t) | \alpha \in \Omega^*\} \in L(\Omega, A)$ be the basis which generates the family of commutative GSOs $\{D_t^\tau | \tau \in \Omega\}$.

Definition 1 *An operator $\mathcal{L}$ for which the equation*

$$\mathcal{L}\varphi_\alpha(t) = \alpha\varphi_\alpha(t) \tag{6}$$

is held for all $\alpha \in \Omega^*$, is called the generalized differential operator (GDO).

Notice that in the general case the generalized differential operator appears as an ordinary differential operator with variable coefficients

$$\mathcal{L} = \frac{d}{dt}, \qquad \mathcal{L} = \frac{d}{dt}t\frac{d}{dt}, \qquad \mathcal{L} = t^2\frac{d^2}{dt^2} + (\lambda + 1 - t)\frac{d}{dt},$$

etc.

Let us now find a connection between the GSOs D_t^τ and GDO $\mathcal{L}$. Almost always this differential operator is of second order:

$$\mathcal{L} = p_2(t)\frac{d^2}{dt^2} + p_1(t)\frac{d}{dt} + p_0(t), \tag{7}$$

where $p_2(t)$, $p_1(t)$, $p_0(t)$ are some variable coefficients.

It can be found on the basis of development into a Taylor series. It is well known that for ordinary shifts holds

$$D_t^s f(t) := f(t+s) = \sum_{k=0}^{\infty} \frac{s^k}{k!} (\frac{d}{dt})^k f(t) = \{\sum_{k=0}^{\infty} \frac{s^k}{k!} (\frac{d}{dt})^k\} f(t). \qquad (8)$$

This expression represents the decomposition of ordinary finite shifts into a series of terms (degrees) of differential operators $\frac{d}{dt}$. Note that the harmonic signals $\mathrm{EXP}(i\alpha t)$ are the eigenfunctions of differential operators:

$$\frac{d}{dt} \mathrm{EXP}(i\alpha t) = i\alpha \mathrm{EXP}(i\alpha t).$$

The eigenfunctions $\mathrm{EXP}(i\alpha t)$ are decomposed into a series of α

$$\mathrm{EXP}(i\alpha t) = \sum_{k=0}^{\infty} \frac{(it)^k}{k!} \alpha^k. \qquad (9)$$

Comparing (8) and (9) one has

$$D_t^s = \sum_{k=0}^{\infty} \frac{s^k}{k!} (\frac{d}{dt})^k = \mathrm{EXP}(is\frac{d}{dt}). \qquad (10)$$

Let us now generalize this construction with harmonic signals and ordinary shifts on an arbitrary generalized harmonic signal and generalized shifts. If a generalized harmonic signal $\varphi_\alpha(t)$ is decomposed into a series of α

$$\varphi_\alpha(,t) = \sum_{k=0}^{\infty} X_k(t)\alpha^k, \qquad (11)$$

then, according to (9) and (10), the expression

$$D_t^s := s(\mathcal{L}) = \sum_{k=0}^{\infty} X_k(s)\mathcal{L}^k$$

defines the family of GSOs $\{D_t^s\}_{s \in \Omega}$, generated by the operator $\mathcal{L}$.

Indeed

$$D_t^s \varphi_\alpha(t) = \sum_{k=0}^{\infty} X_k(s)\mathcal{L}^k \varphi_\alpha(t)$$

$$= \sum_{k=0}^{\infty} X_k(s)\alpha^k \varphi_\alpha(t)$$

$$= \varphi_\alpha(s)\varphi_\alpha(t) = \varphi_\alpha(t \oplus s).$$

Definition 2 *Expressions of the form*

$$\sum_{i=0}^{n} a_i \mathcal{L}^i y(t) = \sum_{j=0}^{m} b_j \mathcal{L}^j x(t), \tag{12}$$

$$\sum_{i=0}^{N-1} a_i \mathcal{D}_t^i y(t) = \sum_{j=0}^{N-1} b_j \mathcal{D}^j x(t), \tag{13}$$

or for the simplicity

$$P_n(\mathcal{L})y(t) = Q_m(\mathcal{L})x(t), \tag{14}$$

$$P_n(\mathcal{D})y(t) = Q_m(\mathcal{D})x(t), \tag{15}$$

*where $x(t)$, $y(t)$ are scalar input and output LDS, $P_n(\mathcal{L}) := \sum_{i=0}^{n} a_i \mathcal{L}^i$, $P_n(\mathcal{D}) :=$
$\sum_{i=0}^{n} a_i \mathcal{D}_t^i$, $Q_m(\mathcal{L}) := \sum_{j=0}^{m} b_j \mathcal{L}^j$, $Q_m(\mathcal{D}) := \sum_{j=0}^{m} b_j \mathcal{D}_t^j$, $\mathcal{D} = D$, R or L are called
generalized $\mathcal{L}$-differential and generalized $\mathcal{D}$-difference equations, respectively.*

These equations describe the input/output behaviour $\mathcal{D}$-stationary LDS. Indeed, substitution of an expression

$$y(t) = \sum_{\alpha \in \Omega^*} Y(\alpha)\varphi_\alpha(t)d\mu(\alpha)$$

$$x(t) = \int_{\alpha \in \Omega^*} X(\alpha)\varphi_\alpha(t)d\mu(\alpha)$$

in (12) gives

$$Y(\alpha)\sum_{i=0}^{n} a_i \alpha^i = X(\alpha)\sum_{j=0}^{m} b_j \alpha^j, \tag{16}$$

i. e. the F-transform carries generalized $\mathcal{L}$-differential equations into algebraic equations. From (16) we obtain expressions of the generalized transition function

$$H(\alpha) = Y(\alpha)/X(\alpha) = \sum_{j=0}^{m} b_j \alpha^j / \sum_{i=0}^{n} a_i \alpha^i = Q_m(\alpha)/P_n(\alpha).$$

Obviously $\mathcal{D}$-stationary LDS belong to systems with variable parameters and hence their behaviour can be described with the help of ordinary differential and difference equations:

$$\sum_{i=0}^{n'} a_i(t)\frac{d^i}{dt^i}y(t) = \sum_{j=0}^{m'} b_j(t)\frac{d^j}{dt^j}x(t), \tag{17}$$

$$\sum_{j=0}^{N-1} a_i(t)y(t-i) = \sum_{j=0}^{N-1} b_j(t)x(t-j), \tag{18}$$

where $n' \geq n$, $m' \geq m$. The set $\{a_i(t), b_j(t)\}$ obviously determines time-structure (time-symmetry) of $\mathcal{D}$-stationary LDS.

2 Space-time structure of $\mathcal{D}$-stationary onedimensional systems

The most simplicity is determined by coefficients $a_i(t)$, $b_j(t)$ of $\mathcal{L}$-differential equations. Indeed, as

$$\mathcal{L} = p_0(t)\frac{d^2}{dt^2} + p_1(t)\frac{d}{dt} + p_0(t),$$

then

$$\sum_{i=0}^{n} a_i(p_2(t)\frac{d^2}{dt^2} + p_1(t)\frac{d}{dt})^i = \sum_{i=0}^{n'} a_i(t)\frac{d^i}{dt^i}.$$

Calculating the left part of the last equation we obtain expressions of $a_i(t)$. Let us now find the connection between coefficients of difference equations (13) and (18). For the beginning we introduce the definitions of D-difference equations (18).

Definition 3 *Expressions of the form*

$$\sum_{i=0}^{N-1} a_i y(t \ominus i) = \sum_{i=0}^{N-1} b_j x(t \ominus j), \tag{19}$$

$$\sum_{i=0}^{N-1} a_i y(t \oslash i) = \sum_{i=0}^{N-1} b_j x(t \oslash j), \tag{20}$$

$$\sum_{i=0}^{N-1} a_i y(i \oslash t) = \sum_{i=0}^{N-1} b_j x(j \oslash t), \tag{21}$$

where $x(t), y(t)$ are scalar input and output LDS are called D^*-, R^*-, L^*- stationary equations with constant coefficients, respectively.

At first we will consider for simplicity the right parts of (19)-(21) as one function $u(t)$. Then the equations (19)-(21) assume the following form

$$\sum_{i=0}^{N-1} a_i y(t \ominus i) = \left[\sum_{i=0}^{N-1} a_i(D_t^i)^+\right] y(t) = P_{N-1}(D_t)y(t) = u(t), \tag{22}$$

$$\sum_{i=0}^{N-1} a_i y(t \oslash i) = \left[\sum_{i=0}^{N-1} a_i(R_t^i)^+\right] y(t) = P_{N-1}(R_t)y(t) = u(t), \tag{23}$$

$$\sum_{i=0}^{N-1} a_i y(t \oslash i) = \left[\sum_{i=0}^{N-1} a_i(L_t^i)^+\right] y(t) = P_{N-1}(L_t)y(t) = u(t), \tag{24}$$

where

$$P_{N-1}(D_t) := \sum_{i=0}^{N-1} a_i(D_t^i)^+, \quad P_{N-1}(R_t) := \sum_{i=0}^{N-1} a_i(R_t^i),$$

$$P_{N-1}(L_t) := \sum_{i=0}^{N-1} a_i(L_t^i).$$

We will solve these equations using the techniques of Green's functions.

Theorem 1 *Suppose that the functions $h(t \ominus \tau)$, $h(t \oslash \tau)$, $h(\tau \oslash t)$ exist such that*

$$P_{N-1}(D_t)h(t \ominus \tau) = \delta(t, \tau), \quad P_{N-1}(R_t)h(t \oslash \tau) = \delta(t, \tau),$$

$$P_{N-1}(L_t)h(\tau \oslash t) = \delta(t, \tau).$$

Then

$$y(t) = \sum_{\tau=0}^{N-1} h(t \ominus \tau)u(\tau) + y_0(t), \tag{25}$$

$$y(t) = \sum_{\tau=0}^{N-1} h(t \oslash \tau)u(\tau) + y_0(t), \tag{26}$$

$$y(t) = \sum_{\tau=0}^{N-1} h(\tau \oslash t)u(\tau) + y_0(t) \tag{27}$$

are the solution of the equations (22)–(24), where

$$P_{N-1}(D_t)y_0(t) = P_{N-1}(R_t)y_0(t) = P_{N-1}(L_t)y_0(t) = 0.$$

Proof. Obviously, the substitution of (25) into (22) gives

$$P_{N-1}(D_t)y(t) = P_{N-1}(D_t)\left[\sum_{\tau=0}^{N-1} h(t \ominus \tau)u(\tau)\right] + P_{N-1}(D_t)y_0(t)$$

$$= \sum_{\tau=0}^{N-1} \delta(t, \tau)u(\tau)x(\tau) = u(t).$$

In analogy the proof for the other equations can be found. ∎

Theorem 2 (LS76) *The variable coefficients $a_i(t)$ of the ordinary difference equation (22)–(24) are expressed through constant coefficients a_i of equations (19)–(24), like in*

$$a_i(t) = \sum_{j=0}^{N-1} a_j \sum_{\alpha \in \Omega^*} \varphi_\alpha(t - i)\bar{\varphi}_\alpha(j)\varphi_\alpha(t), \tag{28}$$

$$a_i(t) = \sum_{j=0}^{N-1} a_j \sum_{\lambda \in \Omega^*} \text{tr}\left[\Phi^\lambda(t - i)(\Phi^\lambda)^+(j)\Phi^\lambda(t)\right], \tag{29}$$

$$a_i(t) = \sum_{j=0}^{N-1} a_j \sum_{\lambda \in \Omega^*} \text{tr}\left[(\Phi^\lambda)^+(t - i)\Phi^\lambda(t)\Phi^\lambda(j)\right]. \tag{30}$$

Similar relations also hold for the coefficients $b_i(t)$ and b_i

$$a_i(t) = \sum_{j=0}^{N-1} a_j D_t^{t-i}(j), \qquad b_i(t) = \sum_{j=0}^{N-1} b_j D_t^{t-i}(j), \tag{31}$$

$$a_i(t) = \sum_{j=0}^{N-1} a_j R_t^{t-i}(j), \qquad b_i(t) = \sum_{j=0}^{N-1} b_j R_t^{t-i}(j), \tag{32}$$

$$a_i(t) = \sum_{j=0}^{N-1} a_j L_t^{t-i}(j), \qquad b_i(t) = \sum_{j=0}^{N-1} b_j L_t^{t-i}(j). \tag{33}$$

The **proof** is given in [LLC92].

The expressions (31)-(33) can be described in the more compact form

$$a_i(t) = a_{t \ominus (t-i)}, \qquad b_i(t) = b_{t \ominus (t-i)}, \tag{34}$$

$$a_i(t) = a_{t \oslash (t-i)}, \qquad b_i(t) = b_{t \oslash (t-i)}, \tag{35}$$

$$a_i(t) = a_{(t-i) \oslash t}, \qquad b_i(t) = b_{(t-i) \oslash t} \tag{36}$$

Hence the knowledge of the matrix representations of the GSO, or the knowledge of basis functions, or finally the knowledge of structure constants of the underlying algebras plays a fundamental role for the application of the described methods.

If characters or irreducible representations of concrete groups are substituted into the three equations of (28)-(30) instead of basis functions, then expressions for variable coefficients will be obtained and its periodicity totally depends on the underlying group structure.

It is interesting to find generalized differentiation operators for arbitrary characters of an arbitrary finite abelian group of m-adic numbers

$$\mathbb{H} = (\mathbf{A}\mathbf{Z}/m) \oplus (\mathbf{A}\mathbf{Z}/m) \oplus \ldots \oplus (\mathbf{A}\mathbf{Z}/m)$$

of order $N = m^n$.

Theorem 3 (Lab80) *Let*

$$\varphi_\alpha(t) := \varphi_{\alpha_1 \alpha_2 \ldots \alpha_n}(t_1, t_2, \ldots, t_n) = \epsilon^{\alpha_1 t_1 + \alpha_2 t_2 + \ldots \alpha_n t_n},$$

where $\epsilon_n = \sqrt[m]{1} \in A$; $\alpha = (\alpha_1, \alpha_2, \ldots, \alpha_n)$ and $t = (t_1, t_2, \ldots, t_n)$ denote the n-digit radix-m-representations of numbers $\alpha, t \in [0, N-1]$. Then

$$\mathcal{L}\{x(t)\} = m^n \sum_{i=1}^{n} \sum_{\tau_i=1}^{N_i-1} \left[x(0) - (m-1)x(t \underset{m}{\ominus} \tau_i m^{i-1}) \right] \left[m^i/(1 - \epsilon^{t_i}) \right]. \tag{37}$$

Analogously one can prove the theorem for an arbitrary abelian group

$$H = (\mathbf{A}\mathbf{Z}/N_1) \oplus (\mathbf{A}\mathbf{Z}/N_2) \oplus \ldots \oplus (\mathbf{A}\mathbf{Z}/N_n)$$

of order $N = N_1 N_2 \ldots N_n$.

Theorem 4 *Let*

$$\varphi_\alpha(t) := \varphi_{\alpha_1, \alpha_2, \ldots, \alpha_n}(t_1, t_2, \ldots, t_n) = \epsilon_1^{\alpha_1 t_1} \epsilon_2^{\alpha_2 t_2} \ldots \epsilon_n^{\alpha_n t_n},$$

where $\epsilon_n = \sqrt[N]{1} \in A$; $\alpha = (\alpha_1, \alpha_2, \ldots, \alpha_n)$ *and* $t = (t_1, t_2, \ldots, t_n)$ *denote the n-digit radix-m-representations of numbers* $\alpha, t \in [0, N-1]$ *in the* $(N_1, N_2, \ldots, N_n)$*-mixed-radix representation. Then*

$$\mathcal{L}\{x(t)\} = N \sum_{i=1}^{n} \sum_{\tau_i=1}^{N_i-1} \left[x(0) - (N_i - 1)x(t \underset{N_i}{\ominus} \tau_i N_{(i-1)!}) \right] \left[N_{i!}^{-1}/(1 - \epsilon^{t_i}) \right], \quad (38)$$

where $N_{(i-1)!} := N_1 N_2 \ldots N_{i-1}$ *and* $N_{i!}^{-1} := N_1^{-1} N_2^{-1} \ldots N_i^{-1}$.

Corollary 1 *If* $N_1 = N_2 = \ldots = N_n = m$, *then*

$$\mathcal{L}\{x(t)\} = m^n \sum_{i=1}^{n} \sum_{\tau_i=1}^{N_i-1} \left[x(0) - (m - 1)x(t \underset{m}{\ominus} \tau_i m^{i-1}) \right] \left[m^i/\{1 - \epsilon^{t_i}\} \right]$$

Corollary 2 *. If* $N_1 = N_2 = \ldots = N_n = 2$, *then*

$$\mathcal{L}\{x(t)\} = 2^n \sum_{i=1}^{n} \left[x(0) - x(t \underset{2}{\ominus} \tau_i 2^{i-1}) \right] 2^i.$$

It is easy to see that the latter equation is a dyadic-difference of the type (22), where $a_0 = 1$, and $a_i(t) = 0$ for all $i \in [1, N-1]$, $b_0 = 1 - 2^{-n}$, $b_1 = 2^{-1}$, $b_2 = 2^{-2}$, $b_\ell = 2^{-3}$, $b_{2^{n-1}} = 2^{-n}$. The other coefficients b_i are equal to 0. The ordinary difference equation

$$y(t) = \sum_{j=0}^{N-1} b_j(t)x(t - j) = \sum_{j=0}^{N-1} b_j x(t \underset{2}{\ominus} j)$$

describing dyadic differentiating unit will have variable coefficients.

The interesting example of Baker-Prigogine dynamical systems was given in [NP87] and was studied in detail in [Pic92] and [PS93].

Let us find out what connection there is between the $\mathcal{D}$-stationary LDS and LDC with ordinary stationery.

Let Ω, and Ω_s be two similar examples of the time moment set. The elements $t \in \Omega$ and $t_s \in \Omega_s$ will be correspondingly called physical (calendar) time and systemic (eigen-)times. Similarly we introduce Ω^* and Ω_s^*, the sets of generalized frequencies $\alpha \in \Omega$ and systemic generalized frequencies $\alpha_s \in \Omega_s$. Let

$$\{\Psi_\alpha(t) \mid \alpha \in \Omega^*, \, t \in \Omega\} \text{ and } \{\varphi_{\alpha_s}(t_s) \mid \alpha_s \in \Omega_s^*, \, t_s \in \Omega_s\}$$

be two bases in the spaces $L(\Omega, A)$ and $L(\Omega_s, A)$, respectively. Let us introduce two families $\mathcal{D}_t^\tau f(t)$, $\mathcal{D}_{t_s}^{\tau_s} f(t_s)$ of GSO, using these bases and LDS, which are invariant to their influence

$$y(t) = \int_{\tau \in \Omega} [\mathcal{D}_t^\tau h(t)] \, x(\tau) d\mu(\tau), \quad y(t_s) = \int_{\tau_s \in \Omega_s} [\mathcal{D}_{t_s}^{\tau_s} h(t_s)] \, x(\tau_s) d\mu(\tau_s), \quad (39)$$

where $d\mu(\tau) = d\mu(\tau_s)$.

Theorem 5 *Suppose* $\alpha_s = \alpha$, $U(t, t_s)$ *is a transfer matrix from the basis* $\Psi_\alpha(t)$ *to the basis* $\varphi_\alpha(t_s)$:

$$\Psi_\alpha(t) = \int\limits_{t_s \in \Omega_s} \varphi_\alpha(t_s) U(t, t_s) d\mu(t_s).$$

Then the mappings

$$y(t) = \int\limits_{t_s \in \Omega} y(t_s) U(t, t_s) d\mu(t_s), \quad x(t) = \int\limits_{t_s \in \Omega_s} x(t_s) U(t, t_s) d\mu(t_s),$$

transformed to LDS, which is invariant to the influence of GSO of another type.

Proof. Indeed

$$y(t) = U y(t_s) = U \int\limits_{t_s \in \Omega_s} \left[\mathcal{D}_{t_s}^{\tau_s} h(t_s) \right] x(\tau_s) d\mu(\tau_s)$$

$$= \int\limits_{\tau_s \in \Omega_s} U \left[\mathcal{D}_{t_s}^{\tau_s} h(t_s) \right] U^{-1} U x(\tau_s) d\mu(\tau_s) = \int\limits_{\tau \in \Omega} \left[\mathcal{D}_t^{\tau} h(t) \right] x(\tau) d\mu(\tau),$$

as $U \left[\mathcal{D}_{t_s}^{\tau_s} h(t_s) \right] U^{-1} = \left[\mathcal{D}_t^{\tau} h(t) \right]$ and $U x(\tau_s) = x(t)$. ∎

For instance in the case where the basis $\{\varphi_\alpha(t_s)\}$ coincides with the Fourier basis any LDS can be reduced to a stationary.

Consequence 1 *All LDS which are invariant to the influence of GSO are reduced to stationary according to Ljapunov.*

2.1 Space-time structure of $\mathcal{D}$-stationary multidimensional systems

So far we have been considering scalar-valued signals and systems. Let us consider now vector-valued signals. Let $\mathcal{L}$ be an arbitrary differential operator, for example,

$$\mathcal{L} = \frac{d}{dt}, \qquad \mathcal{L} = \frac{d}{dt} t \frac{d}{dt}, \qquad \mathcal{L} = t \frac{d^2}{dt^2} + (\lambda + 1 - t) \frac{d}{dt}$$

etc. Let $\chi(\alpha, t)$ be the right eigenfunctions of the operator $\mathcal{L}$: $\mathcal{L}\chi(\alpha, t) = \alpha\chi(\alpha, t)$. For example, the operator $\mathcal{L} = \frac{d}{dt}$ has the eigenfunctions $\chi(\alpha, t) = \mathrm{EXP}(i\alpha t)$, since

$$\frac{d}{dt}(\mathrm{EXP}(i\alpha t)) = i\alpha \mathrm{EXP}(i\alpha t)$$

for all α. The eigenfunctions $\chi(\alpha, t)$ are decomposed into a series by α

$$\chi(\alpha, t) = \sum_{k=0}^{\infty} X_k(t) \alpha^k.$$

For example,

$$\text{EXP}(i\alpha t) = \sum_{k=0}^{\infty} \frac{(it)^k}{k!} \alpha^k.$$

In this case the expression

$$D_t^s := \sum_{k=0}^{\infty} X_k(t)\mathcal{L}^k$$

is the family of commutative GSO, generated by the operator $\mathcal{L}$. For instance,

$$D_t^s = \sum_{k=0}^{\infty} \frac{(is)^k}{k!} \left(\frac{d}{dt}\right)^k = \text{EXP}\left(is\frac{d}{dt}\right)$$

is a usual shift operator, since

$$D_t^s f(t) = \sum_{k=0}^{\infty} \frac{(is)^k}{k!} \left(\frac{d}{dt}f(t)\right)^k = f(t+s).$$

Let $\mathbf{A}$ be an arbitrary linear $(n \times n)$-matrix operator. If it replaces α in

$$\chi(\alpha, t) = \sum_{k=0}^{\infty} X_k(t)\alpha^k$$

a convergence matrix series is obtained

$$\chi(\mathbf{A}, t) = \sum_{k=0}^{\infty} X_k(t)\mathbf{A}^k.$$

For example,

$$\text{EXP}(i\mathbf{A}t) = \sum_{k=0}^{\infty} \frac{(it)^k}{k!} \mathbf{A}^k.$$

It is known that $\text{EXP}(i\mathbf{A}t)$ is the solution of the equation

$$\frac{d}{dt}(\mathbf{X(t)}) = \mathbf{AX}(t),$$

where $\mathbf{X}(t)$ is an $(n \times n)$-matrix operator. Analogously $\chi(\mathbf{A}, t)$ is the solution of the matrix equation

$$\mathcal{L}(\mathbf{X}(t)) = \mathbf{AX}(t). \tag{40}$$

It is evident that LDS the functioning of which is described as

$$\begin{cases} \mathcal{L}\mathbf{x(t)} = \mathbf{Ax}(t) + \mathbf{Bu}(t), \\ \mathbf{y}(t) = \mathbf{Cx}(t) \end{cases} \tag{41}$$

are systems with variable parameters. Let us find the solutions for such equations. Before doing it we should point out the following feature of the matrix function $\chi(\mathbf{A}, t)$

$$\mathcal{L}\chi(\mathbf{A}, t) = \mathbf{A}\chi(\mathbf{A}, s)$$

which follows from a similar feature of the eigenfunctions of the operator $\mathcal{L}$

$$\mathcal{L}\chi(\alpha, t) = \alpha\chi(\alpha, s).$$

A direct test gives the solution of the homogeneous equation (52), i.e. of the equation $\mathcal{L}\mathbf{x}(t) = \mathbf{A}\mathbf{x}(t)$:

$$\mathbf{x}(t) = \chi(\mathbf{A}, t)\chi(\mathbf{A}, t_0)^{-1}\mathbf{x}(t_0) = \chi(\mathbf{A}, t \ominus t_0)\mathbf{x}(t_0).$$

Indeed,

$$\mathcal{L}\mathbf{x}(t) = \mathcal{L}\chi(\mathbf{A}, t \ominus t_0)\mathbf{x}(t_0) = \mathbf{A}\chi(\mathbf{A}, t \ominus t_0)\mathbf{x}(t_0) = \mathbf{A}\mathbf{x}(t).$$

Thus the fundamental matrix $\Phi(t, t_0)$ of this equation is $\Phi(t, t_0) = \chi(\mathbf{A}, t \ominus t_0)$. According to Cauchy's formula we write the solution

$$\mathbf{x}(t) = \chi(\mathbf{A}, t \ominus t_0)\mathbf{x}(t_0) + \int_{t_0}^{t} \chi(\mathbf{A}, t \ominus \tau)\mathbf{u}(\tau)d\tau \tag{42}$$

of the non-homogeneous equation $\mathcal{L}\mathbf{x}(t) = \mathbf{A}\mathbf{x}(t) + \mathbf{u}(t)$. For the equation (53) the solution is written in the following way:

$$\begin{aligned}
\mathbf{x}(t) &= \mathbf{C}\chi(\mathbf{A}, t \ominus t_0)\mathbf{x}(t_0) + \int_{t_0}^{t} \mathbf{C}\chi(\mathbf{A}, t \ominus \tau)\mathbf{B}\mathbf{u}(\tau)d\tau \\
&= \mathbf{C}\chi(\mathbf{A}, t \ominus t_0)\mathbf{x}(t_0) + \int_{t_0}^{t} \mathcal{H}(\mathbf{A}, t \ominus \tau)\mathbf{u}(\tau)d\tau,
\end{aligned} \tag{43}$$

where $\mathcal{H}(t \ominus \tau, \mathbf{A}) = \mathbf{C}\chi(\mathbf{A}, t \ominus \tau)\mathbf{B}$ is the impulse response matrix of the systems (53). In connection with its dependence on $\mathcal{L}$-difference of time arguments this LDS is invariant with respect to the influence of GSO.

Theorem 6 *(Generalized Ljapunov's-Yerugin's theorem). Let*

$$\begin{cases} {}_1\mathcal{L}_t\mathbf{x}(t) = {}_1\mathbf{A}\mathbf{x}(t) + {}_1\mathbf{B}\mathbf{u}(t), \\ \mathbf{y}(t) = {}_1\mathbf{C}\mathbf{x}(t), \end{cases} \quad \begin{cases} {}_2\mathcal{L}_t\mathbf{x}(t) = {}_2\mathbf{A}\mathbf{x}(t) + {}_2\mathbf{B}\mathbf{u}(t), \\ \mathbf{y}(t) = {}_2\mathbf{C}\mathbf{x}(t), \end{cases} \tag{44}$$

are ${}_1D$- and ${}_2D$-stationary LDS and ${}_1\mathcal{H}(t_1 \ominus t_0)$, ${}_2\mathcal{H}(t_1 \ominus t_0)$ their impulse transfer matrices. Then a ${}_1D$-stationary LDS is reducible to a ${}_2D$-stationary LDS if and only if the transition operator $\mathbf{U}(t)$ in the state space ($\mathbf{z}(t) = \mathbf{U}(t)\mathbf{x}(t)$) exists and can be represented in the form

$$\mathbf{U}(t) = \chi_2({}_2\mathbf{A}, t_2 \ominus t_0)\mathbf{U}(t_0)\chi_1({}_1\mathbf{A}, t_1 \ominus t_0),$$

where $\mathbf{U}(t_0)$ is some constant matrix.

The **proof** is given in [LLC92].

Consequence 2 *(Yerugin's theorem). A D-stationary LDS is reducible to stationary in the classical sense if and only if its fundamental matrix is represented as*

$$\chi({}_2\mathbf{A}, t \ominus t_0) = \mathbf{U}(t)\exp\left[-{}_1\mathbf{A}(t - t_0)\right]\mathbf{U}^{-1}(t_0).$$

Generally speaking the systems (56) must not necessarily be $_1D$- and $_2D$-stationary. Theorem 6 can be used in the most general case of non-stationary systems. Here is a certain example of a homogeneous system:

$$\frac{d}{dt}\mathbf{x}(t) = \mathbf{A}(t)\mathbf{x}(t).$$ (45)

If $\mathbf{A}(t) \equiv \mathbf{A}$, then $\Phi(t, t_0) = \exp[\mathbf{A}(t - t_0)]$. The necessary and sufficient [WS76] conditions for representations of the fundamental matrix in similar form

$$\Phi(t, t_0) = \exp\left[\int_{t_0}^{t} \mathbf{A}(\eta)d\eta\right]$$

are known also for the cases when $\mathbf{A}(t)$ depends on time.

In this case the matrix should be represented as

$$\mathbf{A}(t) = \sum_{i=1}^{M} \mathbf{A}_i \alpha_i(t),$$

where $\mathbf{A}_i \mathbf{A}_j = \mathbf{A}_j \mathbf{A}_i$ and $\alpha(t)$ are arbitrary functions. Then

$$\Phi(t, t_0) = \exp\left[\sum_{i=1}^{m} \mathbf{A}_i \int_{t_0}^{t} \alpha_i(\eta)d\eta\right] = \exp\left[\sum_{i=1}^{m} \mathbf{A}_i \beta_i(t, t_0)\right]$$

$$= \exp[\mathbf{A}_j \beta_j(t, t_0)] \exp\left[\sum_{i=1;\, i \neq j}^{m} \mathbf{A}_i \beta_i(t, t_0)\right],$$

where

$$\beta_i(t, t_0) = \int_{t_0}^{t} \alpha_i(\eta)d\eta.$$

Let be the transform coordinates in the state space be

$$\mathbf{z}(t) = \exp\left[\sum_{i=1;\, i \neq j}^{m} \mathbf{A}_i \beta_i(t, t_0)\right] \mathbf{x}(t).$$

We define the system into which the original system maps after this transform. Obviously, the transformed system possesses the following fundamental matrix

$$\Phi_j(t, t_0) = \exp[\mathbf{A}_j \beta_j(t, t_0)].$$

Differentiating this fundamental matrix we find the differential equation which it satisfies

$$\frac{d}{dt}\Phi_j(t, t_0) = \mathbf{A}_j \beta_j(t, t_0)\Phi_j(t, t_0).$$

Consequently, the transformed system satisfies the analogous equation

$$\frac{d}{dt}\mathbf{z}(t) = \mathbf{A}_j \beta_j(t, t_0)\mathbf{z}(t).$$

Let us now introduce the systemic time $t_{s_j} = \beta_j(t, t_0)$. Let $\mathbf{w}(t_{s_j})$ be a new variable such that $\mathbf{w}(t_{s_j}) = \mathbf{z}(t)$. Then

$$\frac{d}{dt_{s_j}}\left[\mathbf{w}(t_{s_j})\right] = \frac{d\mathbf{z}(t)}{d\beta_j(t, t_0)} = \frac{d\mathbf{z}(t)}{\alpha_j(t)dt} = \mathbf{A}_j \mathbf{z}(t) = \mathbf{A}_j \mathbf{w}(t_{s_j})$$

and

$$\frac{d}{dt_{s_j}}\left[\mathbf{w}(t_{s_j})\right] = \mathbf{A}_j \mathbf{w}(t_{s_j}), \quad j = 1, \ldots, m. \tag{46}$$

that is a LDS with classical stationarity. Hence, m different reductions of the basis system (57) to m different stationary (in a classical sense systems) (58) exist, which functioning is described in m different systemic times t_{s_j}, $j = 1, \ldots, m$.

3 Space-time structure of $\mathcal{D}$-stationary continuous and discrete distributed systems

The given considerations do not change if we pass over from multidimensional LDS to discrete and continuously distributed systems or to linear cellular automata.

Let $\mathbf{t} = (t_0, t_1, \ldots, t_{n-1}) = (t, \mathbf{r})$, where $t_0 = t$ denotes the calendar time and $\mathbf{r} = (t_1, t_2, \ldots, t_{n-1})$ are spatial coordinates.

Suppose the continously distributed LDS (CDLDS) operation is described as:

$$\mathcal{L}_t \mathbf{x}(t, \mathbf{r}) = \mathcal{L}_r \mathbf{x}(t, \mathbf{r}) + \mathbf{B}\mathbf{u}(t, \mathbf{r}), \tag{47}$$

where $\mathcal{L}_t$ is the time differential operator, $\mathcal{L}_r$ is usually the space coordinate differential operator, $\mathbf{u}(t, \mathbf{r})$ is the input driving of continuously distributed LDS. The eye's retina is a typical example of such a systems, in which the input effect is caused by the flow of light that changes in time. In some special practical cases the equation (59) is reduced to wave equation, diffusion equation, Laplace equation, Schrödinger equation, Klein-Gordon equation etc.

To estimate the fundamental matrix of the equation (59) eigenfunctions of the operator $\mathcal{L}_r$ can be used. Let $\mathcal{L}_r \varphi(\alpha, \mathbf{r}) = \lambda(\alpha)\varphi(\alpha, \mathbf{r})$, where $\lambda(\alpha)$ is eigenvalue $\mathcal{L}_r$, which satisfies eigenfunction $\varphi(\alpha, \mathbf{r})$. Suppose solution $\mathbf{x}(t, \mathbf{r})$ can be decomposed into the following series:

$$\mathbf{x}(t, \mathbf{r}) = \int_\alpha \mathbf{X}(t, \alpha)\varphi(\alpha, \mathbf{r})d\alpha, \tag{48}$$

where

$$\mathbf{X}(t, \alpha) = \int_r \mathbf{x}(t, \mathbf{r})\bar{\varphi}(\alpha, \mathbf{r})d\mathbf{r}. \tag{49}$$

After substituting equation (61) into the homogeneous form

$$\mathcal{L}_t \mathbf{x}(t, \mathbf{r}) = \mathcal{L}_r \mathbf{x}(t, \mathbf{r})$$

we get

$$\int\limits_{\alpha} [\mathcal{L}_t \mathbf{X}(t,\alpha) - \lambda(\alpha)\mathbf{X}(t,\alpha)]\, \varphi(\alpha,\mathbf{r})d\alpha = 0,$$

and then

$$\mathcal{L}_t \mathbf{X}(t,\alpha) = \lambda(\alpha)\mathbf{X}(t,\alpha) \tag{50}$$

which is a special case of (52). That is why

$$\mathbf{X}(t,\alpha) = \chi(t \ominus t_0, \lambda(\alpha))\mathbf{X}(t_0,\alpha)$$

$$= \chi(t \ominus t_0, \lambda(\alpha)) \int\limits_r \mathbf{x}(t_0,\mathbf{r})\bar{\varphi}(\alpha,\mathbf{r})d\mathbf{r} \tag{51}$$

After substituting (63) into (60) we obtain

$$x_h(t,\mathbf{r}) = \int\limits_r \int\limits_\alpha \chi(t \ominus t_0, \lambda(\alpha))\varphi(\alpha, \mathbf{r} \ominus \mathbf{s})x(t,\mathbf{s})ds d\alpha$$

$$= \int\limits_r \Phi(t \ominus t_0, \mathbf{r} \ominus \mathbf{s})x(t,\mathbf{s})ds,$$

where

$$\Phi(t \ominus t_0, \mathbf{r} \ominus \mathbf{s}) = \int\limits_\alpha \chi(t \ominus t_0, \lambda(\alpha))\varphi(\alpha, \mathbf{r} \ominus \mathbf{s})d\alpha.$$

According to Cauchy's formula we write, finally, the solution of the equation (60)

$$\mathbf{x}(t,\mathbf{r}) = \mathbf{x}_h(t,\mathbf{r}) + \int\limits_{t_0}^{t} \int\limits_{s_0}^{r} \Phi(t \ominus \tau, \mathbf{r} \ominus \mathbf{s})\mathbf{u}(\tau,\mathbf{s})d\tau ds \tag{52}$$

However the fundamental matrix $\Phi(t \ominus \tau, \mathbf{r} \ominus \mathbf{s})$ which is used here must be given a physical meaning, i.e. it must reflect disturbance propagation in the LDS environment from the instantly operating point source and with final speed, and therefore it must be centered in the cone $\{t, \| \mathbf{r} \| \mid t \geq \| \mathbf{r} \| /v\}$, where $\mathbf{v}$ is used for disturbance propagation speed in the LDS environment. It is now of multidimensional LDS can be easily used for the case with continuously distributed LDS. Reduction in the case with all independent coordinates $t \to t_s$, $\mathbf{r} \to \mathbf{r}_s$ is particularly interesting, since non-stationary, non-homogeneous and non-isotropic CDLDS are reduced to stationary, homogeneous and isotropic low "time", and "space" frequency filter. However the possibility of logical and consistent realization of this idea in its general outline must be thoroughly studied.

4 Conclusion

A review of recently developed symmetric models for signals and systems shows that these models can be considered from the general (unified) point of view and investigated by general (unified) methods, taking into consideration the specific features of every concrete signals theory, because a lot of concrete theories are based on models with definite spatio-temporal symmetries. Hence, their integral representations in a basis of orthogonal functions with the same type of spatio-temporal symmetry, as the analyzed signals, can be adapted as a mathematical basis of signal theory. These conclusions allow us to formulate the main purpose of [CLL92] and this paper:

1. on the basis of specific signals and systems theories analysis a general definition of abstract signals and abstract systems is to be given in such a way that known signals and systems are special cases;
2. to make the foundations of such description theoretical scheme of abstract systems work laws and abstract signals processing methods, which would allow specialists not to lose touch with frequency notions and representations and at the same time to investigate signals and systems from the common point of view on the base of unique approach;
3. to conduct classification of signals and systems by considering the complete set of abstract signal and abstract system realizations;
4. to work out an abstract basis of orthogonal functions for signals and systems including:
   - the determination of that generalized stationarity (symmetry) type, which is produced by the chosen orthogonal basis,
   - the proofs of theorems, analogous to Fourier harmonic analysis theorems,
   - a theory of signal transforms by systems with general stationarity
   - a theory of codes based on generalized spectral techniques by *Blahut*.

References

[Ber29]   BERGSON, H.: *Durce et Simultancite*. Paris 1929

[Bet84]   BETH, T.: *Verfahren der schnellen Fourier-Transformation*. Teubner: Stuttgart 1984

[Bet89]   BETH, Th.: *Algorithm engineering a la Galois (AEG)*. Proc. AAECC-7 (1989)

[CT85]   CREUTZBURG, R. - TASCHE, M.: *F-Transformation und Faltung in kommutativen Ringen*. Elektr. Informationsverarb. Kybernetik **EIK-21** (1985), pp. 129-149

[CT86]   CREUTZBURG, R. - TASCHE, M.: *Number-theoretic transforms of prescribed length*. Math. Comp. **47** (1986), pp. 693-701

[CS88]   CREUTZBURG, R. - STEIDL, G.: *Number-theoretic transforms in rings of cyclotomic integers*. Elektr. Informationsverarb. Kybernetik **EIK-24** (1988), pp. 573-584

[CT89] CREUTZBURG, R. - TASCHE, M.: *Parameter determination for complex number-theoretic transforms using cyclotomic polynomials.* Math. Comp. **52** (1989), pp. 189-200

[CLL92] CREUTZBURG, R.; LABUNETS, V. G. - LABUNETS, E. V.: *Towards an "Erlangen program" for general linear systems theory.* Proceed. EURO-CAST'91 (Krems, Austria), (Ed.: F. Pichler), Lecture Notes of Computer Science **585**, Springer: Berlin 1992, pp. 32-51

[CM93] CREUTZBURG, R. - C. MORAGA: *Spectral Techniques - Theory and Applications.* North-Holland: Amsterdam 1993 (in print)

[Get75] GETHÖFFER, H.: *Algebraic theory of finite systems.* Progress in Cybernetics and Systems Research. (1975), pp. 170-176

[Har69] HARMUTH, H. F.: *Transmission of Information by Orthogonal Functions.* Springer: Berlin 1969

[Hol90] HOLMES, R. B.: *Signal processing on finite groups.* MIT Lincoln Laboratory, Lexington (MA), Technical Report 873 (Febr. 1990)

[Kar76] KARPOVSKY, M. G.: *Finite Orthogonal Series in the Design of Digital Devices.* Wiley: New York 1976

[Kar85] KARPOVSKY, M. G.: *Spectral Techniques and Fault Detection.* Academic Press: New York 1985

[LS76] LABUNETS, V. G. - SITNIKOV, O. P.: *Generalized harmonic analysis of VP-invariant systems and random processes (in Russian)* in: *Harmonic Analysis on groups in abstract systems theory.* Ural Polytechnical Institute Press: Sverdlovsk: 1976, pp. 44-67

[LS76b] LABUNETS, V. G. - SITNIKOV, O. R.: *Generalized harmonic analysis of VP-invariant linear sequential circuits.* In: *Harmonic Analysis on Groups in Abstract Systems Theory (in Russian).* Ural Polytechnical Institute Press: Sverdlovsk 1976, pp. 67-83

[Lab84] LABUNETS, V. G.: *Algebraic Theory of Signals and Systems - Computer Signal Processing (in Russian).* Ural State University Press: Sverdlovsksk 1984

[Lab89] LABUNETS, V. G.: *Theory of Signals and Systems - Part II (in Russian).* Ural State University Press: Sverdlovsk 1989

[LLC91a] LABUNETS, V. G.; LABUNETS, E. V. - CREUTZBURG, R.: *Algebraic foundations of an abstract harmonic analysis based on a generalized symmetry principle. Part I: Analysis of signals.* Preprint, Karlsruhe 1991

[LLC91b] LABUNETS, V. G.; LABUNETS, E. V. - CREUTZBURG, R.: *Algebraic foundations of an abstract harmonic analysis based on a generalized symmetry principle. Part II: Analysis of systems.* Preprint, Karlsruhe 1991

[LLC92] LABUNETS, V. G.; LABUNETS, E. V. - CREUTZBURG, R.: *Algebraic foundations of an abstract harmonic analysis based on a generalized symmetry principle, Part III: Space-time of D-stationary systems.* Preprint, Karlsruhe 1991

[Lev73] LEVITAN, B. M.: *Theory of Generalized Shift Operators (in Russian).* Nauka: Moscow 1973

[MC92] MINKWITZ, T.; CREUTZBURG, R.: *A new fast algebraic convolution algorithm.* in: *Signal Processing VI: Theories and Applications.* (Eds.: J. Vandewalle; R. Boite; M. Moonen; A. Oosterlinck) Elsevier Science Publ.: Amsterdam 1992, pp. 933-936

[NP87] NICOLIS, G. - PRIGOGINE, I.: *Die Erforschung des Komplexen.* Piper: München 1987

[Pic70]   PICHLER, F.: *Walsh-Fourier-Synthese optimaler Filter.* Archiv Elektr. Übertragung (1970) No. 24, pp. 350-360

[Pic72]   PICHLER, F.: *Zur Theorie verallgemeinerter Faltungssysteme: dyadische Faltungssysteme und Walshfunktionen.* Elektron. Informationsverarb. Kybernet. **8** (1972) No.4., pp. 197-209

[PM92]   PICHLER, F. - MORENO DIAZ, R.: *Computer Aided Systems Theory - EUROCAST'91* LNCS **585**, Springer: Berlin 1992

[PS92]   PICHLER, F. - SCHWÄRTZEL (Eds.): *CAST Methods in Modelling.* Sprimger: Berlin 1992

[Pic92]   PICHLER, F.: *Realisierung der Λ-Transformation von Prigogine mittels dyadischer Faltungsoperatoren.* Research Report of the Austrian Society for Cybernetic Studies. Vienna, December 1992

[PS93]   PICHLER, F.: *Visualization of Baker-Prigogine dynamical systems* Proceed. EUROCAST'93 (Las Palmas, Canary Islands)

[Sit76]   SITNIKOV, O. P.: *Harmonic analysis on groups in abstract systems theory.* In: *Harmonic Analysis on Groups in Abstract Systems Theory (in Russian).* Ural Polytechnical Institute Press: Sverdlovsk 1976, pp. 5-24

[TK85]   TRACHTENBERG, E. A. - KARPOVSKY, M. G.: *Filtering in a communication channel by Fourier transforms over finite groups.* in: Karpovsky, M. G. (Ed.): *Spectral Techniques and Fault Detection.* Academic Press: Mew York 1985, pp. 179-216

[TT75]   TRACHTMAN, A. M.; TRACHTMAN, V. A.: *The Principles of Discrete Signals on Finite Intervals Theory (in Russian).* Soviet Radio: Moscow 1975

[Ver32]   VERNADSKY, V. N.: *Time problem in modern science (in Russian).* Izvest. Akad. Nauk SSSR, Mathematics and Natural Sciences 1932, pp. 511-541

[Vin69]   VINITSKY, A. S.: *Modulated Filters and Tracking Reception of Frequency Modulated Signals (in Russian).* Moscow 1969

[WS76]   WU, M. Y. - SHERIF, A.: *On the commutative class of linear time-varying systems.* Int. J. Contr. **23** (1976), pp. 433-444

The Shape of Complex Systems

Charles Rattray

Department of Computing Science, University of Stirling,
Stirling, Scotland FK9 4LA, UK
cr@cs.stir.ac.uk

Abstract. Categorical modelling is a useful tool in the study of systems. To motivate this, interpretation of elementary category theory ideas are illustrated in terms of time-varying complex systems. A brief introduction to the notion of categorical shape theory to model approximating situations is given.

1 Introduction

A complex system is defined to be an inter-connection of objects, where objects are of arbitrary character and no commitment is made to a particular system configuration. Such generality might seem to lead only to "abstract nonsense". This is not the case; it is through this generality that conceptual, methodological and organisational problems, associated with the evolving complex system, can be described and reasoned about. In a number of recent papers ([19, 20]), we have harnessed this generality by exploiting a categorical meta-model of software development process models [22] to investigate improved ways for looking at CAST system structure development and software re–usability in the CAST frame. In this paper, we review some of the mathematical concepts used in this *categorical modelling* framework and provide interpretations of the ideas in the system context.

A familiar process in mathematics and engineering is the creation of complex systems from given, more elementary, systems. The elementary systems can be considered as models of the process, and one usually says that the complex or *global* objects are formed from the elementary or *local* objects by "pasting together" models [1]. This provides a more *geometric* view of systems and allows us to consider the shape [6] of such systems. Note that it is the structure of a complex system, its specification, design and development, that we are looking to model: its components, their inter-relations, their complexity level, their evolution in the course of time. The functional aspect is recovered from the synchronic and diachronic interactions among components, and from the fact that a complex component's behaviour is determined by that of its constituents and their links.

The novelty of the work outlined lies in the great flexibility of the mathematical framework. Already it has been used to model the software process [22], blackboard systems [18], software development environments [20], and neural networks with memory and a learning capability [10, 11]. A simplified version has given useful insights into the data–flow diagram representation technique for software engineering design. The framework has great potential in modelling any information system development [7, 14].

Development processes may be evaluated, compared, composed and decomposed by standard categorical methods. Using the notion of time incorporated in the model, design changes and evolution dependent on being completed within a time period, whether at the micro or macro level, are dealt with. Feedback of such changes allow the model to learn; propagation delays are naturally described in the model; hierarchical memory structure regulated by an evolving regulatory subsystem further enhances the overall approach. Precision can be given to qualitative descriptions of development methods, processes, and objects. The model provides a very useful common ground for technology transfer of design notions.

Of course, systems theory means different things to different people. Here, we mean the mathematical study of abstract representations of systems, keeping in mind the problems that seem most important for real systems and the aspects of the theory which seem most amenable to mathematical development. As a result, the approach is more abstract than is normal for the CAST view.

The first step is the mathematical formulation of the basic concepts. In engineered software systems, for instance, many basic concepts have previously been formulated with insufficient generality or have not been formulated at all. We are trying here to be completely general in our explications of the basic concepts yet, at the same time, provide a framework in which qualitative notions may be precisely stated and in which quantitative features can be considered.

A mathematical explication takes a vaguely defined concept from the everyday language of the system under study and gives a precise meaning which includes all or nearly all important related situations; it therefore *clarifies* the concept. Our choice of concepts takes *system, object, interconnection* or *link*, and *behaviour* as basic, and provides the explications within category theory.

The application of the category approach in systems theory is not new. Rosen [24, 25], in 1958, used it to discuss a relational theory of biological systems; Mesarovic and Takahara [16] published their well-known book on abstract systems theory , in which category theory played an important role. More recently, Ehresmann and Vanbremeersch [8, 9] have used category theory in modelling evolving systems in the medical field. The time may be ripe now for CAST researchers to reconsider category theory to help structure their systems problems. Historically, category theory played an important role in mathematics during the period approximately from 1945 to 1975. Four achievements of this development can be singled out: (i) concepts were clarified by phrasing them in categorical terms; (ii) uniform definitions of constructions were given in categorical terms. The specialisations of these constructions in particular categories showed how seemingly ad hoc constructions were "really" the same as well understood constructions in other contexts; (iii) abstraction led to simplification and generalisation of concepts and proofs. By dissecting out the crucial ingredients of a proof and providing categorical descriptions of the constructions involved in the proof, one could obtain a much more general categorical proof which could then be applied in quite different circumstances to yield new results; (iv) complicated computations were guided by general categorical results. A general construction often makes transparent a lengthy computation in a particular category. These same achievements can be had in CAST by using the categorical approach.

Our needs in this paper are altogether much simpler than those above. Rather than concentrate on categories of mathematical structures we shall concentrate on a category *as a mathematical structure*.

Most of the work covered can be understood with only five "basic doctrines" aimed at providing an initial intuition of the elementary categorical concepts [12, 13]:

1. any (species of) mathematical structure is represented by a *category*;
2. any mathematical construction is represented by a *functor*;
3. any natural correspondence of one construction to another is represented by a *natural transformation*;
4. any construction forming a complex structure from a pattern of related simple structures is represented by a *colimit*;
5. given a species of structure, then a species of structure obtained by enriching the original is represented by a *comma category*.

This paper introduces a meta-model suitable for describing development processes for constructing complex systems. The approach involves interpreting elementary category theory concepts in the systems context and is guided by application of some of the basic dogmas identified by Goguen, and his colleagues, in the computer science literature. Section 2 introduces systems as categories, already a common notion, and provides a possible categorical definition for categories of time-varying complex systems. A closer look at a category in Section 3 suggests some internal structure which models typical systems features. Section 4 provides the basis for a categorical framework for discussing the notion of approximating systems. The final section, Section 5, briefly reviews the point reached in the paper. By collecting together these various categorical ideas, perhaps others in the CAST community will be encouraged to think about them in the context of their own problem areas.

2 Systems as Categories

Many representations of systems are possible. For instance, von Bertalanffy [4] gives the view of a system as a set of units with relationships among them. The units themselves may be complex objects and the relationships between them should respect the structure of the objects. Such structure-preserving relations are normally called *morphisms*. With a minimum of conditions on the morphisms the collection of complex objects and morphisms form a *category*.

A category is a directed graph together with a composition operation on composable edges (or arrows). To each arrow a is assigned a pair of graph nodes $dom(a)$ and $cod(a)$, called the *domain* and *codomain* of a, such that a is the arrow from $dom(a)$ to $cod(a)$. Each pair of arrows a and b, which are *composable* ($cod(a) = dom(b)$), is assigned an arrow called the *composition* of a and b, written $b \circ a$. Composition is associative. Composition requires identities, one for each graph node, so that if A is a graph node then there is an arrow $id_A : A \to A$ such that if $a : A \to B$ then

$$a = a \circ id_A = id_B \circ a$$

The nodes of a category are usually called *objects* and the arrows are called *morphisms*. For any category, $\mathbf{G}$, there is an obvious underlying graph, $U_{\mathbf{G}}$.

A *diagram* in a category $\mathbf{G}$ is a set of objects of $\mathbf{G}$ together with a set of arrows between these objects; these sets may be empty.

A *path* in a diagram D is a finite sequence of arrows $(a_1, a_2, \ldots, a_n)$ of D such that $cod(a_i) = dom(a_{i+1})$, $1 \leq i \leq n - 1$. The *length* of the path is n. Diagram D commutes if, for any paths $(a_1, a_2, \ldots, a_n)$, $(b_1, b_2, \ldots, b_n)$ such that $m \geq 2$ or $n \geq 2$ (or both), $dom(a_1) = dom(b_1)$ and $cod(a_n) = cod(b_m)$,

$$a_n \circ a_{n-1} \circ \ldots \circ a_1 = b_m \circ b_{m-1} \circ \ldots \circ b_1 \ .$$

The category **Set** of sets, for which the collection of objects is the collection of all sets and the collection of morphisms is the collection of all functions between sets, is perhaps the most common category. To any directed graph G we can associate a category of directed paths $\mathbf{P}_G$ in G; the objects of $\mathbf{P}_G$ are just the nodes of G, the paths in G are the morphisms, composition is concatenation of paths (which is associative), and identities are zero-length paths associated with each node.

The nature of many systems representations is diagrammatic/graphical. Design representations for software systems, for instance, may be in the form of flowcharts, structure charts, data flow diagrams, SADT "blueprints", statecharts, etc. These are all essentially (directed) graphs with various properties or peculiarities; for information systems, all of the above have been used as have *entity-relation-attribute* diagrams (ERA). Obviously, none of these repesentations form a category as they stand. For each, one must "embed" the representation in a suitable category. Rosen [25] in developing his relational theory of biological systems discusses $(\mathbf{M}, \mathbf{R})$-systems; from these he constructs *abstract block diagrams* which are diagrams in the category **Set**. A similar approach was taken by Rattray and Phua [21] in considering Data-Flow-Diagrams (DFD). Ehresmann and Vanbremeersch [11], in developing a categorical view of general systems, give an illustrative neural network example in which the category in question is the path category of the network. More interestingly, but still following the same general line, Dampney, Johnson and Munro [7] and Johnson and Dampney [14] construct the category associated with ERA digrams as the *classifying category* or *theory* for the information system being analysed. Generating the category requires that the ERA diagram be normalise, ie. many-many relationships be transformed into a pair of many-one relationships by the introduction of a new entity. The resulting diagram is a directed graph with entities and attributes as nodes and arrows a relations, directed in the "many" to "one" direction, together with arrows from entities to attributes. The many-one relations have real world counterparts so that real world compositions must appear in the diagram. Particular attention must be paid to the paths with common domains and common codomains to determine if the diagram commutes as these represent system constraints. Johnson and Dampney make the point strongly that current ERA-modelling technologies ignore the question of commuting diagrams, yet identifying these often leads to modification of the ERA diagram to produce a better model. The category constructed is just the presentation (specification) for the canonical classifying category. For an information system, certain basic "exactness" properties need to be satisfied. In particular, the classifying category will be *lextensive* [5].

Thus, we can see that a system can be modelled by a suitable category.

Systems evolve with time. For instance, a software system goes through a number of phases: conception, requirements, specification, design, implementation, and use. This latter phase is sometimes referred to in the literature as software evolution or software maintenance. It represents a period in which the system , as constructed, is modified to match the changing environment and the changing needs of the user.

The categorical representation of a system is in no sense absolute, rather it represents the state of the system at some time t. The evolving system then moves through a sequence of states to reach some goal state, determined by the requirements specified for the modified system to match its new environment. Unfortunately, not only does this process move the system to a new state but, because of the complexity of the internal structure of the state, it also changes the system structure and its categorical representation. That is, we may be constructing a new category or, at the very least, an incremental change to that of the previous state. The determination of an appropriate strategy to achieve the new goal state may be undermined by timing constraints. reaching a goal state may have to accomplished within a given time period. The chosen strategy will affect the internal structuring and functioning of the system state. The interaction between the external time period available to achieve the goal state and the internal timescales for system component changes can be quite subtle.

To reflect state transitions, these are represented as *functors*.

Let **G** and **H** be categories. A functor $F : \mathbf{G} \to \mathbf{H}$ is essentially a graph homomorphism which preserves composition and identities. That is, F is a function which assigns to each object A in **G** an object $F(A)$ in **H**, and to each arrow $a : A \to B$ in **G** an arrow $F(a) : F(A) \to F(B)$ in **H** such that

- if $A \xrightarrow{a} B \xrightarrow{b} C$ in **G** then $F(b \circ a) = F(b) \circ F(a)$
- $F(id_A) = id_{F(A)}$ for all objects A in **G**.

The collection of all system states and transitions is a category **Cat**, the category of categories anf functors. The sequence of states and their transitions representing the evolution of a system is just a diagram in **Cat**.

To capture this more fully, we need to define a category of time. Consider a total order on the set of positive reals. This defines a category: objects are the real numbers representing instances of time, and the unique arrows (t, t') from instance t to instance t', $t \leq t'$, represents the period of time between t and t'. This is the category **Time**.

A **time-varying complex system** (TVCS) is a functor $S : \mathbf{T} \to \mathbf{Cat}$, where $\mathbf{T}$ is a subcategory of **Time**. The category $S(t) = \mathbf{S}_t$ over the instance t is the category representing the state of the system at time t. In fact, S induces a *fibration* [3, 2] over $\mathbf{T}$ and $\mathbf{S}_t$ is a *fibre* over t.

One of the "basic doctrines" suggests that to study any structure we should determine the category in which it resides. Indeed, in understanding the development process of any complex system, we need to compare possible developments of the system, compare different development processes, or simply determine invariant properties of a single process. Modelling such processes as TVCSs identifies the need for the notion of morphisms between them. A possible way of dealing with this is to compare states of the systems at "corresponding" times. Thus, the timescales of each system must be compared.

Let S and R be two TVCSs over $\mathbf{T}$ and $\mathbf{U}$, respectively, where $\mathbf{T}$ and $\mathbf{U}$ are subcategories of **Time**. A *morphism from S to R* is a pair (F, ϕ) such that

- $\phi : \mathbf{T} \to \mathbf{U}$ is a *change of time* functor
- $F : S \to R \circ \phi$ is a *natural transformation* between S and R such that
  - for each object t in $\mathbf{T}$, $F(t)$ (usually written as F_t) is an arrow in **Cat**

$$F_t : \mathbf{S}_t \to \mathbf{R}_{\phi(t)}$$

  - for each period (t, t^*) in $\mathbf{T}$, the diagram

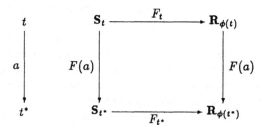

commutes.

Composition of morphisms requires composition of the time change functors, and the correspondences at relevant time instances. With this defined, we now have the *category of time-varying complex systems*.

3 System Fragments and Approximations

For many existing complex systems, it is unlikely that any individual "knows the system"; at best, the individual has a view of the system and only understands certain aspects of it. Communicating individuals have their own systems views but can pass on aspects of their view to a colleague. "Composition" of individual views contribute to a better understanding of the system but such a composition is merely an approximation to, or distortion of, that system. The same problem persists in the development of complex systems which normally require team effort, or the effort of multiple teams. Assuming that individuals and teams of individuals are part of the complex systems of which we speak, the systems is known solely through some particular individuals ("agents") which interact with it, ie. these agents gain only partial quantitative and qualitative information of their environment, through observation. Of course, an external observer can compare the agents' composed view with the full system and identify the level of distortion between them.

A *control centre* (CC) of a TVCS S is represented by a sub-TVCS with its own timescale. The objects of CC are what we have called *agents*.

To model the notion of an agent's view of the system state G and a system approximation over a control centre, we introduce a *slice category*, a special case of a *comma category* [15].

If G is a category and A any object of G, the slice category $G \downarrow A$ is described in terms of

- an object of $G \downarrow A$ is an arrow $b : B \to A$ of G for some object B
- an arrow of $G \downarrow A$ from $b : B \to A$ to $c : C \to A$ is an arrow $f : B \to C$ such that

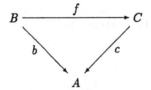

commutes
- the composition of $f : b \to c$ and $g : c \to d$ is $g \circ f$.

The *fragment* for agent A at time t is the category $G_t \downarrow A$ of objects over A in G_t, with $b : B \to A$ an *aspect* of G_t recognised by A. Any ("communication") arrow $\alpha : A \to A'$ between agents in CC induces a functor $\alpha^* : G_t \downarrow A \to G_t \downarrow A'$ between fragments. The collection of fragments over the control centre CC is a *fibration of fragments at t*.

A *system approximation* for the control centre CC at t is a *colimit* of this fibration: its objects are equivalences of aspects b linked by a *zig-zag* of communications between agents. This approximation supports a distortion functor to G_t. For a TVCS, a corresponding system approximation forms a TVCS.

A colimit construction provides a means of forming complex objects from patterns (diagrams) of simpler objects, where a pattern can be considered as a graph homomorphism P from graph G (its "shape") to the underlying graph of some category, G. Pattern P defines a *pattern of linked (related) objects* in G. A *collective link* of the pattern P to the object C of the category is a family of arrows f_A, indexed by the nodes of the graph G, where f_A is an arrow from the object P_A to C, which satisfies the *compatibility condition*: if a is an arrow from

A to B in the graph, then f_A is the composition of $P(a)$ and f_B, ie

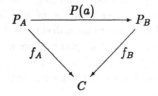

commutes; pattern P and the collective link from P to object C is called a *cone*.

An **colimit** of pattern P is an object C' of the category **G** such that, for any object C, the arrows from C' to C are in one-one correspondence with the collective links from P to C. The unique arrow f associated with the collective link (f_A) is said to *bind* the f_As.

Thus, the colimit object binds together the component objects according to their internal organisation determined by the corresponding pattern. The object P_A of the pattern P is called a component object of the colimit C'. The properties of an object depend on the number and nature of the arrows which link it to other objects of the category **G**. It is natural to compare the properties of the complex object C' with those of its components.

The category **G** models the environment of the pattern P. Modification of the environment category may take various forms. For instance, enlarging **G** to **H** or blurring the distinction between two **G** objects in **H** may make retaining the limits difficult. It may be that a pattern P cannot be bound to a colimit in **G** but can be forced to have a colimit in an extended environment **H**.

By using colimits, a hierarchical structure can be imposed upon the system states. This provides a convenient abstraction concept to make the understanding and development of complex systems more amenable to analysis and construction. We should be conscious of the fact, though, that such abstract objects as modelled by a colimit do not necessarily exist in real systems. We may define a *hierarchical system* to be a category **G** in which the objects are dis tributed on levels $(0, 1, \ldots, p)$, such that each object of level $n + 1$ $(n < p)$ is the limit in **G** of a pattern P of linked objects on level n. In such a hierarchical system, the system components are associated with levels corresponding to increasing complexity of their internal organisation. Any object at level $n + 1$ is the colimit of a pattern of linked objects at level n but it may form part of a pattern of linked objects whose colimit is at level $n + 2$.

Control centres may occur on any level, either singly or as multiple CCs. Higher level CCs will normally be colimits of patterns of CCs at lower levels.

4 The Shape of Complex Systems

The basic idea of categorical shape theory is that in any approximating situation, the approximations are what encode the only information that the system can analyse (Porter [17]). In the previous section, the notion of systems approximations were introduced in relation to control centres. Such approximations are very important in using TVCSs to model software developments. The strategy for constructing a new state is determined by the system approximation of the previous state. That strategy is, in fact, a distortion of the true strategy applicable to the system state. This is interesting and it reflects what actually happens in practice in the development of complex systems.

However, in all engineering disciplines, re-usability (of components, design, manufacturing processes, ...) is of fundamental importance. In software engineering too this same concept is a major research topic. Rather that re-usability of software components, re-usability of software processes is perhaps more likely to lead to the necessary breakthroughs needed to deal with the well-known "software crisis". For, given an unfamiliar system to modify or maintain, a programmer must first understand that system before he can hope to provide the required changes without introducing further errors. If the programmer had complete documentation of the design decisions made by the originator of the system, he would be in a much better position to carry through changes error-free.

In Rattray and Marsden [23], categorical shape theory was introduced as a mathematical framework within which to consider approximation in relation to re-usability of software specifications. Now we wish to consider re-usability of software development processes. Having defined the category of time-varying complex systems, with the express purpose of modelling development processes, we need to consider approximation in this category.

Assume we are given a system of *objects of interest* (eg. TVCSs) and a system of basic objects or *archetypes*. Assume further that such systems are categories (section 2) and a functor $Q : \mathbf{A} \to \mathbf{B}$ allows archetypes to be compared with objects. Given an object B of $\mathbf{B}$ we wish to compare it with an archetype. An *approximation* to B is a pair (f, A) where A is an object of $\mathbf{A}$ (ie. an archetype), and $f : B \to QA$. A morphism between approximations $\alpha : (f, A) \to (g, A')$ is an arrow $\alpha : A \to A'$ in the category of archetypes, $\mathbf{A}$, such that

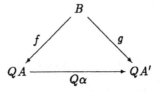

commutes. This defines a category of "approximations" to B, namely, the comma category $B \downarrow Q$.

Note that $B \downarrow Q$ contains information available to the system about B. For α above, the approximation (f, A) may be considered in some way "better" than (g, A'). That is, (f, A) and (g, A') can be compared if there is an arrow from one to the other. The codomain functor δ_B from $B \downarrow Q$ to $\mathbf{A}$ is given by $\delta_B(f, A) = A$, ie. δ_B defines a pattern of archetypes associated with the approximations of B. The limit (dual of colimit) of this diagram, if it exists, would identify an archetype from $\mathbf{A}$ giving a better approximation to B than any other. Even though such a limit may not exist in $\mathbf{A}$ we may formally extend $\mathbf{A}$ to include this abstract archetype.

If $f : B \to B'$ is a morphism in $\mathbf{B}$, any approximation (g, A) to B' gives an approximation $(g \circ f, A)$ to B by composition of the arrows g and f. This induces a functor

$$f^* : B' \downarrow Q \to B \downarrow Q$$

satisfying $\delta_B \circ f^* = \delta_{B'}$. That is, comparing two objects is equivalent to comparing their approximations.

The **shape category**, $\mathbf{Sh}_Q$, of the system Q has as objects the objects of $\mathbf{B}$ and, for $f : B \to B'$ in $\mathbf{C}$, the morphisms are the functors $F : B' \downarrow Q \to B \downarrow Q$. If (f, A) is in $B' \downarrow Q$, then $F(f, A)$ has the form (g, A) for the same A in $\mathbf{A}$, and some $g : B \to QA$ in $\mathbf{C}$.

Two objects have the same Q-shape if they are isomorphic in $\mathbf{Sh}_Q$.

Shape theory has an application in pattern-matching, in re-usability through the use of repositories of specifications, in dealing with semantic notions of learning using the time-varying complex system model, and in modelling partial evaluation. Its role in approximating software processes is still speculative.

5   Conclusions

Category theory can be quite seductive. It appears to be close in form and content to the needs of the systems theorist, and even the software engineer, but few practical applications are available to encourage its further use. This is now being rectified with the work on information systems modelling, software process modelling, and Ehresmann and Vanbremeersch's studies of neural activity in the brain and aging.

From the simple notion of a system as a category we have shown connections with real world problems for some elementary categorical ideas such as functor, colimit, and comma category. There is much more to it than this.

References

1. H Appelgate, M Tierney: "Categories with Models", Lecture Notes in Mathematics, **80**, Springer–Verlag, 1969.
2. M Barr, C Wells: *Category Theory for Computing Science*, Prentice Hall Publ., 1990.
3. J Bénabou: "Fibered Categories and the Foundations of Naive Category Theory", J. of Symbolic Logic, **50**, 1, 1985.
4. L von Bertalanffy: *Les Problèmes de la Vie*, Gallimard, Paris, 1956.
5. A Carboni, S Lack, R F C Walters: "Introduction to Extensive and Distributive Categories", Technical Report, Pure Mathematics 92-9, University of Sydney, 1992.
6. J-M Cordier, T Porter: *Shape Theory: categorical approximation methods*, Ellis Horwood Ltd., 1990.
7. C N G Dampney, M Johnson, G P Munro: "An Illustrated Mathematical Foundation for ERA", in **The Unified Computation Laboratory** (Editors: C Rattray, R G Clark), Oxford University Press, 1992.
8. A C Ehresmann, J-P Vanbremeersch: "Systemes Hierarchique s Evolutifs: une modelisation des systemes vivants", Prepublication No 1, Universite de Picardie, 1985.

9. A C Ehresmann, J-P Vanbremeersch: "Systemes Hierarchique s Evolutifs: modele d'evolution d'un systeme ouvert par interaction avec des agent", Prepublication No 2, Universite de Picardie, **1986**.

10. A C Ehresmann, J-P Vanbremeersch: "How do heterogeneous levels with hierarchical modulation interact on system's learning process?", Proceedings of 3rd International Symposium on Systems Research, Informatics and Cybernetics, Baden-Baden, **1991**.

11. A C Ehresmann, J-P Vanbremeersch: "Outils Mathematiques Utilises pour Modeliser les Systemes Complexes", Cahiers de Topologie et Géométrie Différentielle Catégoriques, **XXX111**, 2, **1992**.

12. J Goguen, J Thatcher, E Wagner, J Wright: "A Junction between Computer Science and Category Theory, I: Basic Concepts and Examples", Technical Report RC 4526, IBM T J Watson Research Centre, NY, **1973**.

13. J A Goguen: "A Categorical manifesto", Mathematical Structures in Computer Science, **1**, **1991**.

14. M Johnson, C N G Dampney: "Category Theory and Information Systems Engineering", in **Algebraic Methodology and Software Technology AMAST'93** (Editors: M Nivat, C Rattray, T Rus, G Scollo), Workshops in Computing Series, Springer-Verlag (to appear), **1993**.

15. S MacLane: *Categories for the Working Mathematician*, Springer-Verlag, **1971**.

16. M D Mesarovic, Y Takahara: *Abstract Systems Theory*, Lecture Notes in Control and Information Sciences **116**, Springer-Verlag, **1989**.

17. T Porter: "Can Categorical Shape Theory Handle Grey Level Images?", *Shape in Pictures* (ed. M Pavel) (to appear), **1993**.

18. C Rattray: "Evolutionary hierarchical systems and the blackboard architecture", Proc. 1st Maghrebin Conference on Software Engineering and Artificial Intelligen ce, Constantine, **1989**.

19. C Rattray, D Price: "Sketching an Evolutionary Hierarchical Framework for Knowledge–Based Systems Design", *Computer Aided Systems Theory – EURO-CAST'89*, Lecture Notes in Computer Science, 410, Springer–Verlag, **1990**.

20. C Rattray: "Systems Factories and CAST", in *Cybernetics and Systems '90* (R Trappl: editor), World Scientific Publ. Co., **1990**.

21. C Rattray, G Phua: "A Meta–Model for Software Processes: Complex System Modelling by Categories — Data Flow Diagrams", Tech Rpt CSET/90.3, University of Stirling, **1990**.

22. C Rattray: "An evolutionary software model", in *Mathematical Structures for Software Engineering* (de Neumann, Simpson, Slater: editors), Oxford University Press, **1991**.

23. C Rattray, M Marsden: "Object Identification and Retrieval in a CAST Library", EUROCAST'91, LNCS 585, Springer–Verlag, **1992**.

24. R Rosen: "A Relational Theory of Biological Systems", Bulletin of Mathematical Biophysics, **20**, **1958**.

25. R Rosen: "The representation of Biological Systems from the standpoint of the Theory of Categories", Bulletin of Mathematical Biophysics, **20**, **1958**.

Polynomial Systems Theory for n-D Systems Applied to Vision-Based Control

Raimo Ylinen

Helsinki University of Technology, Control Engineering Laboratory
SF-02150 Espoo, Finland

Abstract

The polynomial systems theory for time-varying and distributed parameter systems is briefly introduced. The theory is applied to design of a vision based control system for exact positioning of a camera above a moving object. The algebraic structure needed here is a ring of skew polynomials with skew polynomial coefficients. Both an observer for estimating nonmeasurable outputs and a stabilizing controller are designed.

1 Introduction

The polynomial systems theory for time-invariant linear differential and difference systems has proven to be an efficient tool for control system analysis and design ([1], [2], [3] , [4]) . The methodology is based on strong algebraic properties of polynomials with real or complex coefficients. In particular, then the ring of polynomials in a differentiation or shift operator on a suitable signal space is a principal ideal domain, or more strongly, an Euclidean domain with respect to the degree function. This means that the division algorithm can be used for finding common factors, manipulating polynomial matrices to suitable canonical forms and so on.

The generalization of the theory to time-varying (c.f. [6],[7]) or multidimensional (n-D) (c.f. [5]) systems is relatively difficult. In time-varying case the polynomials are not commutative, but they are so called skew polynomials. The other is, however, that in many cases the coefficients (parameter functions) are not invertible, which is necessary in order to obtain a (noncommutative) principal ideal domain. The same problem occurs in n-D systems. The algebraic structure needed there is the ring of polynomials in two or more operators with real or complex coefficients. This is not a principal ideal domain but only a Noetherian domain [6].

n-D systems are usually used as models of dynamic linear distributed parameter systems. This means that one of the operators is the differentiation or shift operator with respect to time. The polynomials can be presented as ordinary polynomials in this operator with polynomial coefficients. If the signal space is suitable, the ring of the coefficients can be extended to its field of fractions. This presumes that the non-zero coefficients are invertible as operators. If this extension can be made, a ring of polynomials with rational coefficients is obtained. This is an Euclidean domain so that all methods developed for ordinary polynomial systems can be applied.

The most difficult case is obtained, if the parameters of a distributed parameter system are dependent both on time and space. Then the structure is a ring of skew polynomials with skew polynomial coefficients. Again, if the signal space and parameters (parameter functions) are suitable, the ring of coefficients can be extended to its (noncommutative) fied of fractions.

Image processing is a typical application area of n-D systems. In continuous vision based robot control an exact position of a possibly moving object is often needed. This means that due to uncertainties in image information and delays in its processing some kind of estimation or prediction is needed. An observer for this purpose can be designed using the polynomial systems theory described above. Unfortunately, in this case the most difficult structure, the skew polynomials with skew polynomial coefficients is needed. However, if the time-invariance of parameters is assumed, a little simpler structure, a ring of ordinary polynomials with skew polynomial coefficients , is obtained.

The methology is also applied to the design of a stabilizing feedback controller for the positioning of camera above a moving object. Two kind of controllers are designed. Due to the complex calculations only the simpler design based on the use of estimated unmeasurable outputs is presented here.

2   Polynomial Systems Theory

A polynomials system description consists of equations of the form

$$\underbrace{(a_0 + a_1\sigma + \ldots + a_n\sigma^n)}_{a(\sigma)} y = \underbrace{(b_0 + b_1\sigma + \ldots + b_m\sigma^m)}_{b(\sigma)} u \tag{1}$$

where $u, y \in X$, which is an additive Abelian group, $a_0, \ldots, a_n, b_0, \ldots, b_m, \sigma \in \text{End}(X) \overset{\triangle}{=}$ the ring of endomorphisms of X.

Often the *coefficients* a_i, b_i belong to a subring of $\text{End}(X)$ so that the *polynomials* $a(\sigma), b(\sigma)$ belong to

$$R\{\sigma\} \overset{\triangle}{=} \{c(\sigma)|c(\sigma) = \sum_{i=0}^{n} c_i\sigma^i, c_i \in R, n \in \mathbf{N_0}\} \tag{2}$$

$R\{\sigma\}$ is closed under addition of $\text{End}(X)$ and it is closed under multiplication of $\text{End}(X)$ if $\sigma a \in R\{\sigma\}$ for every $a \in R$. For example:
(i) If

$$\sigma a = a\sigma \tag{3}$$

then $R\{\sigma\} \triangleq R[\sigma] \triangleq$ the ring of ordinary polynomials in σ over R.

(ii) If

$$\sigma a = \alpha(a)\sigma + \delta(a) \tag{4}$$

where α is a ring endomorphism of R and δ is an α-derivation of R i.e. satisfies

$$\begin{aligned}
&\delta(a+b) = \delta(a) + \delta(b) \\
&\delta(0) = 0 \\
&\delta(ab) = \delta(a)b + \alpha(a)\delta(b)
\end{aligned} \tag{5}$$

then $R\{\sigma\} \triangleq R[\sigma; \alpha, \delta] \triangleq$ the ring of skew polynomials in σ over R determined by α and δ.

Examples

(i) Time-invariant ordinary differential systems

Let $X = C^\infty(T)(\triangleq C^\infty \text{in short}) \triangleq$ the complex valued infinitely differentiable functions on $T \subset \mathbf{R}$ ($\triangleq$ the real numbers), $\sigma = p \triangleq$ the differentiation operator and $R = \mathbf{C}(\triangleq$ the complex numbers). Then

$pa = ap$ for all $a \in \mathbf{C}$ and $R\{\sigma\} = \mathbf{C}[p] \triangleq$ the (commutative) ring of ordinary polynomials in p over $\mathbf{C}$.

(ii) Time-invariant ordinary difference systems

Let $X = \mathbf{C}^{\mathbf{Z}} \triangleq \{\ldots, x(-1), x(0), x(1), \ldots\}$, $\sigma = r \triangleq$ the unit delay operator on $\mathbf{C}^{\mathbf{Z}}$ and $R = \mathbf{C}$. Then

$ra = ar$ for all $a \in \mathbf{C}$ and $R\{\sigma\} = \mathbf{C}[r]$

. (iii) Time-varying ordinary differential systems

Let $X = C^\infty$, $\sigma = p$ and $R = C^\infty$. Then

$pa = ap + \frac{da}{dt}$ for all $a \in C^\infty$ and $R\{\sigma\} = \mathbf{C}[p; 1, p] \triangleq$ the ring of skew polynomials in p over C^∞ determined by $1, p$.

(iv) Time-varying ordinary difference systems

Let $X = \mathbf{C}^{\mathbf{Z}}$, $\sigma = r$ and $R = \mathbf{C}^{\mathbf{Z}}$. Then

$ra = r(a)r$ for all $a \in \mathbf{C}^{\mathbf{Z}}$ and $R\{\sigma\} = \mathbf{C}^{\mathbf{Z}}[r; r, 0]$.

(v) Time-invariant constant parameter dynamic partial differential systems

Let $X = C^\infty(D), (D \subset \mathbf{R}^2)$, $\sigma = p_t \triangleq \frac{\partial}{\partial t}$ and $R = \mathbf{C}[p_x], p_x \triangleq \frac{\partial}{\partial t}$. Then $p_t a = ap_t, a \in \mathbf{C}$, $p_t p_x = p_x p_t$, $p_t a(p_x) = a(p_x)p_t$, $a(p_x) \in \mathbf{C}[p_x]$ and $R\{\sigma\} = \mathbf{C}[p_x][p_t](\cong \mathbf{C}[p_x, p_t])$.

(vi) Time-invariant variable parameter dynamic partial differential systems

Let $X = C^\infty(D)(\triangleq C^\infty)$, $\sigma = p_t$ and $R = C^\infty[p_x; 1, p_x]$. Then $p_t a = ap_t + \frac{\partial a}{\partial t}$, $p_t p_x = p_x p_t$ and $p_t a(p_x) = a(p_x)p_t + \frac{\partial a(p_x)}{\partial t}$ where $\frac{\partial a(p_x)}{\partial t} \triangleq \sum_i \frac{\partial a_i}{\partial t} p_x^i$, $a(p_x) \in C^\infty[p_x; 1, p_x]$

Thus $R\{\sigma\} = C^\infty[p_x; 1, p_x][p_t; 1, p_t]$.

The skew polynomial ring $R[\sigma; \alpha, \delta]$ has the following properties:

(i) If for every $a(\sigma) \in R[\sigma; \alpha, \delta]$

$$a(\sigma) = a_0 + a_1 \sigma + \ldots + a_n \sigma^n = 0 \Rightarrow a_0 = a_1 = \ldots = n = 0 \tag{6}$$

then the *degree-function* $d(a(\sigma)) \triangleq \max\{n | a_n \neq 0\}$ is well-defined.

(ii) If R is a field, then $R[\sigma; \alpha, \delta]$ satisfies the *(left) division algorithm*: for $a(\sigma), b(\sigma) \neq 0$ there exist $q(\sigma), r(\sigma)$ such that

$$a(\sigma) = q(\sigma)b(\sigma) + r(\sigma), d(r(\sigma)) < d(b(\sigma)) \tag{7}$$

(iii) If $R[\sigma; \alpha, \delta]$ is an integral domain and satisfies the *(left) Ore condition*: for $a(\sigma), b(\sigma) \neq 0$ there exist $a_1(\sigma), b_1(\sigma)$ such that

$$b_1(\sigma)a(\sigma) = a_1(\sigma)b(\sigma) \neq 0 \tag{8}$$

then $R[\sigma; \alpha, \delta]$ can be extended to the *field of (left) fractions*

$$R(\sigma; \alpha, \delta) \triangleq \{b(\sigma)/a(\sigma) | a(\sigma), b(\sigma) \in R[\sigma; \alpha, \delta], a(\sigma) \neq 0\} \tag{9}$$

(iv) X is a *(left) module* over $R[\sigma; \alpha, \delta]$ with scalar multiplication $(a(\sigma), x) \mapsto a(\sigma)x$. If every $a(\sigma) \neq 0$ is an automorphism of X, then X is a vector space over $R(\sigma; \alpha, \delta)$ with scalar multiplication

$$(b(\sigma)/a(\sigma))x = a(\sigma)^{-1}b(\sigma)x \tag{10}$$

The *polynomial system (model)* generated by Eq.1 is the relation

$$S = \{(u, y) | a(\sigma)y = b(\sigma)u\} \tag{11}$$

where (u, y) is an *input-output pair* of the system. In multivariable case

$$S = \{(u, y) | A(\sigma)y = B(\sigma)u\} \tag{12}$$

where $A(\sigma), B(\sigma)$ are matrices over $R[\sigma; \alpha, \delta]$ and S is said to be *generated* by $[A(\sigma) \vdots - B(\sigma)]$.

Let $\tilde{S}$ be another system generated by $[\tilde{A}(\sigma) \vdots - \tilde{B}(\sigma)]$. Then under some general conditions

$$S \subset \tilde{S} \Leftrightarrow [\tilde{A}(\sigma) \vdots - \tilde{B}(\sigma)] = L(\sigma)[A(\sigma) \vdots - B(\sigma)] \text{for some} L(\sigma) \tag{13}$$

If in addition $S = \tilde{S}$, then $L(\sigma)$ is *unimodular* i.e. $L(\sigma)$ is invertible and $L(\sigma)^{-1}$ is a matrix over $R[\sigma; \alpha, \delta]$. This means that the generators of the system S are *row equivalent* to each other.

The matrices related to *elementary row operations*:
(i) add the jth row multiplied by $a(\sigma)$ to the ith row (the matrix $T_{ij}(a(\sigma))$),
(ii) interchange the ith and jth rows (U_{ij})
are unimodular and the matrix $V_{ii}(c(\sigma))$ related to the operation
(iii) multiply the ith row by $c(\sigma)$ is unimodular if $c(\sigma)$ is an invertible skew polynomial ($c(\sigma) = c_0$ with $c_0^{-1} \in R$).
Therefore, using the unimodular row operations the generators of systems can be brought to suitable forms without changing the systems themselves.

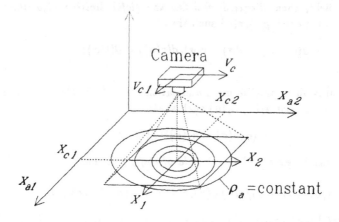

Figure 1. Vision based positioning system

3 Vision Based Control System

Consider the system depicted by Fig.1. The problem is to keep the camera above a moving object using visual information only. The velocity v_c of the camera can be used for manipulating its position x_a in the absolute coordinates x_a. Image coordinates are $x = x_a - x_c$. The object is described by mass density distribution $\rho_a(x_a, t)$ and its surface is supposed to be such that the intensity distribution in the image $\rho(x, t) \approx \rho_a(x_a, t)$.

The system is governed by the equations

$$\rho_a(x_a, t)\frac{\partial v_a}{\partial t}(x_a, t) + \rho_a(x_a, t)\frac{\partial v_a}{\partial x_a}(x_a, t)^T v_a(x_a, t)$$
$$= f(x_a, t) \text{ (force balance)} \tag{14}$$

$$\frac{\partial \rho_a}{\partial t}(x_a, t) + \frac{\partial \rho_a}{\partial x_a}(x_a, t)^T v_a(x_a, t) + \rho_a(x_a, t)(\frac{\partial v_{a1}}{\partial x_{a1}}(x_a, t)$$
$$+ \frac{\partial v_{a2}}{\partial x_{a2}}(x_a, t)) = g(x_a, t) \approx 0 \text{ (mass balance)} \tag{15}$$

where $f(x_a, t)$ is an external force affecting the camera and $g(x_a, t)$ describes the disturbances in the movement of the object.

Putting $x_a = x + x_c$, $v_a(x_a, t) = v(x, t) + v_c(t)$ and $\rho_a(x_a, t) = \rho(x, t)$ the model (14,15) can be brought to the form

$$\rho(x, t)(\frac{\partial v}{\partial t}(x, t) + \frac{\partial v_c}{\partial t}(t)) + \rho(x, t)\frac{\partial v}{\partial x}(x, t)^T(v(x, t)$$
$$+ v_c(t)) = f(x + x_c, t) \tag{16}$$

$$\frac{\partial \rho}{\partial t}(x, t) + \frac{\partial \rho}{\partial x}(x, t)^T(v(x, t) + v_c(t)) + \rho(x, t)(\frac{\partial v_1}{\partial x_1}(x, t)$$
$$+ \frac{\partial v_2}{\partial x_2}(x, t)) = g(x + x_c, t) \tag{17}$$

Suppose then that the nominal solution $\bar{\rho}(x,t), \bar{v}(x,t)$ corresponding to $\bar{v}_c(t)$, $\bar{f}(x + x_c, t)$ and $\bar{g}(x + x_c, t)$ is known and model is linearized near that solution.

$$\bar{\rho}(x,t)(\frac{\partial v}{\partial t}(x,t) + \frac{\partial v_c}{\partial t}(t)) + \rho(x,t)(\frac{\partial \bar{v}}{\partial t}(x,t) + \frac{\partial \bar{v}_c}{\partial t}(t))$$
$$+\bar{\rho}(x,t)\frac{\partial \bar{v}}{\partial x}(x,t)^T(v(x,t) + v_c(t)) + \bar{\rho}(x,t)\frac{\partial v}{\partial x}(x,t)^T(\bar{v}(x,t) \qquad (18)$$
$$+\bar{v}_c(t)) + \rho(x,t)\frac{\partial \bar{v}}{\partial x}(x,t)^T(\bar{v}(x,t) + \bar{v}_c(t)) = f(x + x_c, t)$$

$$\frac{\partial \rho}{\partial t}(x,t) + \frac{\partial \bar{\rho}}{\partial x}(x,t)^T(v(x,t) + v_c(t)) + \frac{\partial \rho}{\partial x}(x,t)^T(\bar{v}(x,t) + \bar{v}_c(t))$$
$$+\bar{\rho}(x,t)(\frac{\partial v_1}{\partial x_1}(x,t) + \frac{\partial v_2}{\partial x_2}(x,t)) \qquad (19)$$
$$+\rho(x,t)(\frac{\partial \bar{v}_1}{\partial x_1}(x,t) + \frac{\partial \bar{v}_2}{\partial x_2}(x,t)) = g(x + x_c, t)$$

4   Polynomial System Model

Denote

$$a_1 \stackrel{\triangle}{=} \bar{\rho}(x,t),\; a_{01} \stackrel{\triangle}{=} \frac{\partial \bar{\rho}}{\partial x_1}(x,t),\; a_{02} \stackrel{\triangle}{=} \frac{\partial \bar{\rho}}{\partial x_2}(x,t)$$
$$b_1 \stackrel{\triangle}{=} \bar{v}_1(x,t),\; b_0 \stackrel{\triangle}{=} \frac{\partial \bar{v}_1}{\partial t}(x,t),\; b_{01} \stackrel{\triangle}{=} \frac{\partial \bar{v}_1}{\partial x_1}(x,t),\; b_{02} \stackrel{\triangle}{=} \frac{\partial \bar{v}_1}{\partial x_2}(x,t)$$
$$c_1 \stackrel{\triangle}{=} \bar{v}_2(x,t),\; c_0 \stackrel{\triangle}{=} \frac{\partial \bar{v}_2}{\partial t}(x,t),\; c_{01} \stackrel{\triangle}{=} \frac{\partial \bar{v}_2}{\partial x_1}(x,t),\; c_{02} \stackrel{\triangle}{=} \frac{\partial \bar{v}_2}{\partial x_2}(x,t)$$
$$d_1 \stackrel{\triangle}{=} \bar{v}_{c1}(t),\; d_0 \stackrel{\triangle}{=} \frac{\partial \bar{v}_{c1}}{\partial t}(t) \qquad (20)$$
$$e_1 \stackrel{\triangle}{=} \bar{v}_{c2}(t),\; e_0 \stackrel{\triangle}{=} \frac{\partial \bar{v}_{c2}}{\partial t}(t)$$
$$p \stackrel{\triangle}{=} \frac{\partial}{\partial t},\; p_1 \stackrel{\triangle}{=} \frac{\partial}{\partial x_1},\; p_2 \stackrel{\triangle}{=} \frac{\partial}{\partial x_2}$$
$$\tilde{f}(x,t) \stackrel{\triangle}{=} f(x + x_c, t),\; \tilde{g}(x,t) \stackrel{\triangle}{=} g(x + x_c, t)$$

Then the model can be presented in the form

$$(a_1 p + (a_1 b_{01} + a_1(b_1 + d_1)p_1)v_1 + (b_0 + d_0 + b_{01}b_1 + b_{01}d_1)\rho$$
$$= -(a_1 p + a_1 b_{01})v_{c1} + \tilde{f}_1$$
$$(a_1 p + (a_1 c_{01} + a_1(c_1 + e_1)p_2)v_2 + (c_0 + e_0 + c_{01}c_1 + c_{01}e_1)\rho$$
$$= -(a_1 p + a_1 c_{01})v_{c2} + \tilde{f}_2 \qquad (21)$$
$$(a_{01} + a_1 p_1)v_1 + (a_{02} + a_1 p_2)v_2 + (p + (b_{01}$$
$$+c_{01} + b_1 p_1 + c_1 p_2))\rho = -a_{01}v_{c1} - a_{02}v_{c2} + \tilde{g}$$

The coefficients of the model depend both on time t and position x. Therefore the model should be considered as a module equation over the ring $((C^\infty[p_1; 1, p_1])[p_2; 1, p_2])[p; 1, p]$ of skew polynomials with skew polynomial coefficients.

Because the coefficient ring $((C^\infty[p_1; 1, p_1])[p_2; 1, p_2]$ is not a field, the algebraic structure of the model is relatively weak. It is for instance not possible to use the division algorithm for manipulating the model etc.

Consider the following special case:

$\bar{v}(x,t) = 0, \bar{v}_c(t) = V_c = $ constant, $\bar{f}(x,t) = 0, \tilde{g}(x,t) = 0$. Then $b_1 = c_1 = 0$, $b_0 = c_0 = d_0 = e_0 = 0$ and $c_{01} = c_{02} = d_{01} = d_{02} = 0$. The model (21) becomes to

$$
\begin{bmatrix}
a_1 p_1 + a_{01} & a_1 p_2 + a_{02} & p \\
a_1 p + a_1 d_1 p_1 & 0 & 0 \\
0 & a_1 p + a_1 e_1 p_2 & 0
\end{bmatrix}
\begin{bmatrix}
v_1 \\
v_2 \\
\rho
\end{bmatrix}
=
\begin{bmatrix}
-a_{01} & -a_{02} \\
-a_1 p & 0 \\
0 & -a_1 p
\end{bmatrix}
\begin{bmatrix}
v_{c1} \\
v_{c2}
\end{bmatrix}
$$

(22)

Suppose further that the boundary conditions in the image are zero, so that the skew polynomials in p_1 and p_2 are automorphisms. Then the coefficient ring $(C^\infty[p_1; 1, p_1])[p_2; 1, p_2]$ can be extended to the field of fractions $(C^\infty(p_1; 1, p_1))(p_2; 1, p_2)$ which gives a stronger structure.

In this special case the coefficients are all time-invariant. Hence the model can be considered also as a module equation over the ring $((C^\infty(p_1; 1, p_1))(p_2; 1, p_2))[p]$ of ordinary polynomials with fractions of skew polynomials as coefficients. This is the class of models used in the following for the analysis of the controlled system and also in the observer and controller design.

5 Analysis

Using the elementary row operations and the division algorithm the model (22) can be brought to an upper triangular form ([1], [5]).

Suppose for simplicity that the nominal movement is in the direction of x_{a2}-axis i.e. $V_{c1} = d_1 = 0$. Then the model can be brought to the form

$$
\begin{bmatrix}
a_1 p_1 + a_{01} & a_1 p_2 + a_{02} & \vdots & p \\
0 & (a_1 p_2 + a_{02}) e_1 p_2 & \vdots & -p^2 \\
\cdots & \cdots & & \cdots \\
0 & 0 & \vdots & p^3 + a_0(p_2) p^2
\end{bmatrix}
\begin{bmatrix}
v_1 \\
v_2 \\
\cdots \\
\rho
\end{bmatrix}
$$

$$
=
\begin{bmatrix}
-a_{01} & -a_{02} \\
-a_1 p_1 p & -a_1 p_2 p \\
\cdots & \cdots \\
a_1 p_1 p^2 + a_0(p_2) a_1 p_1 p & a_1 p_2 p^2 - a_0(p_2) a_{02} p
\end{bmatrix}
\begin{bmatrix}
v_{c1} \\
v_{c2}
\end{bmatrix}
$$

(23)

where

$$
a_0(p_2) \overset{\Delta}{=} (a_1 p_2 + a_{02}) e_1 p_2 \frac{1}{a_1 p_2 + a_{02}}
$$

(24)

or briefly

$$
\begin{bmatrix}
A_1(p) & \vdots & A_2(p) \\
\cdots & & \cdots \\
0 & \vdots & A_4(p)
\end{bmatrix}
\begin{bmatrix}
v \\
\cdots \\
\rho
\end{bmatrix}
=
\begin{bmatrix}
B_1(p) \\
\cdots \\
B_2(p)
\end{bmatrix}
v_c
$$

(25)

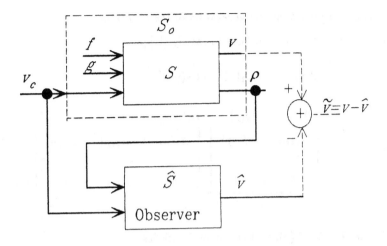

Figure 2. Observer design problem

The system with input v_c and output ρ is generated by the equation

$$\underbrace{\left(p^3 + a_0(p_2)p^2\right)\rho}_{A_4(p)}$$
$$= \underbrace{[a_1 p_1 p^2 + a_0(p_2)a_1 p_1 p \quad a_1 p_2 p^2 - a_0(p_2)a_{02} p]}_{B_2(p)}\begin{bmatrix} v_{c1} \\ v_{c2} \end{bmatrix} \qquad (26)$$

Because $A_4(p)$ and $B_2(p)$ have a common left factor $L(p) = p$, the system is not *controllable*. Uncontrollable mode associated with p is not, however, unstable.

The matrix

$$A_1(p) = \begin{bmatrix} a_1 p_1 + a_{01} & a_1 p_2 + a_{02} \\ 0 & (a_1 p_2 + a_{02})e_1 p_2 \end{bmatrix} \qquad (27)$$

is unimodular so that v is *observable* from ρ and v_c.

6 Observer Design

The image gives only the intensity distribution ρ. The other output, the velocity v, would obviously be also useful in controlling the position of the camera. If it could be estimated continuously, the unknown velocity could be replaced by the estimated one. There are many different solutions to the estimation problem. The *observer* type estimators are based on the system model so that the observer model and the system model belong to the same class of systems. The observer design problem is depicted by Fig.2.

The problem is to design a system $\hat{S}$, an *observer* with two inputs ρ and v_c so that its output $\hat{v}$ estimates the velocity v i.e. the error $\tilde{v} = v - \hat{v}$ is as small as possible and stable irrespective of the input v_c.

It can be shown that the generators of the observers $[C(p) \vdots -D_1(p) \; -D_2(p)]$ have to satisfy ([1],[5]

$$
\begin{bmatrix} C(p) & \vdots & -D_1(p) & -D_2(p) \\ \cdots & & \cdots & \cdots \\ 0 & \vdots & A_4(p) & -B_2(p) \end{bmatrix}
$$
$$
= \underbrace{\begin{bmatrix} T_1(p) & \vdots & T_2(p) \\ \cdots & & \cdots \\ 0 & \vdots & I \end{bmatrix}}_{T(p)} \begin{bmatrix} A_1(p) & \vdots & A_2(p) & -B_1(p) \\ \cdots & & \cdots & \cdots \\ 0 & \vdots & A_4(p) & -B_2(p) \end{bmatrix} \tag{28}
$$

for some $[T_1(p) \vdots T_2(p)]$ and the error $\tilde{v}$ satisfies

$$
T_1(p)A_1(p)\tilde{v} = 0 \tag{29}
$$

The behaviour of the observer should be robust with respect to parameter variations which usually means that it has to be proper.

Left multiplication of $T(p)$ by another matrix of the same type results again in a matrix of the same type. Therefore the condition (28) can be used repeatedly for constructing a suitable $T(p)$. In this case the choice

$$
[T_1(p) \vdots T_2(p)] = \begin{bmatrix} (p+\alpha)(p+\beta) & 0 & \vdots & -1 \\ 0 & (p+\gamma)(p+\delta) & \vdots & p+\gamma+\delta-a_0(p_2) \end{bmatrix} \tag{30}
$$

with constants $\alpha, \beta, \gamma, \delta > 0$ gives a 4 th order proper and robust observer with

$$
\begin{aligned}
C(p) &= T_1(p)A_1(p) \\
&= \begin{bmatrix} (a_1p_1+a_{01})(p+\alpha)(p+\beta) & (a_1p_2+a_{02})(p+\alpha)(p+\beta) \\ 0 & (a_1p_2+a_{02})e_1p_2(p+\gamma)(p+\delta) \end{bmatrix}
\end{aligned} \tag{31}
$$

$$
\begin{aligned}
D_1(p) &= -T_1(p)A_2(p) - T_2(p)A_4(p) \\
&= \begin{bmatrix} (-\alpha-\beta+a_0(p_2)p^2 - \alpha\beta p \\ -((\gamma+\delta-a_0(p_2))a_0(p_2)+\gamma\delta)p^2 \end{bmatrix}
\end{aligned} \tag{32}
$$

and correspondingly

$$
D_2(p) = T_1(p)B_1(p) + T_2(p)B_2(p) \tag{33}
$$

(too long to be written here).

7 Feedback Controller Design

Consider finally the feedback controller design for the system. Suppose first that only the output ρ is used for control.

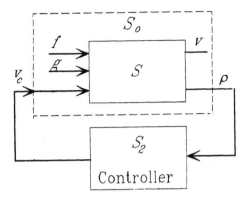

Figure 3. Feedback control system

The problem is to design a system S_2, a *feedback controller*, with input ρ and output v_c such that the overall system depicted by Fig.3 behaves satisfactorily, is stable, robust etc. In what follows, the feedback controller is assumed to belong to the same class of systems than the controlled system S. Then the robustness usually implies that the controller should be proper.

It can be shown (([1],[5])) that the generators $[C(p) \vdots - D(p)]$ of feedback controllers satisfy

$$
\begin{bmatrix}
A_{41}(p) & \vdots & -B_{21}(p) \\
\cdots & & \cdots \\
-D(p) & \vdots & C(p)
\end{bmatrix}
=
\underbrace{\begin{bmatrix}
I & \vdots & 0 \\
\cdots & & \cdots \\
T_3(p) & \vdots & T_4(p)
\end{bmatrix}}_{T(p)}
\underbrace{\begin{bmatrix}
A_{41}(p) & \vdots & -B_{21}(p) \\
\cdots & & \cdots \\
X(p) & \vdots & Y(p)
\end{bmatrix}}_{P(p)}
$$

$$(34)$$

for some $[T_3(p) \vdots T_4(p)]$.

Here $[A_{41}(p) \vdots - B_{21}(p)]$ represents the controllable part of the controlled system i.e.

$$[A_4(p) \vdots - B_2(p)] = L(p)[A_{41}(p) \vdots - B_{21}(p)] \tag{35}$$

where $L(p)$ $(= p$ in this case) is a greatest common left factor of $A_4(p)$ and $B_2(p)$. The *first candidate* $P(p)$ is unimodular, satisfies

$$[A_{41}(p) \vdots B_{21}(p)] = [I \vdots 0]\, P(p) \tag{36}$$

and can be constructed using *elementary column operations* to $[A_{41}(p) \vdots - B_{21}(p)]$.

The closed loop behaviour of the overall system is determined by $T_4(p)$ and the

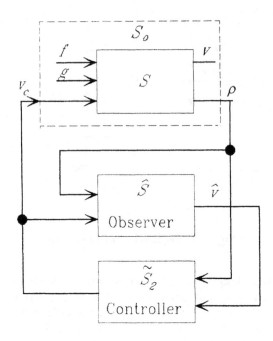

Figure 4. Feedback control with observer

uncontrollable part $L(p)$. Analogously to the condition (28) also the condition (34) can be used repeatedly.

The method applied to the system generated by

$$\underbrace{\left(p^3 + a_0(p_2)p^2\right)}_{A_4(p)} \rho$$
$$= \underbrace{[a_1 p_1 p^2 + a_0(p_2)a_1 P_1 p \vdots a_1 p_2 p^2 - a_0(p_2)a_{02}p]}_{B_2(p)} \begin{bmatrix} v_{c1} \\ v_{c2} \end{bmatrix} \tag{37}$$

leads to relatively complicated calculation and is not presented here.

If the estimated velocity v is also used for control the resulting controller becomes much simpler. The situation is depicted by Fig.4.

Now the controller can be designed most easily if the model (22) is first written to the form

$$\begin{bmatrix} p & a_1 p_1 + a_{01} & a_1 p_2 + a_{02} \\ 0 & a_1 p + a_1 d_1 p_1 & 0 \\ 0 & 0 & a_1 p + a_1 e_1 p_2 \end{bmatrix} \begin{bmatrix} \rho \\ v_1 \\ v_2 \end{bmatrix} = \begin{bmatrix} -a_{01} & -a_{02} \\ -a_1 p & 0 \\ 0 & -a_1 p \end{bmatrix} \begin{bmatrix} v_{c1} \\ v_{c2} \end{bmatrix} \tag{38}$$

After cancelling the greatest common factor

$$\begin{bmatrix} 1 & 0 & 0 \\ 0 & 1 & 0 \\ 0 & 0 & p \end{bmatrix}$$

the first candidate can be constructed easily

$$\begin{bmatrix} p & a_1p_1 + a_{01} & a_1p_2 + a_{02} & \vdots & a_{01} & a_{02} \\ 0 & 1 & 0 & \vdots & 1 & 0 \\ 0 & 0 & p + e_1p_2 & \vdots & 0 & p \\ \cdots & \cdots & \cdots & \cdots & \cdots \\ 1 & 1 & 0 & \vdots & 1 & 0 \\ 0 & 0 & 1 & \vdots & 0 & 1 \end{bmatrix}$$

The controller

$$[\tilde{C}(p) \vdots -\tilde{D}(p)] = \begin{bmatrix} 1 & 0 & \vdots & 1 & 1 & 0 \\ 0 & 1 & \vdots & 0 & 0 & 1 \end{bmatrix} \tag{39}$$

obtained from the first candidate gives a stable closed loop system and is proper. It does not however guarantee the robustness but a more proper controller is needed. This kind of controller is obtained if $[T_3(p) \vdots T_4(p)]$ is chosen as

$$\begin{bmatrix} -1 & -p - \alpha + a_1p_1 + a_{01} & 0 & \vdots & p + \alpha & 0 \\ 0 & 0 & -p - \beta - \gamma + e_1p_2 & \vdots & 0 & (p + \beta)(p + \gamma) \end{bmatrix}$$

with constants $\alpha, \beta, \gamma > 0$. The resulting controller is

$$\begin{aligned} & [\tilde{C}(p) \vdots -\tilde{D}(p)] \\ & = \begin{bmatrix} a_1p_1 & -a_{02} & \vdots & \alpha & 0 & -a_1p_2 - a_{02} \\ 0 & e_1p_2p + \beta\gamma & \vdots & 0 & 0 & (\beta - e_1p_2)(\gamma - e_1p_2) \end{bmatrix} \end{aligned} \tag{40}$$

8 Concluding Remarks

The distributed parameter systems with parameters varying with respect to time and space are very difficult to analyze and design. In this paper a methodology based on generalization of polynomial systems theory of ordinary time-invariant linear systems has been presented. The basic structure was the ring of skew polynomials with skew polynomial coefficients.

The analysis and design of a vision based control system proved the applicability of the methodology. On the other hand, although the methods are relatively straightforward and simple in principle, they are very laborious to carry out with pencil and paper. Therefore it would be useful to develop a program package for symbolic manipulation of matrices over skew polynomials with skew polynomial coefficients. The development of a corresponding program package for ordinary time-varying systems has already been started (c.f.[7]).

References

[1] Blomberg,H.,Ylinen,R.(1983). *Algebraic theory for multivariable linear systems*. Academic Press.

[2] Kučera,V.(1979). *Discrete linear control: The polynomial equation approach*. Wiley.

[3] Rosenbrock,H.H.(1970). *State-space and multivariable theory*. Nelson.

[4] Wolovich,W.A.(1974). *Linear multivariable systems*. Springer.

[5] Ylinen,R.(1975)."On the algebraic theory of linear differential and difference systems with time-varying or operator coefficients". Helsinki University of Technology, Systems Theory Laboratory,**B23**.

[6] Ylinen,R.(1980)."An algebraic theory for analysis and synthesis of time-varying lineal differential systems". Acta Polytechnica Scandinavica **Ma32**.

[7] Ylinen,R., Zenger,K.(1991). "Computer aided analysis and design of time-varying systems".In Pichler,F., Moreno Díaz (Eds.). *Computer Aided Systems Theory-EUROCAST'91*. Lecture Notes in Computer Science 585, Springer.

A Representation of Software Systems Evolution Based on the Theory of the General System

José Parets, Ana Anaya, María J. Rodríguez, Patricia Paderewski

Dept. de Lenguajes y Sistemas Informáticos. Universidad de Granada
Facultad de Ciencias. 18001 - GRANADA (Spain)

Abstract. The more commonly used software development methods assign the activities and results of the process of modification and evolution of software systems to maintenance. In this paper we present an approach to the representation of the evolution of these systems, based on Le Moigne's theory of the General System, and a prototype developed in an object-oriented programming language which implements the concepts used.

To do this we distinguish, in the evolution of software systems, between the Functional and Structural History of the system. These concepts have allowed us to elaborate a recursive-evolutive model of the software development process, which permits us to implement the concepts of the theory and to make prototypes of software systems.

Perhaps the most important contribution comes from the uniformity provided by a well founded epistemological background, the System Theory, which allows us to treat the design, functioning and modifications of software systems in an homogeneous way.

1 Introduction

Software development is based on the elaboration of reality models which are finally transferable to a computer program. The normal way to carry out this modelling process consists in using requirement specification and design methods inserted into a life cycle model. In general, neither of these methods includes the evolution between each performed model, nor the active role that the modeller has in this transition between models.

Using one of the more operational views of General System Theory, the Theory of the General System (TGS) stated by Le Moigne, we are looking for a way to represent the global history of software systems. With this purpose we distinguish the Functional History and the Structural History in the evolution of a software system which correspond, respectively, with the functioning representation (cinematic) and the structural evolution of the system (dynamic).

The concepts we are now using in this representation are those of system, processor, action, event, processed object and decision. Focusing on these, we are developing a class structure in an object-oriented programming language which allows us to obtain elementary prototypes of software systems. This class structure, which implements the previous concepts, has shown us that it is possible to use them both in the Functional and the Structural History representation.

This research is the core part of the development of a software requirements specification, a design and evolution method (SSDEM), and an associated tool (SSDET). The paper presents an abstract of the systemic concepts used and of the prototype carried out in the first phase of research. This has allowed us to check the validity of the concepts and to reinforce our belief that it is possible, in using these, to make useful prototypes of software systems and to reflect upon the active role of the developer in the development and evolution of software systems.

The paper is divided into three main parts. Firstly, a general view of the systemic concepts used as a foundation for the work is given. Secondly, the features of the modelling methods used in software development, from the point of view of the first part, are considered. Finally, the software development model with which we are working and the prototype which we have built based upon the model are explained.

2 Le Moigne's Theory of the General System[1]

Le Moigne represents a strong operational attitude amongst systemic authors. He describes a TGS as that which "allows the identification of the basic concepts, the verification of their coherence and the revelation of their conditions of use in the practice of object modelling... The TGS is *the object modelling theory*, with the aid of this artificial object gradually modelled by human thinking, which L. von Bertalanffy will propose to name *the General System*" [8, p.59-60].

Basically TGS is a modelling theory. General System (GS) is a *process* within an environment used to conceive phenomena as systems by some Representation System (RS) (the modeller conceived as system by means of the GS). This RS prepares the models of phenomena which he himself perceives and conceives as systems. To conceive a phenomenon as system Le Moigne proposes the formation of a matrix between the three traditional modelling conceptions: *functional, organic and historic*, and the three main GS features: *active, structured and evolutive*.

Dealing with the modelling process the distinction made by Le Moigne between models of the cinematic of a system and those of dynamic is interesting. We call CINEMATIC MODELS those models which model an active system, reflecting its structure and function. Structure is conceived as a net of processors in interaction, function is defined as a state equation in a space of states. Cinematic modelling expresses how the system works. DYNAMIC MODELS reflect the trajectory in the space of structures. What distinguishes the state space radically from the structure space is that whilst the first can be determined synchronically, the second will only be determined in the future. To clarify the application of these concepts to software system modelling we will further introduce the concept of System Functional History (SFH), and that of System Structural History (SSH).

---

[1] It is very difficult to reflect exactly the spirit of an author by condensing his ideas. We recommend the reader to see [8,9].

Two important features of the TGS which we use as a support for our work are the classification of processors into nine levels of increasing complexity and the notion of finality. (For a clearer comprehension see [8,9,15]).

3 Software Development Methods in the Light of TGS

By means of our professional and educational experience a number of software requirements specification methods are known. All these methods try, to some extent, to model the real world, and have the following in common: a set of basic concepts in order to abstract reality, a language to represent abstractions and some rules (formal or heuristic) to transform these abstractions into a design and, subsequently, into an implementation. Each of these methods has its concepts, language and rules. However, each lacks important features, i.e.:

1) A well founded epistemological background that allows the integration of modelling concepts and the modelling process.

2) A model evolution representation through time. Roughly speaking, big models are built which are very difficult to modify in practice.

3) The role of the modeller, which forms the active part of the RS, incorporated into the modelling process.

4) Software evolution along with software development evolution are not considered in an integrated and homogeneous way.

The methods of software requirements specification and design normally used in software development allow the elaboration of cinematic models of systems, but do not include dynamic modelling (see [15] for a comparative study of these methods). Dynamic features are not included, and in addition:

1) System finalities are not considered.

2) Modelling the RS is avoided, which produces certain schizophrenia in their practical application since these methods do not model either the modelling activity or SR finalities. In practice, this produces a dissociation between methods and life cycle models, which proliferate in an independent way, as if they dealt with completely separate realities.

4   A Recursive-Evolutive Model of Software Development

From this discussion we deduce the need to adopt a more extensive view of the software development process. A view that joins together the development process and the models made during this process.

4.1 Epistemological Assumptions: the Framework

Software systems are developed by teams of analysts, designers and programmers who elaborate successive models finally implemented in a computer. We will call these teams collectively DEVELOPER or modeller. We assume that we are working in a prototypical situation of a software system development in which the developer conceives the system recursively[2] and represents it in successive models. This conception of the software development process needs some prior suppositions which should be explained:

1) Assuming the validity of the Le Moigne General System and his modelling conception, the principle role of the developer is to make phenomenon models as part of an RS isomorphic with the GS.

2) It does not matter if reality is continuous or not. To study and represent the software system evolution we are obliged to make the history of the system discrete in the same way as when representing the action, we appeal to discrete events which occur in a moment in time. We consider the system *conception* as a discrete event or as a discrete series of discrete events, placed between perception and representation:

reality *perceived as* mental representation *conceived as* system *represented as* model.

The model (the final product) is capable of modifying the representation, conception and perception. Even for software systems we can think that it modifies reality. Since it is unpredictable and unexplorable, for the moment, just how the models will modify reality, we will consider, as hypothesis, that this is autogenerated and autogenerative (a white lie, but interesting because something else would be assumed as being able to predict the impact of a model on the next one: to know the state function *a priori!*) We assume the modeller (observer) as capable of perceiving that the reality and models are no longer tuned in. We attribute the capacities of reconception and re-representation to the modeller.

3) The concepts to be developed should be capable of allowing conception, reconception, re-presentation, and re-re-presentation.

4) The objective which could be considered as ideal would be that the software itself was capable of re-conceiving and re-representing itself. This would imply that a software system has a finalization subsystem (level 9 processor). In the framework within which we are working autofinalization should depend on the information obtained from the environment by the system, at least whilst we accept that only living beings have creative capacities. In the present context of a generalized ignorance of the software evolution process, we are incapable of such high goals and therefore establish our objectives more modestly:

a) The observer reflects on his perceptions and conceives a system. The system is represented using a representation language[3]. We need to provide the observer with a language isomorphic with the GS, which allows him to conceive and represent, i.e. to provide him with a representation language

---

[2] Recursively in the sense that, frecuently, we talk about iterativity. We prefer the term recursivity because each new elaboration serves as a basis for the next. Further, this concept has a strong formalization, which could be interesting to tackle.

[3] Language in a very broad sense.

which allows him to make software requirements specification models and software design models.

b) We will provide the observer with software tools which help him to reconceive and re-represent, that is to say, mechanisms which make the *evolution* from one conception to the next and from one representation to the next, i.e. from one model to another easier. Definitively, to allow an easier transition between different software system models, during initial development as well as during maintenance.

We accept, as a working hypothesis, that it is possible to reach both objectives using the same representation language. It is not a capricious hypothesis, but rather a conceptual and practical need. If we assume that we are going to use the GS as a tool in order to know the reality in which software is inserted, it seems reasonable that the RS of software systems and software production systems be as uniform as possible. The practical justification comes from the object-oriented paradigm, especially Smalltalk. Uniformity, with which objects and classes are treated, leads us to think that the event that associates the SR with the object action is perfectly comparable with the event associated with the incorporation of a new class or new behaviour into the system.

5) The main assumption we have derived from this position is to try to conceive and represent the developer as a METASYSTEM of the successive software systems developed by him. Another position which implies providing the software systems with some deduction rules and an inference engine which allows them to evolve, exists. However, we prefer the first position because development systems have a high degree of complexity. This cannot be reduced to one logic and thus we believe that it is more reasonable to leave the creative aspects in the hands of the developer and help him in formalizable aspects of software development[4].

6) To clarify the application of the concepts of cinematics and dynamics to software system modelling we will introduce the concept of System Functional History (SFH), which here means the memory of actions developed by the system between two structural stability periods, and that of System Structural History (SSH) whisch identifies the memory of the actions developed by the RS to transform one system model into another. SFH reflects the system cinematics, i.e. its synchronic work, SSH reflects dynamics, i.e. the system diachrony.

4.2 The Model

Working within the aforementioned framework, we can conceive the software development process as shown in figure 1. The developer conceives the system recursively and represents it in successive models. During this process the system lacks FH or has an FH irrelevant to the system, normally used by the developer to

---

[4]This perspective, which implies a helpful attitude in design and creativity more than a substitutive aim, has been adopted by some authors in different contexts. See, for example, the design of strategic decision systems [1].

verify the correct functioning of his model. SSH is, at this point in time, very rich and changing; the developer modifies the system structure frequently (Structural History 1 in the figure). When the model is finished the authentic SFH begins in the

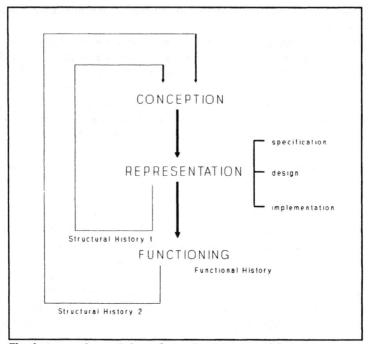

Fig. 1. A recursive-evolutive software development model

user's hands. At any moment the system structure could be modified (Structural History 2 in the figure). These structural modifications will thus be more significant than in the initial development period since SFH must be reinterpreted in the light of the new structural events which have been produced during these modifications.

With this perspective of the software development process we are trying, in particular, to represent SFH, SSH1 and SSH2 with a two-fold objective:

1) Research into software systems evolution and model software systems evolution from the perspective of TGS.

2) Design tools to aid developers in the modelling of software systems.

The goal of this paper is mainly to present the results we have obtained related to our first objective, but there is a clear relationship between both objectives, because our work tries not only to go into the epistemological and methodological fields, but also to capture the concepts using an object-oriented programming language (Smalltalk). One of our main aims has been to elaborate a representation of software systems which can be implemented in a computer. We have therefore used a cyclic prototyping to refine theoretical elements and their representation in Smalltalk. In the next sections the results obtained in SFH and SSH1 representation are shown.

5 Functioning Representation: the System Functional History

5.1 Basic Representation Concepts

The representation we have chosen uses the following concepts:

- PROCESSOR: a level 6 processor. It corresponds to the identification in the real world of an active agent capable of carrying out actions on agents which behave, for the RS, as passive. It is not a problem if a processor acts upon another processor.

- PROCESSED OBJECT: a level 1 processor. It does not carry out any significative action for RS. It is processed by a processor.

- ACTION: identifiable process carried out by a processor upon another processor or processed object.

- EVENT: memorizable representation of an action conducted by a processor. This concept is based on the Le Moigne [8] definition: "Identifiable phenomenon by which the RS knows the object intervention upon the flows it processes or in the fields that process it " (p.97).

- DECISION: permission to carry out an action based on a history of events and on a logic of decision. Each processor has its own decision subsystem, which determines its finalities.

- SYSTEM: level 6 processor. Net of processor and processed objects with a memory.

5.2 Implementation

The standard design decisions adopted to represent these concepts in Smalltalk have allowed us to associate a Smalltalk class to system, processors, events and processed objects (fig.2):

EventManager: Represents a System. It has a memory which remembers the system events, the processors and processed objects that are part of the system, i.e. it stores its FH.

EventBrowser: Presents a system (an EventManager) interactively. It uses a philosophy similar to the Smalltalk Class Browser. [5]

Event: Abstract superclass. It defines the common event protocol.

---

[5] The MVC design model widely used in Smalltalk.

RealObjects: Abstract superclass. Contains the common behaviour of all the objects capable of being processed.

         subclasses: Each one defines the characteristics of a kind of processed object.

         Processor: Abstract superclass. Contains the features common to all the processors.

Actions are defined as processor methods whose execution produces an event stored in the system memory.

Decision about the execution of an event belongs to each processor and can be expressed in terms of any Smalltalk message, obviously regarding that the previous history of events is available in the EventManager associated with each system. This implies that one sole method contains the decision elements and that this method will invoke others that execute the actions which will be reflected as system events.

With this philosophy a *kind of system* is defined as a subclass of EventManager. This abstract superclass provides the system FH storage management which has the ability to remember processors, events and processed objects. All of these are instances of the corresponding subclasses. Each RealObject subclass has its own creation model, and each Event and RealObject subclass has its own comparison model, a model which can be used to take decisions.

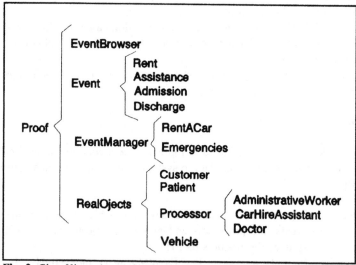

Fig. 2. Class Hierarchy used to implement the main representation concepts. There are two kinds of systems: RentACar and Emergencies.

A prototype of a system is defined associating an EventBrowser instance with an instance of the class that represents a system. This EventBrowser allows us to work with processors, processed objects and events. In [15] a Hospital Emergencies Information System example is showed.

The main advantages working this way are:

- It is very easy to change creation models, and thus the information complexity associated with each processor or processed object.
- The logic to carry out actions is concentrated for each processor and thus is easy to modify. In addition, Smalltalk dynamic binding allows each RealObject and Event subclass to have its own instance comparison model. This allows the decision to adopt patterns of great complexity, which do not blur the action algorithms (in general very simple) (fig. 3 shows the general pattern for decision. Each processor has a method **execute:with:** which decides the action to be taken and the conditions needed for each action. When an action is carried out the system memory will remember the event produced).
- Processors simulate active elements from the real world which are capable of manipulating objects and carrying out events. Both can be selected by the user. Some kind of protection affecting objects and events that a particular processor can work with can be incorporated. This allows the prototyping of the protection mechanism and the privileges of each processor subclass instance to manipulate objects or carry out events.

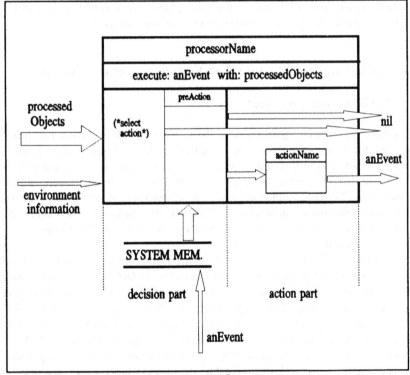

Fig. 3. General pattern of decision and action for a processor

5.3 Criticisms

Using these concepts and their implementation we have learnt some interesting points which have influenced our present work:

1) The design, i.e. the translation of the basic representation concepts to an implementation, is not very efficient and is probably not the most adequate. Now, however, we are less interested in this issue than in refining the concepts and in learning about possible design problems.

2) Decisional logic only works before the execution of an event (like an event precondition). We will need a more complex decisional logic both during and after.

3) The representation of finality with which we are working is the decisional logic separated from algorithmic elements. We need a more complex finality representation to tackle SSH representation [1,10,11].

4) We need a more complex representation of events. Le Moigne has proposed a concept of event to us quite different from the previous one: "expresses the representation of a timed transaction, a transaction between two systems or processes, identifiable and intentionally identified by the RS. This transaction can be a retroaction (transaction of a processor with itself)". This way of conceiving an event expresses not only the intentionality of the RS in the recognition of events and their granularity in a system, but a deep reflection on the conception of time in a system. We are now researching into the representation of the granularity of the events and time.

6 Evolution Representation: The System Structural History

6.1 What is software evolution?

As we have seen above, the activity of the software developer is not traditionally accounted for in software development methods. The SSH representation and its utility have not been considered in any way. Furthermore, in traditional software development there is a conceptual gap between the software product before and after its functioning. The modification activities carried out after delivering the product are called, and considered as, maintenance. This way of considering modifications does not take into account the fact that the activities are very similar to those of the development period.

Authors more involved in software evolution do not consider the role played by software development methods in evolution directly. In fact the concept of software evolution has been considered in very different ways during the last fifteen years. Belady and Lehman [2], at the beginning of Software Engineering (1976), talked about software evolution as 'Program Evolution Dynamics', identifying this evolution with the continuous changes suffered by computer programs. Changes which were seen as a source of problems and were considered as a cause of damage to programs. In a later paper, Lehman [10] established the relationship between evolution, an intrinsical software property, and the continuous backward step in the software development process throughout the software life cycle. Actually, these works can be viewed as a justification for the need for a software production discipline, Software Engineering. It has been here where research into software evolution proliferated in the late eighties. Ramamorthy et al.[17] focus their interest on the control of versions of the development products, management configuration of the software architecture and management configuration of the software life cycle, integrating them all into an aiding development system called ESE. Another point of view is that of Luqi [13], who relates software evolution with the maintenance of integrity and consistency in a software system when these changes. She gives a model of software evolution which integrates configuration management activities and the evolution of non (automatic)-derivable software products. Madhavji [14] works on the evolution of politics, resources and products involved in the development process. He defines the model of change called Prism which reflects the changes in people, politics, laws, resources, processes, results and their consequences in the environment of development.

Authors related with the object-oriented paradigm (OOP) are interested in software evolution at a level closer to programming and design. Gibbs et al. [6] interpret software evolution as the changes made in the class definitions and provide the suitable techniques to control these changes. These modifications in classes could involve modifications in the programs which use these definitions, modifications in the existing instances of these classes (change propagation) or a reorganization of the class structure [5,3]. In the object-oriented database field, software evolution is seen as the changes of the database schema, which can be translated into a basic set of transformations on the object definition [4,7,11,12,16].

Our work tries to join together a development method with the representation of software evolution and its use in aiding development. We conceive software evolution as a representation of the changes in time that the RS recognizes in the system and in its development, that is to say, the SFH, the SSH1, and the SSH2 which relates past functioning with new structural changes. To some extent our conception is related more to that of the OOP authors, but has the intention of being integrated into a method and a tool, of searching for a more general abstraction level and of regarding evolution as a process more than as a result.

6.2 Implementation

A prototype developed with the philosophy stated in paragraph 5.1 has shown us that it is possible to find a representation of SSH1 founded upon the same concepts used to represent SFH, i.e., the system of software development can be conceived as a system, which we call MetaSystem, and the developer as the one sole kind of processor involved in the development process. As a processor, the developer will be able to carry out actions and to have a decision subsystem. In short, the SSH1 of a developed system will be included in the SFH of its MetaSystem. Because of the uniformity of the concepts used, it has been possible to define a primary prototype of the MetaSystem using exactly the same implementation as shown above:

> MetaSystem: an EventManager subclass. Its memory constitutes the SSE of the systems included in it.
> Developer: subclass of Processor.

System, Processors, ProcessedObjects: RealObjects subclasses
All the events which constitute the memory of actions conducted by a developer are
subclasses of Event. Some of these are define, modify, release any ProcessedObject,
define, modify or release System(s),...
Figure 4 shows a MetaSystem prototype obtained by associating an instance of the
MetaSystem class with one of the EventBrowser class. Instances of Developer can be
defined by working in the upper right window. The middle and bottom windows allow
one to work with ProcessedObjects and Events.

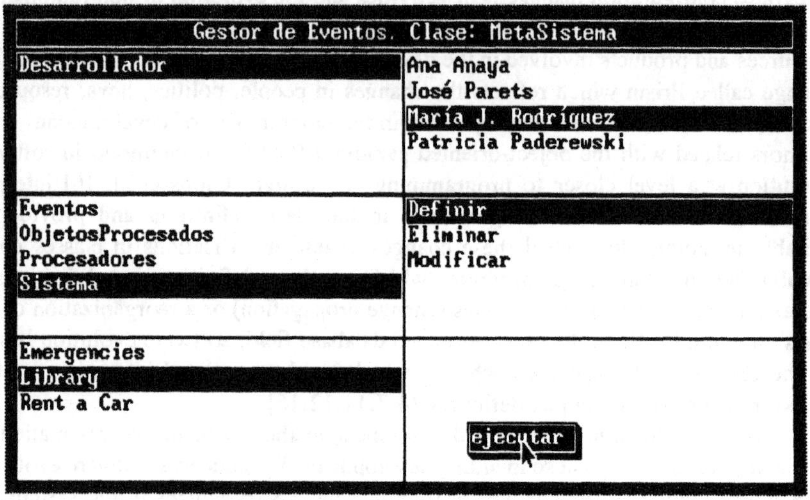

Fig. 4. A MetaSystem prototype

6.3 Criticisms

Obviously the MetaSystem is based on the same concepts as the System(s)
which we can define in it. The detectable problems are the same as those we have
stated above in the SFH representation. At present, parallel to the study of SFH, we
are researching into the following point:

- The problem of complex actions and events and their split in atomic actions
and events, referred to the events carried out by the Developer
- The representation of actions as operations upon the class structure,
considering the operations listed by Kim [7] and considered in the evolution of
class structure by Casais [5].
- The decision of the processors involved in the development process (in this
case our Developer).

5 Conclusions and Further Research

The incorporation of TGS ideas to represent systems and their evolution allow us to treat the representation of systems and development systems (MetaSystem) and their evolution in an homogeneous way.

In addition, the programming paradigm selected, i.e. object-oriented programming, has allowed a relatively simple implementation of systemic concepts, due to the active role that objects play in this paradigm and the uniformity with which these languages treat the classes and their instances.

Finally, our model of software development allows us to use the same concepts to represent software systems and to represent structural evolution, to incorporate maintenance activities as part of the system evolution and to build basic prototypes of systems and metasystems easily.

The concepts we use and the prototype in Smalltalk show that it is possible to make new system prototypes easily, defining the processors, events, actions and decision. The same concepts used to represent software systems are very useful to represent structural evolution, but, as we have stated, criticizing our current SFH and SSH representation, the concepts and their implementation need further refining and development which we centre around the following items:

1) Decision and finality representation.

2) Events and system time conception and representation.

3) Translation of MetaSystem transformations into a Smalltalk class structure representation.

References

1. J.R. Alcaras, F. Lacroux: Towards Intelligent E.I.S.:The Case of Strategic Design. Proceed.CECOIA III, 291-294 (1992)

2. L.A. Belady, M.M. Lehman: A Model of Large Program Development. IBM System Journal 3, 225-252 (1976).

3. P. Bergstein: Object-Preserving Class Transformations. Proceed. OOPSLA'91. ACM SIGPLAN Notices 26,11,299-313 (1991).

4. A. Bjornerstedt, C. Hultén: Version Control in an Object-Oriented Architecture. In:H. Kim, F.H. Lochovsky (Eds.): Object-Oriented Concepts, Database, and Applications. Addison-Wesley/ACM Press, pp.451-485 (1989).

5. E. Casais: Managing Class Evolution in Object-Oriented Systems. In: D. Tsichritzis (Eds.): Object Management. Geneve: Centre Universitaire D'Informatique, pp. 133-195 (1990).

6. S. Gibbs, D. Tsichritzis, E. Casais, O. Nierstrasz: Class Management for Software Communities. CACM 33, pp. 90-103 (1990).

7. H. Kim: Algorithmic and Computational Aspects of OODB Schema Design. In: R. Gupta, E. Horowitz (Eds.): Object-Oriented Databases with applications to CASE, networks and VLSI CAD. NJ. Prentice-Hall. pp. 26-61 (1991)

8. J.L. Le Moigne: La théorie du système général. Théorie de la modélisation. Paris. Presses Universitaires de France. (1977, 1983, 1990).

9. J.L. Le Moigne: La modélisation des systèmes complexes. PARIS. Dunod. (1990)

10. M.M. Lehman: Programs, Life Cycles, and Laws of Software Evolution. Proceedings of the IEEE 68,1060-1076 (1980).

11. B.S. Lerner, A. N. Habermann: (1990). Beyond Schema Evolution to Database Reorganization. Proceed. ECOOP/OOPSLA'90. ACM SIGPLAN Notices 21,25,67-76 (1990).

12. Q. LI, D. McLeod: Conceptual Database Evolution Through Learning. In: R. Gupta, E. Horowitz (Eds.): Object-Oriented Databases with Applications to CASE, Networks, and VLSI CAD. NJ. Printice Hall. pp. 62-74 (1991)

13. Luqi: A Graph Model for Software Evolution. IEEE Trans. Soft. Eng. 16,8, pp. 917-927 (1990).

14. N.H. Madhavji: Environment Evolution: The Prism Model of Changes. IEEE Trans. Soft. Eng. 18,5,380-392 (1992).

15. J. Parets et al.: An Object-Oriented Prototyping Framework Based on General System Theory. Dept. L.S.I. Working paper 93-1 (1993).

16. D.J. Penney, J. Stein: Class Modification in the GemStone Object-Oriented DBMS. Proceed. OOPSLA'87. ACM SIGPLAN Notices 22.12.111-117 (1987).

17. C.V. Ramamoorthy et al.: The Evolution Support Environment System. IEEE Trans. Soft. Eng. 16,11,225-1234 (1990).

Theoretical Considerations About Subset Descriptions

Gabriel Fiol   José Miró-Nicolau

University of the Balearic Islands
Departament de Ciències Matemàtiques i Informàtica
Crra. de Valldemosa, Km. 7.5
07071 Palma de Mallorca, Spain

Abstract. In this paper we present a new theoretical approach about intensional desciptions of subsets extensionally defined. In this approach some subsets of a certain domain are initially defined in an extensional way, in such a way that for each element of the subsets some attributes are known (the attributes are the same for each element). Obtaining an intensional description of the defined subsets consist of finding a property based on the attributes associated with the elements that allows classifying all the elements of the domain in their corresponding subset.
The efficacy of this approach has been proved, setting up as a powerful tool to reduce the complexity of some problems, particularly classification problems.

1 Introduction

It is well known that the complexity of some problems is due to the great amount of information to be manipulated to solve them. The classification problems are typical in these cases. Their complexity may be due to several aspects, such as the great amount of tests to be achieved to classify the elements of a certain domain, the existence of redundant decisions which cause unnecessary tests, the high cost of some decissions to be taken into account, etc.... *Inductive Acquisition of Knowledge from Examples* constitutes a powerful tool to reduce the complexity of some classification problems, particularily those ones whose solution consist of finding intensional descriptions of some equivalence classes (subsets) extensionally defined.
A brief reference to some of the many techniques that have been proposed, follows:
The AQ11 Method, proposed in [8], creates a general description from a given set of positive examples and another set of negative examples, using the language VL1 [7, 8]. It uses the detailed description of properties of the first set not in the second one. The result is given in the form of decision rules.
The ID3 (Iterative Dicotomizer 3) Method, described in [14], generates a *decision tree* from a set of elements characterized by some attributes, using an *entropy* function to select the attributes on which to basis the decision. This method has an elegant central concept and an easy implementation.
The concept of *Rough Set*, described by Pawlak in [12, 13] was developed through a study of approximate description of a subset, starting a whole area of research. The concept of *core*, defined in [11], is useful in finding the *reduct* of attributes to be used in the description of the subset. Although, the concept of *reduct* [11] does not always yield an optimal intensional description of the subsets [1].

The above mentioned methods work well for some problems. For instance, the AQ11 system obtained once a better solution than the expert that provided the data. However, all these methods seem have been conceived without taking into account the nature of the problem to be solved. It is our view that when trying to find an optimal description of a subset, the particular characteristics of the particular problem in hand must be looked into. None of the above methods can always obtain an optimal description in every case.

In this paper we present a theoretical approach about the subset descriptions, based on the theory presented in [9]. In this approach some subsets of a certain domain are initially defined in an extensional way, by a set of examples or specific cases of knowledge about their elements. Obtaining an intensional description of the subsets consists of finding a property that allows classifying all the elements of the domain in their corresponding subset. Two concepts are essential in this approach: the *Object Attribute Table* (OAT) which defines the specification model of the specific elements of knowledge and the concept of *Basis of Attributes* in terms of which the general solution is expressed.

2 The Problem

The problem studied here can be formally described as follow:

Let $D = \{d_1, d_2, ..., d_m\}$ be an extensionally defined set of elements or examples and $R = \{r_1, r_2, ..., r_n\}$ an extensionally defined set of attributes (binary or multivalued), such that for each $d_j \in D$ the values of these attributes are known. Given the extensional subsets $D_i = \{d_a, d_b, ..., d_c\}$, $i = 1, 2, ..., k$, $D \supseteq D_i$ we want to find an intensional description P_i, $i = 1, 2, ..., k$, of these subsets, $D_i = \{d_j \in D / P_i\}$, $i = 1, 2, ..., k$, that is *optimal* under some criterion; moreover, P_i will be expressed in terms of a subset R_x, $R \supseteq R_x$ of the attributes. Before to find an optimal intensional description P_i, one must find an optimal subset R_x of the attributes such that P_i can be expressed as a function of R_x.

The information from the examples will be stored in the *Object Attribute Table (OAT)*, defined as follows:

Definition 1. The OAT is defined as a four-tuple as follows:

OAT = <D, R, V, F>, where

$D = \{d_1, d_2, ..., d_m\}$ is a set of elements or concrete objects of knowledge.

$R = \{r_1, r_2, ..., r_n\}$ is a set of qualities or attributes.

$V = \{V_1, V_2, ..., V_n\}$ is a family of sets, such that V_i is the set of values of the attribute r_i adopted by the elements of D. In database literature V_i is called the domain of r_i.

$F = \{f_1, f_2, ..., f_n\}$ is a set of functions that defines extensionally the values that each $d_j \in D$ takes for each attribute $r_j \in R$, that is, $f_i: D \times \{r_i\} \to V_i$, $i = 1...n$.

The Subset Definition Table (SDT), establishes, in a format compatible with that of the OAT, the subsets (or concepts) whose intensional description is desired, by means of their Characteristic functions $f: D \to \{0, 1\}$.

Definition 2 The SDT is defined as a three-tuple as follows:

SDT = <D, C, f> where

D has the same meaning that in definition 1, C is a set of w subsets of D, $1 \le w \le m$, and the function f assigns to each element of D its corresponding subsets, that is, $f: D \to \Pi(C)$, where $\Pi(C)$ denote the set of parts of C.

The discussion of the problem becomes easier by representing graphically the OAT and the SDT, as in figure 1. We will refer to the Enlarged Object Attribute Table (Enlarged OAT); where $t_i^k \in V_k$, $k = 1...n$, $i = 1...m$, is the value of the attribute r_k associated to the element d_i through the function f_k, $C \supseteq C_i$ is the subset of C associated at the element $d_i \in D$, $i = 1...m$.

$$
\begin{array}{c|cccc|c}
 & \multicolumn{4}{c|}{\overbrace{\qquad\qquad\qquad}^{R}} & \\
D & r_1 & r_2 & \cdots & r_n & C \\
\hline
d_1 & t_1^1 & t_1^2 & \cdots & t_1^n & C_1 \\
d_2 & t_2^1 & t_2^2 & \cdots & t_2^n & C_2 \\
\vdots & \vdots & \vdots & & \vdots & \vdots \\
d_m & t_m^1 & t_m^2 & \cdots & t_m^n & C_m
\end{array}
$$

Fig. 1. Enlarged OAT

3 Intensional Descriptions

The choice of the subset R_x, $R \supseteq R_x$, of attributes to be used to describe intensionally the subsets of C constitutes one of the main stages of the inductive acqyisition of knowledge process.

A subset R_x, $R \supseteq R_x$, of attributes is able to describe the subsets of C if it provides adequate information to describe these subsets. The term adequate indicates that the attributes of R_x allow the description of the subsets of C without any kind of *confusion* (contradiction). This aspect is discussed below.

Consider an arbitrary enlarged OAT. Each row contains the complete information about one example, represented by a (n+2)-tuple as follow: $(d_i, t_i^1, t_i^2, ..., t_i^n, c_i)$, where $d_i \in D$, $i = 1...m$, is the name or identifier of the example, $t_i^j \in V_j$, $j = 1...n$, are the values of the n attributes $r_1, r_2, ..., r_n$ respectively, and $C \supseteq C_i$ is the subset associated at the example. This information can be interpreted in the following way: ' the example d_i has the value t_i^1 for the attribute r_1 and the value t_i^2 for the attribute r_2 and... and the value t_i^n for the attribute r_n and belongs to the concept C_i'. Since the mentioned problem in section 2 is focused exclusively on describing the concepts in terms of the attributes without taking into account the identifier of the examples, then one can forget the first term of the (n+2)-tuple, resulting the following (n+1)-tuple: $(t_i^1, t_i^2, ..., t_i^n, C_i)$.

Consider the following definitions:

Definition 3 Let $R_x = \{r_i, r_j,..., r_k\}$, $R \supseteq R_x$, be a subset of x attributes of the OAT. A (x+1)-tuple $(t_s^i, t_s^j, ..., t_s^k, C_s)$, $1 \leq s \leq m$, made up of the values of the attributes of R_x in the same row 's' in the enlarged OAT and the corresponding concept, is called an *instance* of R_x respect to C. Observe that this particular instance corresponds to $d_s \in D$.

Definition 4 Let $(t_s{}^i, t_s{}^j, ..., t_s{}^k, C_s)$ be an arbitrary instance of R_x respect to C. $(t_s{}^i, t_s{}^j, ..., t_s{}^k)$ is called the *left side* of the instance and (C_s) is called the *right side* of the instance.

Definition 5 Two instances of R_x respect to C are *contradictory* instances if and only if their left sides are identical whereas they have different right sides.

Definition 6 The set of all the instances of R_x respect to C is called the *Instance Set of R_x respect to C* and denoted by $ISR_x{}^C$.

Possibly, the enlarged OAT will contain some identical instances of R_x respect to C, in such a case it is enough to consider only one representative instance of the class, thus $ISR_x{}^C$ will be considered a set of equivalence class of elements of D.

The complete information provided by the n attributes of R respect to C is ISR^C and it is considered adequate to describe the concepts of C if ISR^C contains no couples of contradictory instances, else it is not adequate. If the information provided by ISR^C is not adequate then it is not possible to describe the concepts of C in terms of the attributes of R. If it is adequate there may exist more than one way to describe them.

Definition 7 For a given subset R_x, $R \supseteq R_x$, of attributes, if $ISR_x{}^C$ contains no couple of contradictory instances then R_x is called a *Basis of Attributes* (or simply *Basis*) of R respect to C.

A basis of attributes is a subset of attributes whose instance set is adequate. Therefore, we say that a subset R_x, $R \supseteq R_x$, of attributes is able to describe the subsets of D described by C if it constitutes a basis. Numerous bases may exist, the selection of a particular one to describe the concepts must depend on the characteristics of the particular case.

Some important properties to speed up the process to finding bases are discussed in the sequel.

Theorem 8 There exist at least one basis in R respect to C if and only if R is a basis respect to C.

Proof. $\rightarrow$) Let $R_y = \{r_i, r_j, ..., r_k\}$, $R \supseteq R_y$, a basis of y attributes respect to C. From definition 8, $ISR_y{}^C$ contains no couples of contradictory instances. Each instance of R_y can be obtained from some instance of R by removing the values of the attributes which are not in R_y. Consequently, since of $ISR_y{}^C$ has no contradictory instances, neither has ISR^C.

$\leftarrow$) It is immediate since R is a basis respect to C.
[e.o.p][1]

Theorem 9 establishes the conditions under which one can obtain directly bases from previously obtained ones.

Theorem 9 Let R_y, $R \supseteq R_y$, be a basis of attributes respect to C. Any subset R_z, $R \supseteq R_z$, such that $R_z \supseteq R_y$, is also a basis of R respect to C.

---

[1] The term [e.o.p] indicates the end of the proof

Proof. Each instance of R_y can be obtained from some instance of R_z by removing the values of the attributes which are not in R_y. Consequently, since $IS_{R_y}{}^C$ contains no contradictory instances, neither does $IS_{R_x}{}^C$.

[e.o.p]

Corollary 10 Let R_y, $R \supseteq R_y$, be a subset of attributes such that R_y is not a basis of R respect to C. Any subset R_x, $R \supseteq R_x$, such that $R_y \supseteq R_x$, is not a basis.

Proof. It is immediate, since each instance of R_x can be obtained from some instance of R_y and so, if two instances of R_y are contradictory instances then the instances of R_x obtained from them are also contradictory.

[e.o.p]

Another properties may be found in [1].

According the characteristics of C, one can distinguish two types of bases:

- Bases for a specific set of attributes of D, or *Specific Bases.*
- Bases for an arbitrary subset of D, or *Universal Bases.*

In the case of Specific Bases, C is a specific set of subsets of D, meanwhile in the last case C can be considered an arbitrary set of subsets of D. Notice that an Universal Basis is also specific, but it is not necessarily true otherwise. The main features and properties of Specific and Universal Bases are presented in [1].

The case i which no basis can be found (it is not possible to describe the subsets) is treated in [1] and some general solutions are proposed.

4 On Obtaining an Optimal Basis

To obtain an optimal basis, two steps must be taken, in the following order:
- To establish a criterion for optimality.
- To generate a basis from the established criterion of the former step.

The first step will establish the characteristics that a basis must exhibit to be considered an optimal one. These criteria can be based on a *cost* associated to the attributes (what this cost is depends on the nature of the problem) or on the structural relations that exist among the attributes such as the one described in [14]. The former are appropiate in cases where cost considerations about the attributes are important. The cost of the basis depends only on the cost of its attributes and the basis is optimal if and only if its total cost is minimum. Several criteria of this family have been established, being the *minimum basis criterion* one of the most significative. The second family of criteria demands the attributes of the basis to satisfy certain *decision-tree relationships* among them. The optimal bases are then defined as those attributes that satisfy a certain optimality function defined over the structure. Among the most significant criteria of this family is the *fast basis* criterion, which enables the classification of the elements of the D with a minimum number of decisions.

From this perspective we have proved [1] that a reduct does not always yield an optimal description of the subsets. Some important aspects to be considered in this stage are:

- The characteristics of the attributes, such as type of the attributes and cardinality of their domains determine the efficiency and the quality of decisions.
- The values of |D| and |R| affect the performance of the system, where |D| and |R| denote the cardinal of D and R respectively.

- The quality of the knowledge provided by the elements of D determines the quality of decisions adopted by the method presented here.

To generate a basis, a general two step procedure must be considered:

- To generate a subset R_X, $R \supseteq R_X$, of attributes.
- To check that the subset R_X obtained in the previous step constitutes a basis respect to C (remember that C can be an specific or arbitrary set of subsets).

The way to generate a subset of attributes is conditioned by the applied resolution technique. In [1] one has been developed a general method, and so a general algorithm based on the *Dynamic Programming* technique [5].

To check that the generated subset of attributes constitutes a specific basis, it is enough to verify that R_X contain no couples of contradictory instances. To do this, a new object called *Confusion Table* (CT) is defined:

Definition 11 A CT is formally defined as a three-tuple as follows:

CT = <P, R, G> where

$P = \{p_1, p_2, ..., p_k\}$ is the set of all the couples of different elements of D. That is to say, $p_i = (d_h, d_j)$, $h \neq j$, $i = 1...k$. Each couple (d_i, d_j) of P has two associated instances of R.

$R = \{r_1, r_2, ..., r_n\}$ if the set of attributes of the OAT.

$G = \{g_1, g_2, ..., g_n\}$ is a set of boolean functions defined over the elements of P and R, in the following way: g_i: P x $\{r_i\} \rightarrow \{true, false\}$ such that g_i assigns the value 'true' respect to the attribute r_i to the couples of elements of P whose associated instances are contradictory instances.

The meaning of the CT is the following: If for a subset $R_X = \{r_a, r_b, ..., r_c\}$, $R \supseteq R_X$, of attributes it occurs that any couple $p_i \in P$ leads to the value 'true' respect to all the attributes of R_X, then R_X is not a basis of attributes else it is a basis. That is to say, if $G_X = \{g_a, g_b, ..., g_c\}$, $G \supseteq G_X$, is such that $g_w(p_i, r_w) = $ 'true', $\forall g_w \in G_X$, then R_X is not a basis else it is a basis.

In [1] are exposed some important aspects about the computational structure of the CT and the algorithms to generate it.

5 Experimental Results

Several experiments [2] over several application environments have been realized using the program UIB-IK-2 (Inductive Acquisition of Knowledge System), which is based on the above considerations. One of them [1, 2] shows how to reduce the test circuitry needed to diagnose the system errors and how to optimize the diagnosis time to detect the errors of an electronic system. The results were compared with those provided by others methods, such as the ID3, AQ11 and ILS [15] systems, showing that the nature of the problem must be considered and an exhaustive analysis of it must be made before solving it. Another experiment [2] has to do with medical diagnosis and it describes how to perform an intensional description of a set of eleven extensionally defined diseases. The information about the diseases has been obtained from real data, provided by Doctors of the Hospital 'Son Dureta', at Palma de Mallorca (Spain).

6 Bibliography

1. Fiol, G.. *Contribución a la adquisición inductiva de conocimiento.* PhD Thesis, Universitat de les Illes Balears, Palma de Mallorca, Spain, 1991.
2. Fiol, G., Miró, J., Canudas, B... Some Experiments About Knowledge Acquisition Using a Subset Description Theory. Technical Report UIB-DMI-A-012. Universitat de les Illes Balears. 1993.
3. Fiol, G.; Miró, J.. S.A.I.C., un sistema de adquisición inductiva de conocimiento. *Revista de Ciència.* Vol. nº 7, 61-78. December, 1990.
4. Fiol, G.; Miró, J.. Theoretical Considerations about the Multivalued Attributes in the Process of Inductive Acquisition of Knowledge from examples. *Fifth International Symposium on Knowledge Engineering.* Sevilla, Spain, October 1992.
5. Horowitz, S. Sahni. Fundamentals of computer algorithms. *Computer Science Press,* 1978.
6. Michalski, R. S.. A geometrical model for the synthesis of interval covers. Report No. 461, Dpt. of Computer Science, University of Illinois, Urbana, Illinois. 1971.
7. Michalski, R. S.. Variable-valued logic: system VL1. Proceeding of the 1974 International Symposium on Multiple-Valued Logic, West Virginia University, Morgantown, West virginia. 1974.
8. Michalski, R. S.; Larson, J. B.. Selection of most representative training examples and incremental generation of VL1 hypotheses: The underlying methodology and the description of programs ESEL and AQ11. Technical Repord 867, Computer Science Department. University of Illinois at Urbana-Champaign, Urbana, IL, 1978.
9. Miró, J.. On defining a set by a property. Technical Repord, Universitat de les Illes Balears, Palma de Mallorca, Spain, 1987.
10. Pawlak, Z.. Information Systems-Theoretical Foundations. *Information Systems,* 6: 205-218. 1981.
11. Pawlak, Z.. On Superfluous Attributes in Knowledge representation Systems. *Bulletin of the Polish Academy of Sciences,* 32(3-4), 1984.
12. Pawlak, Z.. Rough Sets and Fuzzy Sets. *Fuzzy Sets and Systems,* 17:99-102, 1985.
13. Pawlak, Z.. Rough Sets. *International Journal on Information and Computer Sciences,* 11:341-356, 1982.
14. Quinlan, J. R.. Induction of Decision Trees. *Machine Learning,* 1:81-106, 1986.
15. Wong, J.H.. An Inductive Learning System-ILS. A thesis submitted to the Faculty of Graduate Studies and Research. University of Regina. 1986.

Sampled Data Passive Systems

P. Albertos

Departamento de Ingeniería de Sistemas,
Computadores y Automática. (DISCA)
Universidad Politécnica de Valencia
e-mail: pedro@aii.upv.es
Apdo.22012, E-46071 Valencia, Spain

Abstract

Advances in computer-based control systems require to consider digital controllers on the base of multirate and unconventional sampling schemes. The purpose of this paper is to extend the concepts of system passivity which are well known ideas in the context of continuous and discrete time systems to the case of sampled-data systems. Initially, the classical sampling pattern is considered. Then, unconventional sampling schemes lead us to the general treatment of periodic systems.

1 Introduction

The theory of linear passive systems, and more generally speaking linear dissipative systems, has been largely studied in the context of system theory and stability analysis. The former approach dealt with finite dimensional time invariant continuous time (CT) systems, as reported in [1], and lately extended to the case of discrete time (DT) systems, (see, for instance,[2, apendix C]). Sampled-data (SD) systems use to be handled by the approximated DT systems, the intersampling behaviour being underconsidered. But dissipativeness is a system property involving the values of the system input and output for all the time instants, and a carefully analysis of the SD systems shows up some differences to the equivalent DT one.

Also, in the case of unconventional sampling, that is, when output sampling and input updating do not happen at regularly spaced time intervals, the dissipativenes property must be tested taking into account the actual input/output values. The problem of dealing with nonlinear or time varying systems is quite complex, but some especial cases can also be studied. In particular, for multirate and periodic systems [3,4], presenting a growing interest in the field of control system implementation, the dissipativeness analysis leads to analytical expressions, easy to check, as shown in the paper.

In the next section, a review of CT and DT models and dissipativeness concepts is presented. These concepts may be interpreted in both time and frecuency domains, leading to different formulas to compute the system properties. Then, the same concepts are applied to SD systems. Initially, a regular sampling period is assumed. The results are then extended to periodic (multirate) systems. Alternative expressions to compute the system passivity are stated in both internal and external representations.

2 Preliminaries

2.1 Models

Let us summarize the representation of the signals and systems we are interested in.

We will consider a finite dimensional linear time invariant continuous time system, **S**, described by

$$
\begin{aligned}
\dot{x}(t) &= A_c x(t) + B_c u(t) \\
y(t) &= C x(t) + D u(t)
\end{aligned}
\tag{1}
$$

where $x(t) \in R^n$ is the state vector, $u(t) \in R^m$ is the control vector, $y(t) \in R^p$ is the output vector, and $A_c, B_c, C,$ and D are real matrices of appropriate dimensions.

The input signals are assumed to be elements of $\mathcal{L}_2^m(R_+)$, the space of square integrable functions $u : R_+ \rightarrow R^m$. Nevertheless, from a practical viewpoint, only finite interval signals will be considered. The truncation operator is defined by:

$$
\begin{aligned}
P_T u(t) &= u(t) \quad \forall t < T \\
&= 0 \quad \forall t \geq T
\end{aligned}
\tag{2}
$$

The enlarged space of square integrable functions on all finite intervals $[0, T]$, for $T \geq 0$, is denoted by

$$
P_T u(t) \in \mathcal{L}_{2e}^m(R_+)
$$

The input signals are elements of $\mathcal{L}_{2e}^m(R_+)$.

The system **S** may be also defined by a linear operator $\bar{G}$ mapping the input space into the output space

$$
(\bar{G} u)(t) = y(t).
\tag{3}
$$

This output space will be represented by $\mathcal{L}_{2e}^p(R_+)$.

It is well known that the transfer function matrix of system **S** is given by

$$
G(s) = C(sI - A_c)^{-1} B_c + D
\tag{4}
$$

where $G(s)$ is a proper matrix.

In the time domain, the system output can be expressed by

$$y(t) = \int_0^t H(t - \tau)u(\tau)d\tau \tag{5}$$

where the impulse response matrix $H(.)$ is given by

$$H(t) = Ce^{A_c t}B_c + D\delta(t) \tag{6}$$

and $\delta(t)$ is the impulse Dirac function.

Let us remind that the system frecuency behaviour is expressed by its frecuency response matrix, given by $G(jw)$, where its complex entries represent the output/input frecuency gain and phase shift.

Discrete systems

For a generic sampling period Δ, the discrete time system, **D** obtained by discretization of the plant model (1) will be:

$$
\begin{aligned}
x[(l+1)\Delta] &= A(\Delta)x(l\Delta) + B(\Delta)u(l\Delta) \\
y(l\Delta) &= Cx(l\Delta) + Du(l\Delta); \qquad l = 0, 1, 2, \ldots
\end{aligned}
\tag{7}
$$

where

$$
\begin{aligned}
A = A(\Delta) &= e^{A_c\Delta} \in R^{n\times n} \\
B = B(\Delta) &= \int_0^\Delta e^{A_c\tau}B_c d\tau \in R^{n\times m}
\end{aligned}
$$

In this case, the input signals are assumed to be elements of $\mathcal{L}_2^m(Z_+)$, the Hilbert space of sequences $\{u_k\}$. To deal with finite time sequences, the truncated sequence over the interval $[0, N]$ is represented by $\{\bar{u}_k\}_N = P_N\{u_k\}$, where

$$
\begin{aligned}
\bar{u}_k &= u_k \quad \forall k < N \\
&= 0 \quad \forall k \geq N
\end{aligned}
\tag{8}
$$

The discrete transfer function is

$$\hat{G}(z) = C(zI - A)^{-1}B + D = G(s)|_{s=\frac{1}{T}\ln z} \tag{9}$$

2.2 Passivity

Let us summarize some basic definitions of dissipativeness and related functions. In the following transfer function square matrices are assumed, i.e., $p = m$.

Definition 2.1.- System passivity. A system **S** is passive iff the following inequality holds

$$J_u(T) = \int_0^T y\prime(\tau)u(\tau)d\tau = < y, u >_T \geq 0 \tag{10}$$

$$\forall T \geq 0; \qquad \forall u \in \mathcal{L}_2^m(R_+)$$

where $\prime$ denotes the matrix transposition operation. By (5), this condition may be expressed by

$$J_u(T) = \int_0^T \int_0^t u\prime(t) H(t - \tau) u(\tau) d\tau \; dt \qquad (11)$$

Definition 2.2.- System passivity (Internal representation). A system **S** is passive iff $\exists \Phi : R^n \to R_+$ such as

$$\Phi(x) \;=\; \frac{1}{2} x\prime P x; \quad P > 0 \qquad (12)$$

$$\frac{d\Phi(x)}{dt} \;\leq\; u\prime y$$

The Kalman-Yakubovich lemma gives an algorithmic procedure to evaluate this condition.

Definition 2.3.- System dissipativeness. Attached to a system **S** and a time interval T, the following Energy Supply Function may be defined:

$$E(u, y, T) = < y, Qy >_T + 2 < y, Su >_T + < u, Ru >_T \qquad (13)$$

leading to the set of definitions, [2],

$$\textbf{Dissipative System:} \quad E(u, Gu, T) \leq 0$$

and, in particular, for spacial matrices

$$\begin{aligned}
\textbf{Passive System} \quad &: \quad Q = R = 0; S = 1/2I \\
\textbf{Input Strict Passive System} \quad &: \quad Q = 0; S = 1/2I; R = -\epsilon I \\
\textbf{Output Strict Passive System} \quad &: \quad Q = -\delta I; S = 1/2I; R = 0 \\
\textbf{Finite Gain Stable System} \quad &: \quad Q = -I; S = 0; R = k^2 I
\end{aligned}$$

Definition 2.4.- System positiveness. A system matrix transfer function $G(s)$ is possitive-real iff

$$\begin{aligned}
i) \qquad & G^*(s) + G(s) \geq 0; \quad \forall s \in \mathcal{C}_+ \qquad (14) \\
ii) \qquad & \text{all poles} \quad [G(s)] \in \mathcal{C}_- \\
iii) \qquad & G\prime(-jw) + G(jw) \geq 0; \quad \forall w \in R;; \quad jw \neq \quad \text{pole} \; [G(s)]
\end{aligned}$$

where $()^*$ stands for the hermitian matrix.

In the SISO case, the above condition is equivalent to

$$Re[G(jw)] \geq 0 \qquad (15)$$

The following theorem, [1], connects both properties:

Theorem 2.1. A system **S** is passive iff its transfer function matrix is positive-real. That is

$$J_u(T) \geq 0 \iff G(s) \text{ positive-real} \qquad (16)$$

In the scalar case this implies $G(jw) \in \mathcal{C}_+$

Definition 2.5.- Discrete time System passivity. A system **D** is passive iff the following inequality holds

$$J_u(N) = \sum_{k=0}^{N} y'_k u_k = < y, u >_N \geq 0 \qquad (17)$$

$$\forall N \geq 0; \qquad \forall u \in \mathcal{L}_2^m(Z_+)$$

Frecuency domains interpretations.

In the scalar case, the dissipativeness properties may be graphically shown attached to the polar plot of the frecuency response function, [2]. In particular, system passivity, (15), will require the polar plot to lay in the right half plane.

2.3 Passivity Applications

Dissipativeness concepts above described are useful to solve a number of interesting control problems.

Feedback systems stability is a key problem in control theory. In [1], the following result has been proved: "Given a feedback system, as shown in fig. 1, under stable operating conditions, the input-output stability is guaranted if G_1 is passive and G_2 is input strict passive".

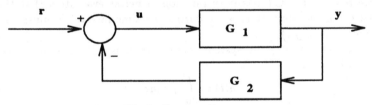

Fig 1. System diagram

This result is a general one, valid for any kind of systems, (CT, DT, non-linear, distributed parameters, and so on). Most of the stability criteria for non-linear systems (circle, Popov, ...) can be proved in the framework of this theory.

The robust control of plants with uncertainties in the model is another key issue in control. Adaptive control is a well known approach to deal with this situation, [2]. One of the most popular schemes is the Model Refernce Adaptive Control, fig. 2. The main problem is to develop parameter tuning rules to stabilize the global system behaviour.

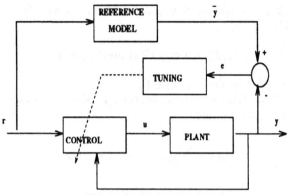

Fig 2. MRAC scheme

It can be shown that if the plant transfer function is positive-real, there is a MRAC system which is Lyapunov stable.

As it will be presented later on, passivity properties may be also useful to analyze the convergence conditions of parameter estimation algorithms.

3   Sampled-data Systems

Basically, a SD system may be considered as a CT system, its input being kept constant in the intersampling time. From an I/O point of view, the input signal is a piecewise constant one, the output being a CT one. In spite of any kind of feedback, in the intersampling periods, the system is behaving like an open loop system, driven by the constant input, with some initial conditions.

If we refer to the CT passivity function, a better evaluation that the one obtained using the product $y_k \times u_k$ during the time interval k, must consider the actual value of $y(t)$.

If we denote by $g(t)$ the system impulse response, the system unity step response will be

$$h(t) = \int_0^t g(\tau)d\tau \tag{18}$$

That means, the process output may be expressed by

$$y_k(t) = \sum_{i=0}^{k} h(t - i\Delta)u(i\Delta); \text{ for } k\Delta \leq t \leq (k+1)\Delta \tag{19}$$

The system passivity condition will be expressed by the following proposition:

Proposition 3.1 *A SD system is passive iff the passivity function, as defined by*

$$J_u(N) = \sum_{k=0}^{N} u(k\Delta) \int_{k\Delta}^{(k+1)\Delta} y_k(t)dt. \tag{20}$$

is positive. ▽▽▽

This expression is quite difficult to handle, even assuming $h(t - i\Delta) =$ *constant* for most i-values.

Some simplifications may be considered. The simplest one is to analyze the DT equivalent system, as previously mentioned. It means to consider the output also constant in the intersampling time.

Let us now study the system frecuency response. In a recent paper, [5], a new frecuency domain sensitivity function is defined. Denoted by *Reference Gain* function, $R(jw)$,

$$Y(jw) = R(jw)U(e^{jw}) \tag{21}$$

it relates the frecuency content of the CT system output, $y(t)$ w.r.t. a DT sinusoidal input u_k. This is, of course, the main frecuency component in the output, but, as it is well known, this input will generate a number of output harmonics.

Using the frecuency domain passivity test (.), the following proposition is stated.

Proposition 3.2 *A SD system could be consider passive if the following inequality holds:*

$$Re[R(jw)] \geq 0. \tag{22}$$

$$\nabla\nabla\nabla$$

For MIMO the equivalent inequality will be

$$R(jw) + R\prime(-jw) \geq 0.$$

If the sampling period is short enough, a simpler option is to make the following approximation:

$$\int_{kT}^{(k+1)T} y(\tau)d\tau \approx T\frac{y_k + y_{k+1}}{2} \tag{23}$$

by means of the trapezoidal integral approximation.

In that case, the passivity function will be

$$\tilde{J}_u(N) = <\frac{y_k + y_{k+1}}{2}, u_k >_N . \tag{24}$$

Based on the DT process model, the next sampled output value may be computed. In fact, all SD system representing a digitally controlled process have at least one period time delay. If the DT transfer function is $G(z)$, for the one-period of time delayed output, the transfer function is $z.G(z)$.

Thus, the following proposition is concluded:

Proposition 3.3 *In the frecuency domain, the positiveness of the passivity function $\tilde{J}_u(N)$ implies the averaged transfer function*

$$\bar{G}(z) = \frac{1+z}{2}G(z) \tag{25}$$

to be positive-real.

$$\nabla\nabla\nabla$$

Remark 3.1. As the interesting range of frecuencies is $0 < w < \frac{\pi}{\Delta}$, the modified frecuency response, $\bar{G}(e^{jw\Delta})$ is always phase led w.r.t. $G(e^{jw\Delta})$. For stable minimum-phase systems, this condition is less restrictive than the previous one, (17).

4 Periodic System

Periodic controllers, leading to periodic system, is a topic of growing interest [3,4]. Again, the issue of intersampling behaviour is not fully considered and some recent results [6] point out the degradation of performances of the controlled process, if this is not taken into account. To precisely evaluate the passivity condition, expressions similar to (20) must be handled.

The same happens if one is dealing with unconventional sampling schemes, where input and output data are not synchronously available, [7]. One possible simplification is to only consider the case of not equally spaced samplig but periodic anyway.

In the sequel, periodic DT systems are considered. The periodicity is assumed to be represented by either one (or a number of them) periodic static gain element or a set of periodic state space matrices.

4.1 State space model

The model matrices, A_K, B_K, C_K, D_K , are assumed to be periodic. That is, for instance, $A(k + K) = A(k)$.

Let us define vectors of output and input blocks:

$$
Y_k = \begin{bmatrix} y_{kK} \\ y_{kK+1} \\ \vdots \\ y_{(k+1)K-1} \end{bmatrix} \qquad U_k = \begin{bmatrix} u_{kK} \\ u_{kK+1} \\ \vdots \\ u_{(k+1)K-1} \end{bmatrix}
$$

Derive

$$
\begin{aligned}
X_{(k+1)K} &= FX_{kK} + GU_k \\
Y_k &= HX_{kK} + LU_k
\end{aligned} \tag{26}
$$

where:

$$
F = \prod_{i=K}^{1} A_i \qquad G = [\prod_{i=K}^{2} A_i B_1 \ldots A_K B_{K-1} \ B_K]
$$

$$
H = \begin{bmatrix} C_1 \\ C_2 A_1 \\ \vdots \\ C_K A_{K-1} \ldots A_1 \end{bmatrix} \qquad L = \begin{bmatrix} D_1 & 0 & \cdots & 0 \\ C_2 B_1 & D_2 & \cdots & 0 \\ \vdots & \vdots & \cdots & \vdots \\ C_K A_{K-1} \ldots A_2 B_1 & \cdots & \cdots & C_K B_{K-1} D_K \end{bmatrix}
$$

According to the previous definitions, the z-transform of (26) is

$$Y(z) = \{H(zI - F)^{-1}G + L\}U(z) = M(z)U(z) \tag{27}$$

The entries of $M(z)$, the block tranfer function are

$$M_{ij}(z) = C_i A_{i-1} \ldots A_1 \delta_z A_K \ldots A_{j+1} B_j + D_{ij}$$

where

$$\delta_z = (zI - F)^{-1} \qquad D_{ij} = \left\{ \begin{array}{ll} D_i & j = i \\ 0 & j \neq i \end{array} \right.$$

The system frecuency response is obtained by:

$$z = e^{-jwT\Delta} = e^{-j\Omega T}$$

The above expressions, combined by (14-17) allow us to state the following result.

The passivity condition for a periodic system is given by the **Periodic system passivity test I:**

$$T_1(\omega) = M(e^{j\omega T\Delta}) + M(e^{-j\omega T\Delta})\prime \geq 0 \quad for\ all\ \omega \tag{28}$$

$$\nabla\nabla\nabla$$

Remark1. The model (27) and the passivity test (28) also stands for MIMO systems.

Remark2. Δ is not a relevant parameter. Instead, $\Omega = \omega\Delta$ can be used.

4.2   Frecuency domain approach.

Let us now consider periodicity in the gain, λ_k, of a scalar element. The following reasoning may be extended to many such a device

$$y_k = \lambda_k.u_k \quad \text{if } \lambda_k = \lambda = constant \quad \rightarrow Y(e^{j\omega\Delta}) = \lambda U(e^{j\omega\Delta})$$

Let us express $\{\lambda_k\} = \{\lambda_0, \lambda_1, \ldots, \lambda_{K-1}\}$ in Fourier series, denoting by $\Omega_s = \frac{2\pi}{K}$,

$$\lambda_k = \sum_{n=0}^{K-1} l_n e^{jk\Omega_s n} \quad \text{where} \quad l_n = \frac{1}{K} \sum_{k=0}^{K-1} \lambda_k e^{-jk\Omega_s n}$$

Then :

$$\lambda(\Omega) = \sum_{n=-\infty}^{\infty} 2\pi l_n \delta(\Omega - n\Omega_s) \quad , \Omega = \omega.\Delta$$

The frecuency response of the periodic device is:

$$Y(\Omega) = y(e^{j\omega\Delta}) = \frac{1}{2\pi} \int_0^{2\pi} \lambda(\eta)u(e^{j(\Delta\omega-\eta)})d\eta = \sum_{n=0}^{K-1} l_n u(e^{j(\Delta\omega-n\Omega_s)})$$

Denote

$$\hat{Y}(\Omega) = \begin{bmatrix} y(e^{j\omega\Delta}) \\ y(e^{j(\Delta\omega-\Omega_s)}) \\ \vdots \\ y(e^{j(\Delta\omega-(K-1)\Omega_s)}) \end{bmatrix} ; \hat{U}(\Omega) = \begin{bmatrix} u(e^{j\omega\Delta}) \\ u(e^{j(\Delta\omega-\Omega_s)}) \\ \vdots \\ u(e^{j(\Delta\omega-(K-1)\Omega_s)}) \end{bmatrix}$$

We can express

$$\hat{Y}(\Omega) = \Lambda\hat{U}(\Omega), \tag{29}$$

where

$$\Lambda = \begin{bmatrix} l_0 & l_1 & \cdots & l_K \\ l_T & l_0 & \cdots & l_{K-1} \\ \ddots & \ddots & \ddots & \ddots \\ l_1 & l_2 & \cdots & l_0 \end{bmatrix}$$

is a circulant matrix.

In a similar way, we can also relate extended I/O vectors of a time invariant linear dynamic system. If the discrete time transfer function is $H(z) = S(z)/E(z)$, the frecuency response is

$$S(e^{j\omega\Delta}) = H(e^{j\omega\Delta})E(e^{j\omega\Delta})$$

For the block of inputs/outputs, we have

$$\hat{S}(\Omega) = \hat{H}(\Omega)\hat{E}(\Omega) \tag{30}$$

Where $\hat{S}(\Omega), \hat{e}(\Omega)$ are defined as $\hat{Y}(\Omega)$ and $\hat{U}(\Omega)$ were, and

$$\hat{H}(\Omega) = diag[H(e^{j\Omega})\ H(e^{(\Omega-\Omega_s)})\cdots H(e^{j(\Omega-(K-1)\omega_s)})]$$

(29) and (30) allow to model any periodic system.

For instance, in the system given by the block-diagram, fig. 3,

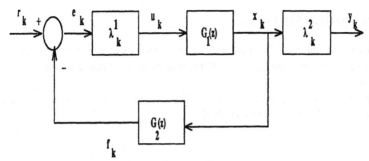

Fig 3. System diagram

the following relationships are straighforward:

$$\left.\begin{array}{rcl}
\hat{Y}(\Omega) & = & \Lambda_2 \hat{X}(\Omega) \\
\hat{F}(\Omega) & = & \hat{H}_2(\Omega)\hat{X}(\Omega) \\
\hat{E}(\Omega) & = & \hat{R}(\Omega) - \hat{F}(\Omega) \\
\hat{U}(\Omega) & = & \Lambda_1 \hat{E}(\Omega) \\
\hat{X}(\Omega) & = & \hat{H}_1(\Omega)\hat{U}(\Omega)
\end{array}\right\} \hat{Y} = \Lambda_2(I + \hat{H}_1\Lambda_1\hat{H}^2)^{-1}\hat{H}_1\Lambda_1\hat{R}$$

$$\hat{Y}(\Omega) = \hat{M}(\Omega)\hat{R}(\Omega)$$

Thus, in that example,

$$\hat{M}(\Omega) = \Lambda_2(I + \hat{H}_1(\Omega)\Lambda_1\hat{H}_2(\Omega))^{-1}\hat{H}_1(\Omega)\Lambda_1$$

The system passivity condition will be expressed by the
Periodic system passivity test II:

$$T_2(\omega) = \hat{M}(\Omega) + \hat{M}(-\Omega)\prime \geq 0 \quad \textit{for all } \Omega(= \omega.\Delta) \tag{31}$$

$T_1(\omega)$ or $T_2(\omega)$ will be more or less suitable to check the passivity condition of the periodic system, according to the available periodic system model. Of course, process in fig. 3 may be also modelled by a state space representation, like (26). In this case, both tests are equivalent.

5 Application

The above lemmas have been applied to prove the convergence of a recursive least square parameter estimation algorithm suitable to incomplete measurement data scenarios, [8].

Fig. 4 shows the parameter estimation lay-out, where a linear plant model $G(z) = B(z)/A(z)$ has been assumed.

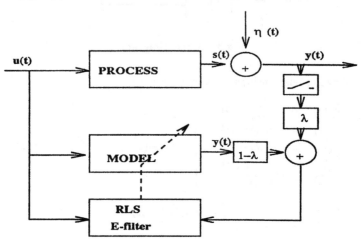

Fig 4. Missing data parameter estimation scheme

The measurement pattern is

$$\lambda \quad = \quad 1 \quad \text{if measurable output} \tag{32}$$
$$= \quad 0 \quad \text{otherwise}$$

and the RLS regressor entries are E-filtered, [2].

In [9], the convergence of the estimation algorithm is directly attached to the passivity of a time-varying system as depicted in fig. 5.

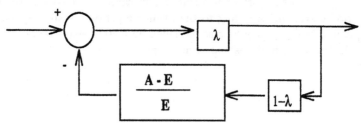

Fig 5. Auxiliar time-varying system

If the measurement pattern is periodic, so this system is.

6 Conclusions

Passivity concepts are useful to deal with many issues of control theory. Although they usually lead to conservative constraints, they can be applied to a great variety of models.

Based on a generalized frecuency domain representation of periodic systems, passivity has been analyzed in this class of systems. The main results are two passivity tests for periodic systems.

7 References

[1] Hill, D.J. and Moylan, P.J. " Dissipative Dynamical Systems: basic input/output and state properties" *J. Franklin Inst.* Vol 309, No 5. 1980.

[2] Goodwin, G.C.G. and Sin, K.S. *Adaptive Filtering, Prediction, and Control* Prentice Hall, 1984.

[3] Francis, B.A. and Georgiou, T.T. "Stability Theory for Linear Time-invariant Plants with Periodic Digital Controllers.*IEEE Trans. on Auto. Control.* Vol AC-33, No. 9. pp 820-832. 1988.

[4] Khargonekar, P. P., Poolla K., and H. Tannebaum, "Robust control of linear time-invariant plants using periodic compensation," *IEEE Trans. on Automat. Contr.*, Vol. AC-30, pp. 1088-1096, November 1985.

[5] Goodwin, G.C. and Salgado, M." Frecuency domain sensitivity functions for continuous time systems under sampled data control". *CICS Internal report.* Newcastle. AUS. 1992.

[6] Feuer, A. and Goodwin, G.C. "Linear Periodic Control: a Fracuency Domain Viewpoint" *Systems & Control Letters* 1992.

[7] Albertos, P., "Block Multirate Input-Output Model for Sampled-data Control Systems". *IEEE Trans. on Auto. Control.* Vol AC-35, No. 9. pp 1085-1088. 1990.

[8] Albertos, P., Goodwin, G.C., and Isaksson, A.I. "Missing-Data Pseudo Linear Regressive Estimation Algorithm" CICS Internal report. August 1992.

Computer-Aided Systems Technology for CAD Environments: Complexity Issues and a Repository-Based Approach

Tuncer I. Ören

University of Ottawa
Computer Science Department
Ottawa, Ontario Canada K1N6N5
oren@csi.uottawa.ca

Douglas G. King

Data Kinetics
Ottawa, Ontario Canada K1S3K5
king@csi.uottawa.ca

Abstract: Design is a model-based activity and is essential in any engineering field. Computer-Aided Design (CAD) can benefit from Computer-Aided Systems Technology (CAST) since systems theories on which CAST is based on, provide powerful bases to tackle complexity issues as well as modelling and model processing formalisms. In computerization, another type of complexity, tool interface complexity arises. For a system where n software tools communicate, the order of the interface complexity is n^2 which may become unmanageable. With the use of a repository, the tool interface complexity is reduced to n. The article describes an example repository-based CAST-CAD environment which uses Finite State Automata-based tools.

1 Introduction

Design is essential in any engineering activity. A *design problem* has two aspects: (1) development of a set of requirements for a system; (2) development of a model (or design) based on which the real system that will satisfy the requirements can be built or developed. Modelling (or development of a model) requires the specification of the component models and their communication interface, that is, their coupling. Component models can be static or dynamic. Behaviors of static component models do not depend on time; therefore, they do not need state transition functions and associated output functions. However, dynamic component models are just the opposite; their behaviors depend on time and they require their internal dynamics to be specified in terms of state transition functions and possibly in terms of the associated output functions [4].

Design, similar to analysis and control problems, is a model-based activity. A model-based activity consists of three parts: specification of a model-based problem, model processing, and processing of the behavior of the model [3].

Specification of a model-based problem consists of the specification of the model (or design) and the specification of the goal-directed solution generation or search conditions. Most often, the model will be a parametric one to be associated with different parameter sets each of which specifying the values of the model parameters. Solution may be a behavior of the model or a function of the model or its behavior. The details of the solution generation conditions determine the nature of the model processing.

Model processing can be done for two purposes: to generate model behavior and to process a model symbolically. Symbolic processing is done for analysis or for transformation purposes; therefore, the outcome can be either another model or information about the symbolically processed model.

Model behavior can be a point, trajectory, or structural behavior; point behavior for static models and trajectory and structural behaviors for dynamic models. Part of the processing is its compression (statistically or analytically) and display (still or animated).

System theories provide powerful formalisms for modelling and model processing [2, 11]. The developments in the Computer-Aided Systems Theory (CAST) [6, 7] warrant Computer-Aided Systems Technology (CAST) which, in turn, has a very promising role in model-based activities and hence, in computer-aided problem solving, in general and in Computer-Aided Design (CAD), in particular.

2 Tool Interface Complexity

In CAST, two categories of complexities exist. One is inherent of the systems under investigation. Klir and Wymore have in depth discussions on this important topic [2, 11]. Another important type of complexity stems from the interface of several software tools needed to solve different aspects of a problem. The software tool interface complexity problem has been covered in several publications [8-10].

In this article, the aim is to highlight the complexity issue raised from the interface of software tools and a repository-based solution to the problem. The repository-based solution decreases the order of complexity from n^2 to n where n is the total number of cooperating CAST-based software tools. Examples are software tools implementing Finite State Automata-based CAST concepts.

3 Integrated versus Integrative Software Environments

The integration of CAST tools can be approached at two different levels; namely,

(1) Integration at the user level. This approach presents the user with a package that provides the functions of the whole toolset.

(2) Integration at the tool level. This approach presents the user with a set of distinct tools that can work together to provide the function of the whole toolset. This means that the tools are able to share data and especially share intermediate results.

Integration at the user level is the most commonplace way of building environments. This approach leads to an *integrated* (or closed) system with a very high level of tool interdependence. This approach is inconvenient when dealing with existing tools because the individual tools need to be re-engineered to use a new data-sharing design.

Integration at the tool level provides the user with a set of tools that can be variable and subject to expansion as improved tools become available. Care should be taken to ensure that the tools are put together to form an *integrative* environment.

The key to the success of the system is the focus on building an *integrative* rather than an *integrated* system. The ability to customize a previously configured system to include a newly available tool is invaluable when one considers the savings of efforts in otherwise re-implementing any existing tools which would like to make use of the new tool through a data sharing mechanism. The savings in human effort and cost as well as the improvement of the efficiency can be tremendous. The configured repository-based system presents the user with a set of strongly coupled components which cooperate through use of the repository as a vital component in the communication interface. By carefully defining the data types of each tool in terms of a shared schema which is used by the repository for storage purposes, the tools gain an advantage through the ability to transparently trigger data transformation routines during the tool-to-tool communication.

4 Repository-Based Approach

The integrative approach is most directly supported using a *repository* as the means of sharing data between tools. In this model, the repository is used as a communication mechanism between tools, in an asynchronous data exchange.

The architecture of a CAST-based design environment which takes advantage of an underlying repository is presented. This stand-alone system is built from existing specialized tools. The reuse of existing subcomponents inside the system allows a sophisticated system to be tailored to the specific needs of a user. The user would otherwise attempt to cope with using a set of individual tools and managing the flow of data and information using manual methods.

The presented repository-based CAST-CAD system is typically configured. The set of tools chosen for the example allows the use of several simple tools to build and analyze models based on Finite State Machine Theory. A verification and validation tool is used for checking the logical consistency of a set of statements of facts about the finite state machine. This provides the system with a high degree of reliability in this rather important CAST-CAD field.

The repository stores shared data in a single internal format by triggering transformations from the source and destination data formats. The triggering of data transformations is transparent to the user, allowing sharing of otherwise incompatible data types. The data transformations are defined at the time the tool is installed into the system. This results in an *integrative* environment in which new tools can be installed as they become available.

The use of a shared schema results in a great reduction in complexity when considering the number of data transformations required to support the communicating tools. This situation is illustrated in the figure directly below.

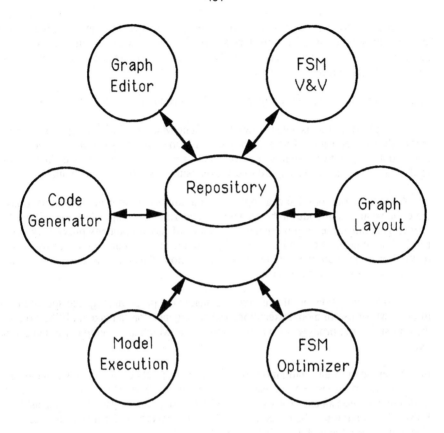

Figure 1. Connectivity in the CAST Toolset

5 CAST Toolset

The repository-based approach has been applied to a pilot project in which a set of simple tools has been gathered to build and analyze models based on Finite State Machine Theory. The tools that were present in the pilot project are shown in Figure 1.

One of the main considerations in the design of the CAST toolset was the need to build a stand-alone system based on existing specialized tools. This reuse of existing components allows creation of a unique sophisticated system tailored to the needs of a particular user or set of users.

The alternative to generating a CAST toolset environment would be for the user to attempt to cope with a set of individual tools and manage the flow and transformation of data among the tools.

As a group, the tools and repository form an *integrative* system. The strongly coupled tools can cooperate using the repository as the vital communication interface reducung the necessity to communicate between themselves. Therefore the complexity of the tool interface is reduced from n(n-1) to 2n.

The toolset in this typically configured CAST environment consists of

(1) A graph editor that is able to draw Finite State Machine (FSM) diagrams and save the results in a repository. FSMs already stored in the repository can be retrieved to allow updates as part of a step-wise refinement of a model. This tool was "repository-ready" in that it had been designed to use a repository as a storage mechanism.

(2) A graph layout tool that can apply one of a number of built-in algorithms to generate the positions of nodes to be drawn in a rendering of a graph. This tool allows the user to generate a pleasing presentation of an FSM based upon any previously stored FSM. The input FSM could have been generated using the graph editor, or by using a text editor to generate the state machine information. This tool was designed to use a repository as a storage mechanism.

(3) A code generation tool that is able to generate the program language specific form of an FSM for use in model execution. The code generator can take an FSM stored in the repository and produce a C program source file which after being compiled can be executed.

(4) A verification and validation tool that is able to check for inconsistencies and problematic structures (e.g., unreachable states) in an FSM. This tool is implemented in Prolog and uses an externally maintained rulebase to check particular aspects of the FSM. The existing tool had used a Prolog fact base as the input format for the FSM. The output from the tool is written to a log file.

(5) A model execution tool that can compile and execute the FSM program. The source of inputs to the FSM is based on a user-supplied parameter. At present, all inputs are taken from a file. This tool is a very simple C shell program that runs on a UNIX platform.

(6) An FSM optimization tool. This tool takes an FSM as input and generates an optimized FSM as output. The previously existing tool used input and output in the form of text data files containing the FSM information. Installation into the repository consisted of defining the transformations between the shared schema and the text data format.

Many other tools to handle several types of FSM formalisms in modelling and model processing would be useful in a CAST toolset. These tools could be introduced into the system without affecting the existing tools. A new tool would be installed into the repository, and thereby gain access to the existing set of FSM models.

Some of the other useful tools we can envision adding to the toolset would include

(1) A knowledge-based system to analyze input/output requirements of new tools for their automatic installation to the repository system.

(2) An expert system to generate test data for exercising the FSM. The expert system could generate the test data based on the user-configured choice of criteria. Such criteria would include entering all states at least once, travelling along each transition at least once, etc. These criteria would be taken from a list of choices. The user would have the ability to combine criteria.

(3) A tool to automatically generate documentation of the FSM model. This could be done in a two-stage approach using a table generating tool and a table-description tool. This approach has been discussed in [1].

(4) Simulation has a natural place in such systems to provide a testing facility for the design by allowing experimentation with dynamic models. Therefore the design environment can become a simulative design environment leading to simulative CAST-CAD.

6 Conclusion

CAST has gaining momentum in providing appropriate bases for fields where modelling and model processing are essential. Several CAST tools and environments are expected to arise similar to CASE tools and environments. To avoid tool interface complexity problem which surfaced in software engineering, one can think repository-based CAST-CAD environments as integrative, that is, open-ended and powerful CAD environments.

References

1. D. King, T.I. Ören, M. Hitz: Automatic generation of natural language documentation for SLAM II programs. In: E.N. Houstis, J.R. Rice (eds.): IMACS Transactions on scientific computing '91. Amsterdam: North-Holland (1992)
2. G.J. Klir: Facets of systems science. New York: Plenum Press (1991)
3. T.I. Ören: Model-based activities: A paradigm shift. In: T.I. Ören, B.P. Zeigler, M.S. Elzas (eds.): Simulation and model-based methodologies: An integrative view. Berlin: Springer, pp. 3-40 (1984)
4. T.I. Ören: Model behavior: Type, taxonomy, generation and processing techniques. In: M.G. Singh (ed.): Systems and control encyclopedia. Oxford, England: Pergamon, pp. 3030-3035 (1987)
5. T.I. Ören, D.G. King, L.G. Birta, M. Hitz: Requirements for a repository-based simulation environment. In: Proc. of Winter Simulation Conf. pp. 747-750 (1992)
6. F. Pichler, R. Moreno-Díaz (eds.): Computer aided systems theory: EURO-CAST'91. Berlin: Springer (1992)
7. F. Pichler, H. Schwärtzel (eds.): CAST methods in modelling: Computer aided systems theory for the design of intelligent machines. Berlin: Springer (1992)
8. R.M. Poston: Proposed standard eases tool interconnection. IEEE Software: 6:11, 69-70 (1989)
9. D. Soufflet: Emeraude V12. PCTE Newsletter: 3 (Feb.): 4 (1990)
10. I. Thomas: PCTE interfaces: Supporting tools in software engineering environments. IEEE Software: 6:11, 15-23 (1989)
11. A.W. Wymore: Model-based systems engineering: An introduction to the mathematical theory of discrete systems and to the tricotyledon theory of systems design. Boca Raton: CRC (1993)

On Requirements for a CAST-Tool for Complex, Reactive System Analysis, Design and Evaluation

Christoph Schaffer and Herbert Prähofer

Johannes Kepler University Linz
Department of Systems Theory
Austria

Abstract. There is a general consensus on the lack of appropriate methods and tools to support complex reactive system design, i.e., design of systems which are, to a large extent, event-driven, continuously having to react to external and internal stimuli. Design methods and tools developed for transformational systems have shown to be inadequate for complex reactive system design. These methods have no means to represent the event-driven behavior of reactive systems and they do not provide any analysis and test methods essential for reactive system design. In this paper we want to give a short statement of the problem of complex reactive system design from our systems theory point of view. We state requirements we pose on a CAST (Computer Aided Systems Theory) tool for reactive system design and discuss approaches in existence. We also evaluate some of the commercial available tools with respect to the requirements.

1 Introduction and Motivation

A *reactive* system, in contrast to a *transformational* system, is characterized by being, to a large extent, event-driven, continuously having to react to external and internal stimuli [14]. There is an unignorable tendency in systems engineering that more and more systems are reactive by nature. While typical reactive systems like communication systems, distributed systems, real-time control systems and others are becoming more important, also typical applications so far built in a transformational manner, like business software, database applications, etc., nowadays work interactively and, in that way, also become reactive. There also is a general consensus on the lack of appropriate methods and tools to support reactive system design. Design methods and tools developed for transformational systems have shown to be inadequate for complex reactive system design. These methods have no means to represent the event-driven behavior of reactive systems and they do not provide any analysis and test methods essential for reactive system design.

In more detail, in our investigations on systems engineering methods we identified the following major problems in reactive system design:

- The majority of tools that are available on the market today are using methods that were developed many years ago to build typically transformational systems and have no means to represent event-driven behavior.
- There are only a few tools that support hardware / software codesign. However in reactive systems design, the collaboration and the integration of hardware and software components is very important.
- There is a lack of available analysis methods and tools for reactive systems.
- The integrated approach is missing in most of the available tools. They only cover a small part of the overall system building process.
- Currently, many safety-critical hard real-time systems are built using methods which can not guarantee that critical timing constraints will be met. However fulfillment of timing constraints is indispensable for correct and safe functioning of hard real-time systems.

Nowadays we also have to deal with more and more *complex* systems caused by the increasing requirements that have to be fulfilled by these systems. What a complex system is, is difficult to define. Just to give two examples of possible definitions we state the definitions of Flood and Carson [10] and Mesarovic and Takahara [18]. Flood and Carson define complexity in the following way:
"In general, we seem to associate complexity with anything we find difficult to understand",
while Mesarovic and Takahara use the following definition:
"... a complex system is a system whose components are systems in their own right. '

To give an exact, all inclusive definition of complexity probably is not possible but, intuitively, we know what a complex system is. We can think of a complex system having many parts with a lot of non-trivial interconnections and being difficult to comprehend. It is also obvious that the design of complex systems is a difficult task requiring a careful design using powerful design methods and tools. In our context of reactive systems design, the complexity especially emerges from the event-driven behavior, from its heterogeneity, from the real time constraints, from the demand of a high system reliability, from its intensive interaction with a complex environment, and, last not least, from the magnitude of current reactive systems.

In this paper we want to give a short problem statement of complex reactive system design from our systems theory point of view. We state requirements we pose on a CAST (Computer Aided Systems Theory) tool for reactive system design. Then, we discuss some approaches in existence today to tackle these requirements. Finally, we evaluate some of the commercial available tools with respect to the requirements stated above. In our exposition we restrict ourselves to hardware/software systems and do not deal with microsystems [24] which also allow the integration of optical and mechanical parts.

2 Requirements on a CAST - Tool

2.1 Hardware / Software Codesign

In the design of embedded systems, which are systems that consist of hardware (HW) and software (SW) and which are embedded within a physical environment, the problem arises that for the development of the hardware and software components different formalism are employed. Such an undertaking necessitates the partitioning of the entire system into hardware and software in the early phases of the design process. However this contradicts with the objectives of a systems engineering approach where the design should be done independently from a particular realization technology and partitioning should not be done e.g. before performance data are available to support respective decisions.

The reason why different design approaches are used for HW and SW design is that these two areas have evolved quite independently and the requirements on the design process have been quite different. For example, hardware designers always had to cope with problems like parallelism, performance, execution speed etc., while in SW design the sequential functionality has dominated. Timing requirements and parallelism with all the related problems, have widely been neglected in SW design for a long time. SW emphasized the sequential execution mode. SW engineers usually only deal with one abstraction level, the level of programming languages, while HW engineers had to learn to deal with several levels to manage the increasing complexity of VLSI design.

Due to the increasing number of distributed and reactive systems coming into being, SW engineering more and more has to deal with issues like parallelism, asynchronous components and timing requirements. This diminishes the difference between SW and HW systems so that similar techniques for their design are needed. The objective should be to bring into being an environment which allows the SW as well as the HW engineer to use the same set of formalisms and solution techniques for their design tasks. Due to the desire to specify the system behavior independently from the realization technology, one also wishes to specify timing requirements in the early phases of the design process. In summary, we can say that systems engineering has to concentrate on a design methodology which is independent of any realization technology.

2.2 Development Life-Cycle

While in former times the top-down design approach was dominant, and for some simple systems this also was justifiable, today a combined top-down, bottom-up, and midside-out design approach is obligatory. This approach often is termed "Yo-Yo" design. One reason to reject the pure top down design is that it does not support the reuse of existing components. Also, it is an illusion, that all the requirements on a really complex system can be comprehended in the very early phases of the design cycle (figure 1). At a particular time $t1$, the designer rather will comprehend some parts of the system in detail while other parts will not be identified at the highest level. He rather will discover the existing circumstances during the design process and put these discoveries in the design. In this context one can speak of a "learning by doing" process.

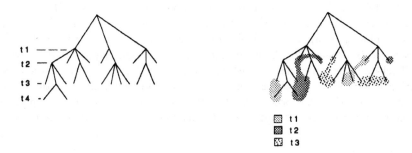

Fig. 1. YO-YO design

In the worst case, errors which have direct effects onto the specification, are detected not before the implementation phase, the system integration phase, or even the operation phase of the system. If a tool is used which does not allow a direct feedback from the later design phases to the specification phase, the usual approach is that changes in the realization are not supplemented in the specification. In current software and hardware development often the system is modified until it reaches a stable state. The modifications are not recorded in the specification and implementation documentation and will be known by the systems implementors only (figure 2 (a)). The situation gets even worse when the system should be extended by integrating additional functionality.

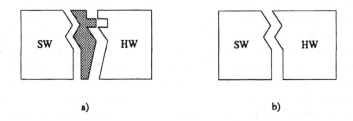

Fig. 2. a) Bad design process b) Well done design process

A development life cycle for complex systems therefore has to support feed-backs between the different phases of the systems design process so that, for example, modifications in the implementation phase also have their effects in all the former phases of the design. Well defined, clear interfaces between the different design phases and between different system components are a basic assumption for a successful design (figure 2 (b)).

Therefore, the waterfall model in its most primitive form without any possibility of feedbacks is not applicable for complex systems design. The requisite that the whole system is understood in all its details already in the requirement phase prohibits its application [35]. Much more appropriate is the spiral model introduced by Boehm [2] where through evolutionary development with several feedbacks and design cycles the risk to detect errors in the integration and operation phases is minimized as much as possible. In that context is the CBSE process model which has been created by the CBSE-IEEE task force group [20] is an interesting approach. It emphasizes the feedbacks and an evolutionary design process.

2.3 Functional Requirements

To meet the functional requirements of a system, one needs a formal, well structured specification technique (ST), which allows one to specify the functionality of the system in a precise, comprehensible, and unambiguous manner. It should be formal because only a formal specification allows formal analysis of the correctness of the specification and design. Besides that, a ST has to support reusability of components, allow a hierarchical description, and, as mentioned above, support Yo-Yo design. For system evaluation, a ST providing a executable system description is desirable since, in that way, the correctness can be tested already during the specification.

Formal techniques however have the disadvantage that they are difficult to comprehend by non-experts. To cope with that problem, a simple, graphical representation is preferred. According to Flood [10] and as generally accepted, graphical representations are better to comprehend by humans than textual equivalents ("a picture says worth than a thousand words").

Regarding the formal specification we have to remark that by formal specification it is not possible to decide if a specification is complete nor is it possible to test if a formal specification fulfills all the requirements. A formal specification only can decide that that what has been specified is correct. Such tests are called validation. Therefore, the graphical representation should reflect the requirements in a way that it can be used for going through the specification with the client and eliminate misunderstandings. Such an approach, to base the view of the customer and the engineer on a common ground, is certainly one of the greatest challenges. A very good survey on formal specification techniques for the hardware design can be found in [13].

Prototyping Prototyping has shown itself to be a very powerful concept. That is, because by prototyping a lot of errors and misconceptions can be detected already in the very early phases of the design cycle. Roes [26] distinguishes between exploratory, experimental and evolutionary prototyping. The objective of *exploratory prototyping* is to provide a communication medium between the customer and the system developer to eliminate misunderstandings in the requirements specification. In contrast, *experimental prototyping* is rather used more by the engineer himself to help him in solving problems of design and implementation. This approach also could be termed as "Learning by Doing". *Evolutionary prototyping* is a technique where a prototype is modified and extended until all the requirements are met. However, by using this type of prototyping one usually get problems with the bad structure of the system, bad maintainability, bad documentation, and, as a result, bad reliability.

To employ the prototyping approach as successfully as possible, the employment of interface prototypers is desirable. Interface prototypers (figure 3) are used to ease the testing of the system. It should be possible to build a fictive interface of the real component by using a simple graphic editor so that it is possible to test the functionality of the system by operating on the interface prototype.

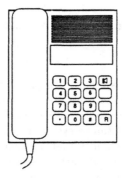

Fig. 3. interface prototype of a telephone set

2.4 Nonfunctional Requirements

The specification of non-functional requirements is supported by the fewest commercially available design tools. This is probably because non-functional requirements have not been recognized to be essential at the early phases of the design process.

According to Dunkel [9], the nonfunctional requirements of a system can be classified into process, product and external requirements.

- *Process requirements* are all the constraints on the development process itself like the particular method or toolset used for designing the system.

- *Product requirements* describe the characteristics (parameters) of the system being developed. Examples for these parameters are the size, performance, availability and reliability of the system.
- *External requirements* describe the overall resource budget, personal and available development time.

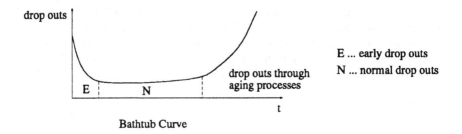

Bathtub Curve

Fig. 4. hardware drop outs

Nowadays, more and more systems, like flight control systems, power plant control systems, chemical process control systems etc., have to care for safety critical tasks. Such systems have to guarantee a high degree of availability, which is a measure for the fraction of time a system is operational, and a high degree of reliability, which is the probability that a system remains operational for a given period of time. These two parameters shall show the importance of non-functional requirements in the early requirements specification phase. Due to the usual breakdowns of hardware components (figure. 4), the probability of system breakdowns is directly proportional to the quantity of HW-components used in the system. Breakdowns of hardware components are due to production errors, aging effects etc. Therefore, breakdowns are - which is a fact often ignored - a "normal" event which should be considered in systems design. The availability A of a system is defined as the ratio

$$A = \frac{MTTF}{MTTF + MTTR}$$

with $MTTR$ denoting the *mean time to repair* and $MTTF$ denoting the *mean time to failure*. As an availability as high as possible is desired (availability rate of 1), one is forced to minimize $MTTR$ (bring it to 0). So localizing and fixing up an error has to be done as quickly as possible. To do so, the system has to be tested periodically according to certain criteria. If continuous operation is required, one is forced to introduce redundancy into the system. For large systems the overhead to introduce safety precautions will make up half of the whole system development efforts.

2.5 Design Knowledge Representation

Systems nowadays are of such a complexity that they are not to comprehend by one single person. Additionally it is a big effort to learn about the functionality and structure of such a system if the documentation is available in written form only. The access to the information is cumbersome and the knowledge is hardly reusable in other projects. It would be much better to incorporate the knowledge gained during the design process into the design itself, i.e., a knowledge base should be created which can be used by the designer for other projects but which also can be used to support colleagues in getting familiar with the system.

3 Proposal of Approaches to Meet the Requirements

Having stated the principal requirements on a CAST tool for complex, reactive systems design, we are in the position to propose some approaches to meet these requirements. In a first step it is very meaningful to evaluate existing methods used for HW and/or SW design to work out a common ground accepted by both societies. So the resulting approaches will be accepted by the practitioners doing systems design more easily.

Since the object oriented paradigm was discovered by the SW community, there is no major difference between HW and SW design methodologies. This is because object oriented design concepts are very similar to the concepts used for designing HW systems. According to Rubin [28] object behavioral analysis sequences through the following five steps:

- setting the context for analysis,
- understanding the problem by focusing on behaviors,
- defining object that exhibit behaviors,
- classifying objects and identifying their relationships, and
- modelling system dynamics.

Such an approach is applicable to systems implemented in hardware too. The different circuit modules can be seen as the objects. Hardware components also fulfill the requirement of data encapsulation very well because a hardware component can communicate via its ports only.

3.1 Modular Hierarchical System Modelling

To cope with the complexity of systems, the hierarchical decomposition of systems has been applied successfully. Systems are built up from *components*, i.e., modular subsystems, which are *coupled* to other components. Coupling is done by connecting the output and input ports of components (figure 5). Coupled systems again can be used as components in larger coupled systems which allows the hierarchical structuring of systems. Communication exclusively can be done via the coupling channels, i.e., also for accessing common data storage, these channels have to be used. Such a restricted form of communication is desired in order to achieve a high degree of reusability. The modular hierarchical modelling has proved itself to be a very powerful concept in HW design for quite a long time. Here circuit modules can communicate through their input and output pins only.

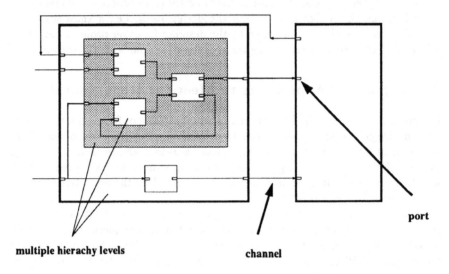

multiple hierachy levels   channel

Fig. 5. modular hierarchical systems modelling

port

Because hardware components work in parallel by nature, the design of parallel heterogeneous systems is accomplished much easier in HW. But pure HW solutions have the disadvantage that the flexibility of SW with respect to later changes and extensions is lost. The usual way to rebuild such a parallel functionality in SW is to introduce one software process for each parallel module. But if there are a lot of processes in parallel controlled by one single processing unit only, the performance of the system decreases drastically due to the operating system overhead for scheduling these processes. A possible solution for this problem is to implement the parallelism of components not by a operating system process but by a different scheme. The code generation module of SDL [30] or the abstract simulator for modular, hierarchical discrete event systems simulation introduced by Zeigler [39] show possible solutions. There the scheduling of the components is done not by the operating system but all the components running on one processor are handled by one scheduling process.

Additionally the realization of the communication principle between the components is not an easy task in SW. In HW communication is done by simply connecting the appropriate pins (ports) by a wire. In SW the whole communication has to be controlled by the SW too. To avoid complex interprocess communication, often common data storages are used. However, this violates the requirement for modularity and is a potential source of errors in the implementation. Through the non-modular implementation, the I/O interfaces of the components are more difficult to recognize which leads to bad reusability of the components.

In a modular, hierarchical SW modelling approach, ports as well as channels should support data typing and automatic type conversion. Type conversion could be realized by the ports. Also, it should be possible to define hierarchies of communication channels and ports. To ease understanding, it should be possible to represent a complex communication system between two components, e.g. a bus system, by one coupling. Additionally to that, it should be possible to model a channel by an autonomous system itself to specify how the communication between the components is realized. Then, on a high design level the communication between two components could be represented in an abstract manner only by a coupling. On such an abstract level it might be irrelevant how the coupling actually is realized, one only wants to model that a communication takes place. In the more detailed level, however, the coupling could be replaced by special communication modules specifying how the communication actually is accomplished.

Additionally we want to stress that equal techniques can be employed for the representation of the *architectural view* and the *design* view. In the architectural view the physical decomposition of the system into its distinct physical components is shown whereas in the design view the logical, functional view of the system is given. The assignment of the components of the design view to the components of the architectural view is done by the so-called *mapping* operation. The resulting relation can be 1:1, 1:n, or n:m.

But there exits another great problem that can be tackled by the modular hierarchical approach, namely documentation. As training of employees, either SW/HW engineers or technicians, in complex, comprehensive systems usually is accompanied only with a lot of effort, multimedia techniques for systems description could be used to increase the training efficiency. For example, if the design engineer or the maintenance technician is interested in the operation of a radio or if he wants to identify the radio within the system because he has never seen it in nature, it could be helpful that by clicking on a radio component a picture of the radio appears or a short video strip which explains the radio is shown. The different components could be identified much easier in that way. Such a multimedia approach would support the knowledge representation in design.

With these points we only want to touch the topic of multimedia systems to show the potential power coming up with this technology [1, 4]. A more detailed exposition of the possibilities of multimedia in system design would go far beyond the scope of this paper.

3.2 System Entity Structure

Conventional systems do not provide much support in finding and accessing reusable components from a component repository. Also, most non-functional requirements cannot be handled so that selection of components based on non-functional requirements, like costs, performance etc., can be done. A means do deal with that kind of problems is offered by the knowledge based system design approach [27] based on the *systems entity structure* (SES) [38] knowledge representation scheme.

The system entity structure is a formal knowledge representation scheme which is intended to support the designer in the modelling task. It is a graph like structure which is used to represent the possible decompositions, the coupling constraints, and the possible variants of systems and system components. By that, it represents the possible configurations for a systems design. An entity node is used to represent the system components which are identified in one or more decompositions. Various decompositions may exist for one entity which are represented by aspect nodes and should serve to model various views one may have of an entity. Besides the various decompositions of an entity, several specializations, i.e., variants, may be identified for one entity. Representation of entity variants corresponds to the class, specialization and inheritance concepts known from object oriented systems. The SES knows inheritance of behavior and data from general entities to their specializations. The SES concept reaches full power by assigning coupling constraints to aspects which restrict the way components of the aspect can be coupled. But besides coupling constraints, there are also synthesis and selection constraints. Selection constraints are assigned to the specialization nodes. They decide how and if variants can be selected and how and if they fit together. Synthesis constraints give rules under which constraints which components and which configuration should be preferred. A rule-based *pruning* process is employed to find one or more adequate configurations out of all configurations represented in the SES. Figure 6 shows an example SES representation of a ship communication system.

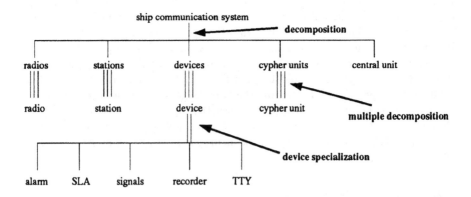

Fig. 6. SES of a ship communication system

In this context the evolution of the SES to the FRASES environment [27] seems to be very interesting. FRASES is an entity-oriented knowledge representation scheme augmented with rules. To each entity an entity information frame (EIF) is assigned which can contain the non-functional requirements, e.g. the costs, of this entity.

3.3 Higraph-Based State Transition Diagrams + Inheritance of Behaviour

As already outlined in section 3.1, systems can be structured and decomposed into subsystems based on the concept of modular, hierarchical modelling. In that way it is possible to define the static structure of a system with its components with their input/output interfaces and couplings. For a complete system description however, also means for the specification of the dynamic behavior of the components are necessary. Finite state automata and their state transition diagram graphical representation are a well known concept for the specification of the dynamic behavior of discrete systems. However, state transition diagrams have shown to be inadequate for complex systems because of their unmanageable, exponentially growing multitude of states (figure 7). To solve this problem, Harel [14], [15] introduced *higraphs* and higraph-based state transition diagrams called *statecharts*. Higraphs are a general extension of conventional graph representations by introducing means for the representation of (1) set enclosure, exclusion and intersection and (2) the parallelism of state machines. This is accomplished by exploiting the area of the diagram similar to the well-known concept of Venn diagrams. Higraphs have a lot of potential applications and have advantageously been employed for the Statechart visual formalism for specification of complex reactive systems which is the basis for the Statemate design environment [16].

Through the parallelism concept, it is possible to represent independent behavior in one component independently as automatons existing in parallel. How advantageously that can influence the number of states to be represented in the diagram is depicted in figure 7.

number of parallel automata	number of states contained in each automaton	number of states in a parallel structure	number of states in a flat structure
2	2	4	4
2	3	6	9
5	3	15	243
10	3	30	59.049
10	10	100	10.000.000.000

Fig. 7. Table of state

The concept of set enclosure, exclusion and intersection provides a means

149

to form clusters of nodes. Such a cluster forms a more abstract concept and can be used to represent equivalence relations of nodes with respect to a particular edge. A edge - see also figure 8 - originating from a cluster means that this edge applies equivalently for all the nodes included in the cluster. This concept is advantageous in state transition diagrams for compact representations of transitions which are equivalent for a set of states.

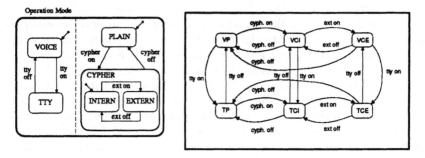

Fig. 8. State charts versus conventional state transitions diagram

Figure 8 gives an example statechart representation for part of a radio. The functionality of the radio can easily be extracted from this representation: The radio can operate in the VOICE or the TTY mode. Switching between these two modes is done through the signal *tty-on* and *tty-off*, respectively. After switching on, the system will find itself in the VOICE mode. This is signified by the arrow with the small circle as source. Independent from the VOICE or TTY mode, the system can encrypt its messages or not. This parallelism is represented by the dotted line separating the two functions. Here the default mode is PLAIN. Switching is signified by the input *cypher-on* and *cypher-off*. The transition *cypher-on* enters the CYPHER cluster node which means that it enters the default, which is INTERN. The transition *cypher-off* is specified for the CYPHER cluster which means that it equivalently applies to the INTERN and EXTERN state, i.e., whatever the state is, when *cypher-off* occurs, then the system will transit to PLAIN. If the system is in CYPHER mode, then there is a difference if a cypher unit internal or external to the radio is used. Switching is done by signals *ext-on* and *ext-off*. The same behavior described by a conventional state transition diagram results in a much more complex graphic. This is due to the fact that in the conventional state transition diagram the cartesian product of the parallel states has to be represented explicitly.

In systems development we usually have to create components which show a lot of similarities in structure and behavior to other components of the system or components we created in former designs and which we can find now in a model base. In that context we speak of component families. To handle such families, to define and represent their common features and differences, the object oriented paradigm can be employed. An object oriented class system is appropriate for defining component types, from which more special ones can be derived. Figure 9 depicts that issue by our radio system. The principal behavior of a radio system will not differ if the actual radio is an UHF, VHF or AM one. Therefore it is possible to define a principal pattern of behavior for a generic radio type which then can be specialized to the particular types.

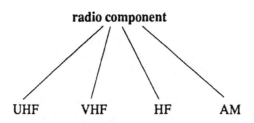

Fig. 9. inheritance of behavior

3.4 (Half)- Automatic Code Generation

A comprehensive design methodology should provide automatic, or at least half automatic code generation, i.e., a design specification should be translated into a program or a digital circuit automatically. The designer could concentrate on the higher levels of design and would not have to invest a lot of time into burdensome low level implementation and tests. Also, an automatic code generation would reduce possible sources of errors usually introduced when translating a design into an implementation by hand. Changes always would be done in the design description only. The code generation automatically propagates these changes down to the implementation. In SW design such an approach could be compared to the developments of high level programming languages and compilers. A design approach with automatic code generation should be regarded as a direct evolution of high level programming languages to higher levels of description. In HW design the situation is similar. Silicon compilers are state of the art in ASIC design where an algorithmic description of the problem situation is automatically realized as an ASIC component.

If automatic code generation is not a feasible approach, then half automatic code generation is applicable, i.e., a code frame can be generated which contains all the routine parts required by the particular programming language and operating system in hand as well as the basic behavior specified in the design. The code frame then has to be complemented by the programmer by code not specified in the design. However such an approach, although feasible and employed in current systems like SDL [30], has its disadvantages. So the complementation of the code frame again is a source of errors. Also for complex systems, it is hard work to get familiar with and understand the code frame.

Hence, an automatic code generation should be preferred but it is an approach which requires further maturation. Code generation techniques have to be improved to generate more efficient code.

3.5 The Analysis of Functional Requirements

The validation of the functional requirements can be done in two ways:

- *automatic tests*
  These are tests which can be done automatically. Such tests include the reachability of states or the localization of deadlocks. The methods are based on petri nets or finite state automaton theory.
- *interactive tests*
  These are tests which are conducted by the programmer by interacting with a running prototype (or executable design specification) of the system. To facilitate such tests, a support by interactive graphics for systems animation is desirable. Also batch simulations can be helpful where, through the realization of a particular system environment, the systems reaction can be studied.

3.6 Analysis of Non-Functional Requirements

As outlined in section 2.5, the performance of the system is one essential parameter of the non-functional requirements. Performance analysis usually is done using discrete event simulation techniques because analytical methods stemming from queuing theory are only applicable to small and simple systems. Simulation allows one to give estimates of mean queue length and utilization of resources. Different design alternatives can be compared by doing simulation runs with equal work loads for different design configurations. The design can be tailored to best fit special requirements.

One major and so far unsolved problem in SW design is worst-case timing analysis, i.e., to make predictions if certain timing constraints can be fulfilled in all cases. Such guarantees are essential for hard real-time systems where the violation of such a timing constraint means a malfunction of the system. Usually, the cases where the timing constraint can not be fulfilled are detected not before run time.

In HW design, worst-case timing analysis is not a problem because through the realization technology or the types of components used, one is in the position to give pretty exact estimates based on the timing of known components. Thus specifications of input/output delay times of components as given in figure 10 are quite common in HW design.

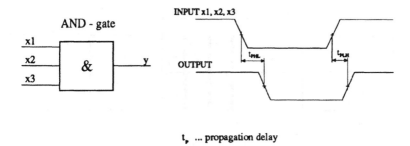

t$_p$ ... propagation delay

Fig. 10. Timing Specification

In SW design such specifications of timing based on input/output definitions of components have not been used so far. This is probably because the execution times of processes usually are not known because they depend on too many uncertain factors, like processes running on one processor.

3.7 Fault Tolerant Systems

The requirement that the mean time of repair of a system discussed in section 2 should approach zero is hard to fulfill. To bring a broken system back to operation as soon as possible, the localization of an error has to be done as quickly as possible. For complex systems in particular, localizing an error can be a time consuming task if the localization procedure is not supported by the system itself. To detect an error in a system as soon as possible, components have to be tested periodically (figure 11).

Localizing and identifying an error in the broken component is the major problem. Most times several error messages are received simultaneously and they have to be combined to make a good judgment on the error. As it is shown in [6] artificial intelligence methods can be used to develop model based diagnosing techniques [7, 25] to reason about errors and to locate them.

In HW, system tests are difficult as chips usually only allow getting information through their input/output pins. Therefore, HW components should have self-test capabilities. Scan-paths or boundary-scan methods have been introduced for that purpose [11].

Additionally, to self-test mechanisms, a design tool also should support the introduction of redundancy for fault tolerance.

4 Evaluation of Some Existing Tools

In the following we shortly mention some commercial available design tools and evaluate them based on the criteria discussed above. Because of the numerous tools and approaches available on the market today, we only can give a brief

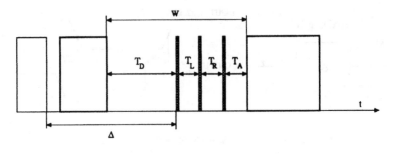

Δ   ...   time interval between two system tests

w   ...   restart time

T_D ...   time interval between break down and the next test of the component

T_L ...   testing time

T_R ...   repair time

T_A ...   start up time

Fig. 11. Cyclic System Tests [3]

overview here. We want to concentrate on such tools that are well known and accepted by software and hardware engineers. Other tools not discussed here are HOOD [31, 17], CHILL [29], Z [36] and VDM [8]. A more detailed comparison of SDL, ESTELLE and LOTOS is given in [5].

4.1 SA/RT (Structured Analysis / Real Time)

SA/RT [34] is an evolution of the usual SA design method. SA (Structured Analysis) supports transformational software systems only. The system is described as a hierarchical network of pure functional components, i.e., each component is described by a function mapping the input values to the output values. As such an approach is not appropriate for real-time reactive system design, SA has been extended to SA/RT by introducing controller components, i.e., state machines, to control the functions of the components. Functional components may be activated, deactivated, triggered, or changed in functionality by events in the controllers. SA/RT supports hierarchical specifications of function networks and controllers.

Several commercial available tools which are based on SA/RT are available. One popular tool is TEAMWORK. It provides a powerful interface prototyper and is also able to generate a code frame for C and ADA, respectively.

4.2 ESTELLE

ESTELLE, which is a formal language, was developed for the specification of distributed systems showing a high degree of intercomponent communication.

ESTELLE itself consists of a language and a model. The model is described by extended finite state machines and can be decomposed into four layers which are the *specification*, the *module*, the *transition*, and the *action* or *process* layers.

A model is set up as a hierarchical, modular coupled structure of communicating modules. Communication is done by passing messages over bi-directional channels which are connected to the ports of the component modules . Interaction primitives may be messages consisting of structured data [33]. A FIFO-queue is associated with each input port to queue the incoming messages for being processed by the component.

4.3   LOTOS (Language of Temporal Ordering Specification)

LOTOS which is also a formal language was designed by the ISO especially for protocol applications. LOTOS is based on two formal concepts which are the *Calculus of Communication Systems* (CCS) [19] and the abstract data type language ACT-ONE. CCS is used to describe process behaviour and their interactions whereas ACT-ONE deals with the description of data types. Processes are the elements the system consists of, and can be refined in a hierarchical way. Communication between these processes is only possible by exchanging events via the event gates of a process [33].

4.4   SDL (Specification and Design Language)

SDL is a formal language and design technique which has been defined by CCITT and which evolved to a standard in 1980. It is a design technique primarily used for telecommunication applications.

An SDL model is a modular, hierarchical coupled model which is called the *block diagram* model. *Process diagrams* are employed as a graphical description technique of the extended finite state machine models of the atomic components. The system itself is communicating with the environment whereas its parts are communicating with each other [30]. Signals which are the packets of information are used to exchange data via channels linked to the different blocks between the environment and the blocks itself. Another concept provided by the SDL method are *sequence charts*, which became a standard in 1992 [12], to graphically describe the dynamic input/output behavior of components and the communication behavior between the components and the environment. They are used to model the input/output behavior of a system to be designed and they can be used to graphically trace the interactive tests of a designed system.

SDL tool implementations are well developed and support the designer in the various design tasks. Commercial tools provide a convenient user interface for model specification and they provide half-automatic code generation.

4.5   Statemate

The Statemate tool [16] is based on an evolution of the SA/RT method. In Statemate the controller components are specified by statecharts described in section 3. Besides statecharts, for the system specification the *activity charts* and *module charts* are employed. The activity charts correspond to the hierarchical functional decomposition of the system as known from SA. The module charts are used to represent the physical decomposition of the system into its HW and SW components and their communication channels.

In our opinion, the Statemate tool is the best developed and most powerful tool for reactive system design available. So it provides powerful methods to test reachability and deadlock properties, it provides an very convenient interface prototyper, and it supports half automatic code generation for C, Ada and VHDL.

4.6 SES/Workbench

Just as one representative of the numerous discrete event simulation tools available we discuss SES/workbench [32] here. SES/Workbench is a new, powerful, state of the art simulation tool usable to validate correctness and evaluate performance of system designs. It belongs to the class of transaction flow oriented simulation systems. System modelling is done by specifying a hierarchical network from mostly predefined building blocks in a graphical, interactive way. The building blocks available can be classified into resources management nodes, transaction flow control nodes, submodel management nodes and miscellaneous nodes (figure 12).

SES/Workbench provides a powerful, interactive user interface for simulation modelling and simulation data output. Simulations can be executed in interaction with the user or in batch mode. In the interactive mode it is possible to observe the transactions flowing through the system. An example of a system modelled with SES/workbench is depicted in figure 13.

5   Summary and Outlook

In this paper, we defined several requirements posed on a CAST-tool for complex, reactive system analysis, design and evaluation. A CAST-tool has to be integrative and comprehensive covering all phases of the development life cycle. A formal specification approach is desirable because it leads to executable specifications and makes formal analysis feasible. An integrative CAST-tool also has provide means to deal with functional and non-functional requirements. Furthermore, the approach taken should cover design of software as well as hardware in a comprehensive manner.

We identified the following existing approaches as being appropriate to meet the requirements:

- A modular, hierarchical modelling approach is appropriate to handle the structuring of complex systems, supports a combined top-down and bottom-up design process and makes the reuse of model components possible.

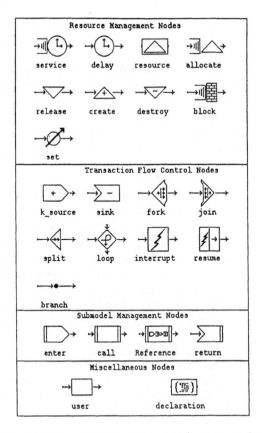

Fig. 12. Different Nodes within SES/workbench [32]

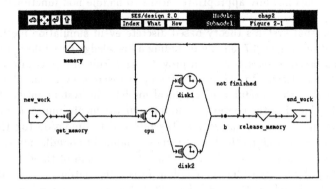

Fig. 13. System modelled with SES/workbench [32]

- The SES knowledge representation scheme and the SES based knowledge-based system design methodology builds a good framework to support a comprehensive design process. It allows the representation of possible design configurations, the allocation of reusable components from a model repository, the specification and treatment of functional and non-functional requirements early in the design process, and the selection of possible designs based on specified constraint knowledge.
- We regard the statechart method being an extension of the conventional state transition diagrams to be the best method for reactive system behavior specification in existence. A statechart specification is formal, executable, graphical, compact and comprehensible.
- Automatic code generation from a formal specification would be most useful. It would shorten the development life cycle drastically and would eliminate one of the sources of errors, viz. the implementation. However, code generation techniques have to mature to produce more efficient code.
- Formal specifications lead to executable specifications which can then be used to test the system already in the very early phases of the design process. Analytical methods based on petri nets and automata theory can be employed to test certain properties of the design, like deadlock or reachability. Simulation runs can be employed to test the behavior of the system interactively.
- When components are augmented with timing parameters, simulation runs can be used to evaluate the performance of the system. In that way, important non- functional parameters can be tested. However, no method for worst-case timing analysis of software systems currently exists.

Currently, none of the commercial tools meets all of the requirements. The commercial tools employ some of the approaches proposed but none takes them all. So, most of the tools take a modular, hierarchical modelling approach. But none of the tools has any appropriate means to handle non-functional requirements.

We think that systems theory based discrete event simulation methodology as developed by Zeigler [37, 38, 39] including a knowledge-based design methodology as discussed in section 3.2 can provide most fruitful inputs into complex, reactive system design. In this methodology, a discrete event simulation program is specified as a modular, hierarchical model. The system entity structure has been developed in the context of modular, hierarchical simulation modelling to facilitate model base organization. Discrete event simulation together with the model- based design methodology is most qualified to handle non-functional requirements missing in most design tools. The extension of the discrete event simulation methodology to handle combined discrete/continuous modelling and simulation developed by Praehofer [22, 23, 21] allows the simulation and test of a reactive control system embedded in the continuous model of the systems environment.

References

1. John A. Adam. Applications, implications. *IEEE Spectrum*, pages 24–31, March 1993.
2. Barry W. Boehm. A Spiral Model of Software Development and Enhancement. *IEEE-Computer*, pages 61–72, May 1988.
3. Mario Dal Cin. *Fehlertolerante Systeme*. Teubner Stuttgart, 1979.
4. Bernard Cole. The technology framework. *IEEE Spectrum*, pages 32–39, March 1993.
5. SPECS Consortium and J. Bruijning. Evaluation and Integration of Specification Languages. *Computer Networks and ISDN Systems*, 13:75–89, 1987.
6. Carla Decker and Susanne Loesken. Ein modellbasiertes Expertensystem fuer die Wartung von Telekommunikationsnetzwerken. In J. Encarnacao, editor, *Telekommunikation und multimediale Anwendungen der Informatik*, number 293 in Informatik Fachberichte, pages 176–187, Darmstadt, October 1991. Springer-Verlag. in German.
7. J. deKleer and B.C. Williams. Diagnosing Multiple Faults. *Artificial Intelligence*, 32:97–130, 1987.
8. Jeremy Dick. VDM Methodology Guide. Technical report, ESPRIT, 1991. Atmosphere Esprit Project 2565.
9. Juergen Dunkel, Kevin Ryan, and Franz-Josef Stewing. A Note on Non-Functional Aspects of CBSE. CBSE-workshop, December 1992.
10. Robert L. Flood and Edward R. Carson. *Dealing with Complexity*. Plenum Press, 1988.
11. Manfred Gerner, Bruno Mueller, and Gerd Sandweg. *Selbsttest digitaler Schaltungen*. Oldenburg Verlag, 1990. in German.
12. Peter Graubmann, Ekkart Rudolph, and Jens Grabowski. Telecommunication System Development Based on Message Sequence Carts and SDL. Technical report, ESPRIT, 1991. Atmosphere Esprit Project 2565.
13. Aarti Gupta. Formal Hardware Verification Methods: A Survey. *Formal Methods in System Design*, 1:151–238, 1992.
14. David Harel. STATECHARTS: A Visual Formalism for Complex Systems. *Science of Computer Programming*, 8:231–274, 1987.
15. David Harel. On Visual Formalisms. *Communication of the ACM*, 31(5):514–529, 1988.
16. David Harel, Hagi Lachover, Amnon Naamad, Amir Pneuli, Michal Politi, Rivi Sherman, Aharon Shtull-Trauring, and Mark Trakhtenbrot. STATEMATE: A Woking Environment for the Development of Complex Reactive Systems. *IEEE Transactions on Software Engineering*, 16(4):403–414, April 1990.
17. Klaus Kronloef. John Wiley and Sons, wiley professional computing edition, 1993.
18. M.D. Mesarovic and Y. Takahara. *Abstract Systems Theory*. Springer-Verlag, 1989.
19. R. Milner. A Calculus of Communicating Systems. *Lecture Notes in Computer Science*, 92, 1980.
20. David W. Oliver. A Tailorable Process Model for CBSE. Technical report, GE Research and Development Center, Schenectady, New York, February 1993.
21. Franz Pichler and Heinz Schwaertzel, editors. *CAST Methods in Modelling*. Springer-Verlag, 1992.

22. Herbert Praehofer. *System Theoretic Foundations for Combined Discrete-Continuous System Simulation*. PhD thesis, Johannes Kepler University of Linz, Linz, Austria, 1991.

23. Herbert Praehofer, Franz Auernig, and Gernot Reisinger. An environment for DEVS-based multiformalims simulation in Common Lisp / CLOS. *Discrete Event Dynamic Systems: Theory and Application*, 1993. (to appear).

24. Herbert Reichl, editor. *Micro System Technologies 92*. VDE Verlag, 1992.

25. R. Reiter. A Theory of Diagnosis from First Principles. *Artificial Intelligence*, 32:57–95, 1987.

26. Henry Roes. *Quality*, chapter 7, pages 189–222. John Wiley and Sons, wiley professional computing edition, 1993.

27. Jerzy W. Rozenblit, Jhyfang Hu, Tag Gon Kim, and Bernard P. Zeigler. Knowledge-based Design and Simulation Environment (KBDSE): Foundational Concepts and Implementation. *Operational Research Society*, 41(6), 1990.

28. Kenneth S. Rubin and Adele Goldberg. Object Behavior Analysis. *Communication of the ACM*, 35(9):48–62, September 1992.

29. W. Sammer and H. Schwaertzel. *CHILL: Eine moderne Programmiersprache fuer Systemtechnik*. Springer Verlag, 1982.

30. R. Saracco and P.A.J. Tilanus. CCITT SDL: Overview of the Language and its Applications. *Computer Networks and ISDN Systems*, 13:65–74, 1987.

31. D. Schefstroem and G. van den Broek. John Wiley and Sons, wiley professional computing edition, 1993.

32. Scientific and Inc. Engineering Software. SES/wokbench; Introductory Overview, January 1991. Product information.

33. Katie Tarnay. *Protocol Specification and Testing*. Plenum Press, 1991.

34. Paul T. Ward and Stephen J. Mellor. *Structured Development for Real-Time Systems*, volume 1. Prentice Hall, 1985.

35. Stuard Whytock. *The Development Life-Cycle*, chapter 4, pages 81–96. John Wiley and Sons, wiley professional computing edition, 1993.

36. J.B. Wordsworth. *Software Development with Z*. Addison-Wesley, 1992.

37. Bernard P. Zeigler. *Theory of Modelling and Simulation* . John Wiley, 1976.

38. Bernard P. Zeigler. *Multifacetted Modelling and Discrete Event Simulation*. Academic Press, 1984.

39. Bernard P. Zeigler. *Object-Oriented Simulation with Hierarchical, Modular Models*. Academic Press, 1990.

Automating the Modeling
of Dynamic Systems*

Heikki Hyötyniemi

Helsinki University of Technology
Control Engineering Laboratory
Otakaari 5 A, SF-02150 Espoo, Finland

Abstract. The current situation in the field of computer aided dynamic systems modeling is syrveyed, and the contemporary tools for computer aided modeling are discussed. It is proposed that there is a need also for special purpose modeling tools that are able to manipulate the models without destructing their structure. The role of symbolic calculation in combination with object-oriented modeling is emphasized. An experimental tool is discussed that is intended specially for implementing the presented ideas. The syntax of the formalism is not elaborated on extensively, only to an extent that is necessary to understand a few illustrative examples.

1 Modeling of dynamic systems

Control engineering work is one of those fields where intuition and real life phenomena intersect. As an consequence of this fact, the analysis and design methods usually are mixtures of heuristics and mathematics. Computer Aided Control Engineering (CACE) is also a challenging application area for Artificial Intelligence (AI) methodologies. One of the central application fields of CACE is the modeling of dynamic systems, where special tools are needed to express the real systems behavior in a compact form.

In control engineering, the models are needed for system formalization, behavior analysis, and finally, for control design tasks. In control design, there are two extreme attitudes to the role of the model:

1. Very little needs to be known of the process, because the feedback control compensates the modeling deficiencies, or
2. Once the accurate model is known, the problem of control design becomes trivial.

Usually, in practical control engineering, the former approach dominates. For example, linearized process models and approximate solutions are regarded as satisfactory. In complex process plants no exact models can usually be found, and this is really the only possible way to proceed. But the interest has been focused on the general purpose tools also because of the software enterprises'

---

* This work has been financed by The Academy of Finland and by the Jenny and Antti Wihuri Foundation.

commercial calculations—the market potential of these tools is large. In this paper, however, the latter approach is emphasized, and the problems and possibilities are discussed, when sticking to the physical *first principles,* the exact laws governing the physical phenomena. It turns out that considerable benefits might be obtained also this way.

1.1 Traditional approaches to modeling

The field of Computer Aided Control Systems Design (CADCS/CACSD) has been an active area of research for a long time. Modern packages and assumed future trends in the field are surveyed, for example, in [10]. In [8], guidelines for constructing CACE environments are given in abstract terms, effectively restating the main objectives of the current development work. Yet, there is need for an alternative approach.

At the moment, there are dozens of packages for doing computer aided modeling. A prototypical modeling tool is, for example, Simulab (Simulink) that is widely used for creating hierarchical models with a graphical user interface [3]. Typically, the computer aided modeling tools offer a visually appealing working environment, where blocks of subprocesses can easily be combined. The behavior of the model can be simulated, and the results can be analyzed, again visually. This is not only the usual scheme of the tool realization, it actually seems to be the standard, and it even seems to have affected the image of models in general: *models are what modeling tools help do!*

Even if the *modular* modeling approach, being the basic idea behind the hierarchical modeling environments, sounds natural, it is only a pragmatic approximation. The internal behavior of the submodels is assumed invariant, no matter what the surrounding system is like, and superposition, linear additivity of the submodel properties is supposed. In many cases this reductionistic view collapses. Even if the modularity has proved successfull in computer programming, for example, it should not be adopted blindly in dynamic systems modeling. The analogy between flow charts and process blocks is not exact. In dynamic systems, the interactions between modules are not only flows of information, because there does not exist pure information flows—the carrier of the information affects the system dynamics, and the properties of the 'sink' also change the behavior of the 'source'. This fundamental coupling problem is not usually critical, and the idealized approach that the standard modeling tools are based on works well. However, in some applications the idealization collapses, and it is dangerous always to neglect these problems directly and apply the tools routinely.

These problems of modern modeling environments, from the point of view of expressing processes in the most useful form, that means, in the most compact or the most accurate form, depending on the application, will be elaborated on in the next sections. Concisely, these deficiencies are the following:

- **Inflexibility.** The modeling tools always assume the same input-output structure of processes.

– **Implicity.** The created models can only be used for simulation, and the only means of analysis is visual inspection.

1.2 'Structural' modeling

The model should not be seen merely as an instrument for simulations. A very useful definition, a *constructive* view of modeling, is given in [12]:

> *Modeling means the process of organizing knowledge about a given system.*

The task of modeling of dynamical systems is to find *structure* to data, dependencies between the structures being dictated by the mathematical properties of the modeled components. Adopting this ideal of *structural modeling*, it is not wise to abandon the structure that is already known.

At the moment, the structureless model representations are used almost exclusively. In the other end, there may be only *quantitative* relationships between model components. For example, an implicit black-box modeling based solely on the input-output relationship carries little information of the internal function. combining two input-output models is again a 'flat' model with no more internal structure. On the other hand, there are the purely *qualitative* models that are the standard way of applying AI in mathematical fields [4]. The exact mathematical structure is forgotten, and only the general trends of the behavior are extracted. No exact calculations can be based on these qualitative models. Both of the above representations capture a fraction of the system behavior, but neither of these approaches alone is good from the structural point view. What is needed is not a compromise, but a combination of the quantitative and the qualitative representations. This can be achieved by *symbolic calculation*.

Symbolic calculation is based on algebraic manipulations of mathematical equations. There are various commercial general purpose tools available for doing symbolic mathematics (Mathematica, Macsyma, and Maple, for example), but there are some fundamental problems that restrict the applicability of this methodology. The symbolic structures usually grow fast, and therefore the *explosion of complexity* often results. From the point of view of practical control engineering, another problem is that no analytic mathematical solutions always exist, and the solutions are based on algorithms that are iterative by nature. In general, numeric methods are more robust, while symbolic methods are prone to deficiencies and the explosion of complexity. Because of these problems, symbolic calculation is often regarded only as an aid for a human, for doing simple manipulations of expressions.

Long symbolic expressions are difficult to handle and maintain as the calculations proceed. Symbolic manipulation cannot be used routinely to all possible tasks, and the application of symbolic methods must be controlled somehow. The conclusion is that there is need for tools that take care of the 'structure of the structures'! Once the structural approach is adopted, there is structure on various levels, not only between the data elements, but also between functional operations. What is needed, is a higher level framework for organizing all the levels in a comprehensible hierarchy. This can accomplished by *object-orientation.*

The object-oriented formalisms are well suited to systems modeling (see, for example, [2]). Often, however, the formalisms are introduced simply for notational purposes, with no emphasis on how the formalism could be implemented. In this paper, the object-orientation is an essential part of the structured operational model.

Summarizing the above discussion, we can state the analogies on different levels of abstraction:

1. **Symbolic vs. numeric calculation**
2. **Object-oriented vs. procedural programming**
3. **Structural vs. simulation based modeling environments.**

Depending on the application area, the analytic mathematical methods are more or less useful. Application of symbolic calculation is always limited to special tasks, where closed form solutions are realistic. That is why, the modeling environment that is based on symbolic calculation, probably differs from one modeling application to another. This means that the modeling tool must be modifiable and extendable to meet various needs. Better still if the system is so compact that the modifications can be carried out by the end-user himself. As compared to the robust, general purpose modeling environments, this kind of specialized tools can be used as a 'surgeons knife', for attacking special problems with high accuracy and efficiency. Within any field, there are plenty of special tasks that can be handled efficiently using tailored and domain-oriented tools—see examples in Sec. 3!

The need of structure and structured environments becomes acute in AI applications, where interdependencies of structures play a central role. Structural models are a step towards *conceptual models* that are central in AI research.

2 Combining AI in systems engineering

Modern control engineering tools are being developed worldwide, and also at the Control Engineering Laboratory of Helsinki University of Technology. The target of the **CES** project[2] is to combine AI and control engineering. The starting point is the hypothesis

> *Intelligence is an illusion caused by a complexity of a large amount of simple operations—Artificial Intelligence is a science for managing this complexity.*

Whether this view of *quantitative* becoming *qualitative* is correct or not, is of philosophical interest. From the engineering point of view, it offers concrete possibilities.

The engineering approach to tackling with complexity is the *decomposition.* The huge step from the conceptual level to the actual data manipulation may

---

[2] Short for 'Coupling Expert Systems on Control Engineering Software'

be divided in steps that can be attacked separately. Systems theory and control engineering is a field of science, where the accumulating simple operations are mathematical. To decompose the AI challenges of control engineering, tools for efficiently manipulating the mathematical structures are essential. In what follows, an experimental approach for constituting an intermediate level between the control theoretical concepts and mathematical data is presented.

2.1 The research tool

The **CES** system constitutes an interface between conceptual and concrete data processing. The system facilitates the *hierarchical* organization of different types of information processing—the mathematical data manipulation takes place below, while the conceptual knowledge processing happens above this interface. The connections between the concepts and corresponding mathematical operations are defined using a specialized, object-oriented formalism.

To guarantee good performance and reliability of the system, the established, ready-to-use mathematical tools are utilized. These lower level tools are only loosely integrated, and they can be changed if needed. Communication between the packages is carried out automatically by the system when necessary. Normally, the basic tools are Mathematica [11] and Matlab, the *de facto* standard program packages for doing symbolic and numeric mathematics. In structural modeling, capability of symbolic calculations is specially important.

The definition of the conceptual 'world models', the application environments, is done off-line, by a domain area expert. This expert needs to understand the application area well enough, because he has to express his knowledge in a reasonable form, giving a mathematical description to those concepts that are relevant in that environment. Because of the powerful manipulation tools and compact task-oriented formalisms, the operations can often be defined in a few lines of code. From the point of view of mastering the complexity of a specific application, it is important that the definitions of concepts are given just in one place. Thus they can easily be modified if needed, and the management of the information becomes simpler. This facilitates not only data, but also knowledge *reuse*. The role of the system as a connection between various software tools is illustrated in Fig. 1.

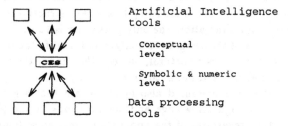

Fig. 1. The role of the system as a connection between tools

After the world model descriptions have been written and compiled, the environment can be utilized on-line as a sophisticated calculator. The user of the environment gives a command, applying the predefined concepts, and the system evaluates it. The end-user need not understand all the details of the underlying world model—this 'user' may be a human, or some kind of an AI application.

To make the **CES** language applicable to varying needs, its syntax can easily be modified—the definitions of the syntax are defined using *metaformalisms* [5].

2.2 Modeling applications

In this paper, we concentrate on the lower levels of the scenario shown in Fig. 1, that means, the **CES** system potential is assessed in creating structural models for dynamic systems. Only the 'sub-intelligence' level is concentrated on, where all manipulations are, in principle, mechanical, even if they may be complex. The mechanical nature of the solutions, together with the powerful tools for mathematical manipulations, facilitates autonomous operation of the system, and it turns out that this partial realization of the scenario can already be beneficial in dynamic systems modeling tasks.

Even if the intermediate level is intended to be used only as an interface between the AI tools and mathematical manipulations, the created models of the application environments can be pretty complex. The model being created is supposed to be the framework for the knowledge based systems, and it should provide the higher level system with all connections to the data that it might need. The application model contains *not only* the dynamic model of the system, but also controller structures, connections of different substructures, design algorithms, etc., whatever is essential in that application environment, and what can be defined exactly and compactly in mathematical terms. In this context, we use the term conceptual model, instead of structural model, to refer to the application environment, together with the actual dynamic process model. Needless to say, all these submodels are structural.

This approach to regarding modeling as a conceptual modeling task is relatively general, and allows many uses to the defined models. Depending on the application environment, the models can be applied not only to simulation, but directly to design tasks, depending on the description of the world model. In Fig. 2, the changing role of the human is illustrated—if the whole application environment is modeled, the simple and iterative tasks can be carried out by the computer. When the output of the modeling tool is raw, structureless time series data, coming from some simulation, the only possibility is to utilize the pattern recognition capabilities of the designer to analyze and modify the resulting model. If the output of the tool is structural, on the other hand, the resulting data structures can be analyzed and modified automatically.

This kind of ideas are presented also in [1], where the targets reach still higher: The whole design knowledge and specifications of the plants would be integrated, so that regardless of the application area, an autonomous, complete modeling environment could always be created. However, it is questionable

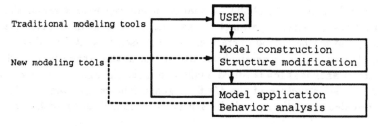

Fig. 2. The new tools can simplify the burden of redundant model analysis and modification. The 'model' consists not only of the dynamic model of the actual process, but also of the model of the application environment

whether complete automation of modeling of plants will ever be possible. One thing against this view is that the mathematical tools cannot be applied with equal success to all tasks, so that this kind of a universal modeling system cannot offer a *homogeneous* view of the environment. Failing in one subtask would collapse the whole structure. This problem is not so severe if simple simulation models are regarded as satisfactory, so that demands of the mathematical machinery are not too great. Another problem is that the large scale plants differ from each other a great deal, and the same design information hardly can be reused. New information is needed for all new applications, so that the burden of creating a new complex environment for a unique task may become unreasonable. Huge data and knowledge bases are necessary, with the maintenance problems of corresponding size.

The approach presented in this paper is less ambitious. Computer can be used as a tool for simplifying or even automating simple subtasks. No universal modeling environment can be constructed, at least not with the tools currently available. To facilitate the realization of a tailored environment for a special task, the flexibility of the tool is a key question. *In some cases it is possible to create complete, automatic modeling environments.*

3   Applications in modeling of flow processes

This far, the **CES** system has been applied to various tasks, like dynamic modeling and analysis of mechanical structures using the Lagrangian method [6]. In this context, the system is applied to different kinds of *flow processes*. The 'world models' in each case consist of the basic process components, their properties, their interconnections, and the dynamic models of the total system. The examples are very reduced and the amount of rules in the descriptions is minimal. The possibilities of manipulating and analyzing processes are rather limited, and the rules in each case can only be used for generating the dynamic models for processes. Additional rules around the kernels are needed if the models are to be used for some further purposes—controller design or parameter optimization, for example.

All calculations are symbolic, so that resulting expressions reveal the connections between system parameters. Mathematica is utilized extensively in the rules, and mathematical operations are expressed in that formalism—to go through the examples in detail, basic literacy of Mathematica is needed. The output from Mathematica is edited slightly to make the expressions more readable. The basic structure of the **CES** language follows the standard ideas of object-oriented formalisms, and the syntax is explained along with the examples.

3.1 Introductory example

The first modeling environment can be used to model ideal mixers connected in series. This application is hardly useful, but the basic syntax of the **CES** language will become more concrete.

The definition of the world model is done off-line, by a more or less experienced application programmer. This expert needs to understand the application area well enough, because he has to express its structure in a reasonable form, and give its constituent properties a mathematical description.

```
DEFINE Process,
  WITH Response == {{ Simplify[InverseLaplace[@TransferFunction
                    *LaplaceTransform[@Input,t,s],s,t]] }}
  AND Simulate: DO {{ Plot[@Response, {t,0,50}] }}
```

The world model is object-oriented and consists of class definitions. The class **Process** above is the prototype to be used as a framework for input-output processes in a transfer function form. For any object of the class **Process**, the **Response** denotes the time function $y(t)$ that can be solved mathematically using Laplace transforms when the **TransferFunction** $G(s)$ of the process and the **Input** $u(t)$ is known:

$$y(t) = \mathcal{L}^{-1}\{G(s)\,\mathcal{L}\{u(t)\}\}.$$

All mathematics is done by Mathematica—the expressions to be evaluated are given between double braces. Because Mathematica expressions are symbolic themselves, a special sign ('@') is needed in front of those symbols that refer to variables or properties defined on the **CES** level. To calculate the **Response**, variables **Input** and **TransferFunction** need to be known, and they have to be defined before **Response** is used. These variables can be either properties defined for the corresponding class, or they can be global variables. Further, operation **Simulate** can be used to plot the calculated **Response** using the Mathematica facilities.

```
DEFINE IdealMixer OF Volume
  IS_A Process,
  WITH TimeConstant == {{ @Volume/@Flow }}
  AND TransferFunction == {{ 1/(@TimeConstant*s+1) }}
```

Above, class hierarchy is utilized. The `IdealMixer` is a *derivation* of the *super class* `Process`. All properties that were defined for general linear processes are also applicable to ideal mixers, but there are also some additional features, like `TimeConstant`, that are not shared by other processes.

```
DEFINE Connection OF Process P1 AND Process P2
  IS_A Process,
  WITH TransferFunction ==
      {{ @P2.TransferFunction*@P1.TransferFunction }}
```

This rule completes our simple world model. The `Connection` of two processes is again a `Process` with the standard properties. The two processes that are given as parameters are given unique local names `P1` and `P2` within the class definition, and references to their properties can be distinguished using the dot ('.') notation. The transfer function of the combined process is calculated by multiplying the transfer functions of the component processes.

After the world model description above has been written and compiled, first to C++ and then to executable code, it can be used during an on-line session. The system is like a sophisticated *calculator:* The end-user writes a command, and the **CES** system evaluates it. The end-user does not need to understand all the details of the underlying world model. A simple session below shows how the world model is utilized. First a tank system is defined, where two ideal mixers with volumes V_1 and V_2, respectively, are connected in series:

```
Tank1 IS_A IdealMixer OF {{ V1 }}
Tank2 IS_A IdealMixer OF {{ V2 }}
System IS_A Connection OF Tank1 AND Tank2
```

The on-line definitions describing the application process are now complete, and ready for use. First, some parameters are defined that are used later:

```
%% Process flow and input concentration:
Flow = {{ Q }}
Input = {{ 1 }}  %% Unit step
```

Properties that were defined for general processes are now fit for use also for the `System`. The transfer function can be calculated as `System.TransferFunction` returning

$$\frac{Q^2}{Q^2 + Q(V_1 + V_2)s + V_1 V_2 s^2}.$$

Because the input function form was defined above to be a unit step, the `System.Response` will be

$$1 - \frac{V_1}{V_1 - V_2}\, e^{-\frac{Q}{V_1}t} + \frac{V_2}{V_1 - V_2}\, e^{-\frac{Q}{V_2}t}.$$

Direct **Mathematica** commands can also be given, if something has not been included in the rules. For example, numerical values can be substituted, using the **Mathematica** notation, as

```
{{ @System.Response /. {V1->1, V2->2, Q->10} }}
```

resulting in

$$1 + e^{-10t} - 2e^{-5t}.$$

In simple cases using Mathematica exclusively is enough, because, despite symbolic manipulations, it can be used also for numerical calculations and for generating graphics—for example, the Response between $0 \leq t \leq 50$ can easily be plotted by typing System.Simulate.

3.2 Free flow

The rules in this section constitute a **CES** world model for defining, linearizing, and discretizing linearizable time-invariant flow processes. Using these simple rules only series connections of fluid vessels can be modeled. There can be any number of vessels in the system, and the shape of the vessels is not constrained. Pathological cases with vessels getting full, etc., are not taken care of.

```
DEFINE NonLinearProcess

DEFINE LinearProcess,
   WITH TransferFunction == {{ Simplify[@MatrixC .
      Inverse[s IdentityMatrix[Dimensions[@MatrixA][[1]]]-@MatrixA] .
      @MatrixB][[1,1]] }}
   AND TimeResponse == {{ Simplify[InverseLaplace[ExpandAll[
      @TransferFunction*Laplace[@Input,t,s]],s,t]] }}
```

The class NonLinearProcess above is a trivial class to be used as a framework for nonlinear processes. A more interesting class LinearProcess is also defined. For all instances of that class, properties TransferFunction and TimeResponse are available. Linear processes are supposed to be in the state space form

$$\begin{cases} \dot{x}(t) = Ax(t) + Bu(t) \\ y(t) = Cx(t) \end{cases}$$

where $x(t) = [x_1(t), \ldots, x_n(t)]^T$, and $y(t)$ and $u(t)$ are scalars. Given the matrices A, B, and C, the transfer function for the process is calculated as

$$G(s) = C \left(sI_{Dim(A)} - A\right)^{-1} B$$

The rule for a LinearProcess is a relatively general definition. The actual matrices A, B, and C, expressed in the rules as MatrixA, MatrixB, and MatrixC, respectively, characterising the specific process, are supposed to be given in some later phase. Further, class Discretization represents the discretized version of the linear process. It has one parameter—when declaring a discretization, the corresponding continuous time linear process has to be given:

```
DEFINE Discretization OF LinearProcess LP,
   WITH MatrixF == {{ Simplify[InverseLaplace
      [Inverse[s IdentityMatrix[Dimensions[@LP.MatrixA][[1]]]
      - @LP.MatrixA],s,t]] /. t->@T }}
```

```
AND MatrixG == {{ Integrate[InverseLaplace
   [Inverse[s IdentityMatrix[Dimensions[@LP.MatrixA][[1]]]
   - @LP.MatrixA],s,t] . @LP.MatrixB, {t, 0, @T}] }}
AND MatrixH == {{ @LP.MatrixC }}
```

The discretized system model is in the form

$$\begin{cases} x(t+T) = Fx(t) + Gu(t) \\ \quad\quad y(t) = Hx(t) \end{cases}$$

where T denotes the sampling time period, and the matrices F, G, and H are defined as

$$\begin{aligned} F &= e^{AT} \\ G &= \int_0^T e^{A\tau} B\, d\tau \\ H &= C \end{aligned}$$

with e^{At} calculated as

$$e^{At} = \mathcal{L}^{-1}\left\{ (sI_{Dim(A)} - A)^{-1} \right\}.$$

Next, we want to express the linearization of a nonlinear process in an explicit form. In the current world model, the state equations of the nonlinear process are supposed to have the following outlook, with $f_i(x, u)$ being differentiable functions:

$$\dot{x}_1(t) = f_1(x, u)$$
$$\vdots$$
$$\dot{x}_n(t) = f_n(x, u)$$
$$y(t) = g(x).$$

Linearization of a nonlinear process is accomplished by first solving for the steady state $x_{,,}$ when the input flow is fixed, and thereafter constructing the Jacobi matrices. The steady state is found by setting $\dot{x}(t) = 0$ and solving the set of nonlinear equations for x. Because only series connections of flow vessels are possible, this set of equations is usually soluble for any number of connected vessels. When the nominal value of the input flow is denoted $u_{,,}$, the matrices characterizing the linearized system can be calculated from the nonlinear model as

$$A = \begin{pmatrix} \left.\frac{\partial f_1(x,u)}{\partial x_1}\right|_{x_{,,},u_{,,}} & \cdots & \left.\frac{\partial f_1(x,u)}{\partial x_n}\right|_{x_{,,},u_{,,}} \\ \vdots & \ddots & \vdots \\ \left.\frac{\partial f_n(x,u)}{\partial x_1}\right|_{x_{,,},u_{,,}} & \cdots & \left.\frac{\partial f_n(x,u)}{\partial x_n}\right|_{x_{,,},u_{,,}} \end{pmatrix}$$

$$B = \begin{pmatrix} \left.\frac{\partial f_1(x,u)}{\partial u}\right|_{x_{,,},u_{,,}} \\ \vdots \\ \left.\frac{\partial f_n(x,u)}{\partial u}\right|_{x_{,,},u_{,,}} \end{pmatrix}$$

$$C = \begin{pmatrix} \left.\frac{\partial g(x,u)}{\partial x_1}\right|_{x_{,,},u_{,,}} & \cdots & \left.\frac{\partial g(x,u)}{\partial x_n}\right|_{x_{,,},u_{,,}} \end{pmatrix}$$

Effectively, the rule below accomplishes this. The steady state input $u_{,,}$ is called InputInSteadyState, the operating point $x_{,,}$ is in SteadyStateValues, and the system matrices A, B, and C are thereafter solved.

```
DEFINE Linearization OF NonLinearProcess NLP
  IS_A LinearProcess,
  WITH States == {{ @NLP.States }}
  AND SteadyStateValues ==
    {{ Join[Solve[Table[@NLP.dx_dt[[i]]==0,{i,Length[@NLP.dx_dt]}],
      @NLP.States][[1]] /. @NLP.Input ->
      @InputInSteadyState,List[@NLP.Input -> @InputInSteadyState]] }}
  AND MatrixA == {{ Table[Table[D[@NLP.dx_dt[[i]],@NLP.States[[j]]],
      {j,Length[@NLP.States]}],{i,Length[@NLP.dx_dt]}]
      /. @SteadyStateValues }}
  AND MatrixB == {{ Table[List[D[@NLP.dx_dt[[i]],@NLP.Input]],
      {i,Length[@NLP.dx_dt]}]  /. @SteadyStateValues }}
  AND MatrixC == {{ List[Table[D[@NLP.Output,@NLP.States[[j]]],
      {j,Length[@NLP.States]}]]  /. @SteadyStateValues }}
```

Some special types of nonlinear processes are next defined. Because the application area is modeling of flows in connected vessels, a prototype for a tank is first needed. For any tank, the following scalar state equation is supposed to hold

$$\begin{cases} \dot{h}(t) = \dfrac{u(t)-k\sqrt{h(t)}}{A(h(t))} \\ y(t) = k\sqrt{h(t)} \end{cases}$$

This is approximately true for fluid vessels with flow out being proportional to the square root of the liquid level, according to the Bernoulli law. Another thing that makes Tank objects nonlinear is the possibility of giving the area of the fluid in terms of any function of h. The level of the fluid is chosen as the state of the process:

```
DEFINE Tank OF Area AND OutFlowFactor
  IS_A NonLinearProcess,
  WITH Input == {{ @NAME"FlowIn" }}
  AND States == {{ List[@NAME"Level"] }}
  AND dx_dt == {{ List[(@Input-@OutFlowFactor*Sqrt[h])/@Area]
      /. h->@States[[1]] }}
  AND Output == {{ @OutFlowFactor*Sqrt[h] /. h->@States[[1]] }}
```

There are some predefined special properties for the classes. For example, NAME always returns the name of the defined object, so that the name can be used to distinguish objects from each other. Further, strings between double quotes together with NAME are concatenated with the object name. For example, names of the states and input signals in various Tank objects must be unique. That is why, these names are constructed by concatenating the unique object name to the strings "Level" and "FlowIn", respectively.

The Connection of two vessels is a natural extension to the world model. The rule below connects two nonlinear processes in series. The Output of the former process is set equal to the Input of the latter one, and the set of states of the combined process is the combination of the states of the component processes.

172

```
DEFINE Connection OF NonLinearProcess P1 AND NonLinearProcess P2
  IS_A NonLinearProcess,
  WITH Input == {{ @P1.Input }}
  AND States == {{ Join[@P1.States,@P2.States] }}
  AND dx_dt ==
      {{ Join[@P1.dx_dt,@P2.dx_dt] /. @P2.Input -> @P1.Output }}
  AND Output == {{ @P2.Output }}
```

This concludes the simple world model definitions. Next this modeling environment is utilized, and two vessels, Tank1 and Tank2 are defined. The first Tank object has a constant area $A_1 = A_{10}$, not dependent of the level in the tank, but the area of the second tank, $A_2(h) = A_{20}h_2$, grows linearly as a function of the fluid level h_2. The rate of flow k out from the tanks is equal for both tanks. The System is constructed by connecting the tanks in series (Fig. 3).

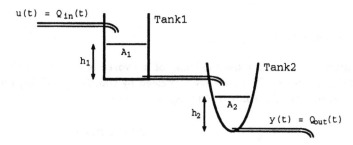

Fig. 3. A system of two vessels

```
Tank1 IS_A Tank OF {{ A10 }} AND {{ k }}
Tank2 IS_A Tank OF {{ A20 h }} AND {{ k }}
System IS_A Connection OF Tank1 AND Tank2
```

After this, the global variable InputInSteadyState is given a fixed value and the tank system is linearized around this operating point. The sampling period T is set to a numerical value of 1/10, and the linearized version of the system is discretized:

```
InputInSteadyState = {{ Flow }}
LinearVersion IS_A Linearization OF System
```

```
T = {{ 1/10 }}
DiscreteVersion IS_A Discretization OF LinearVersion
```

The process definitions are now complete and ready for use. For example, the transfer function of the linearized process, LinearVersion.TransferFunction, is available:

$$\frac{k^6}{(k^2 + 2A_{10}Flow\ s)(k^4 + 2A_{20}Flow^3\ s)}$$

Further, `DiscreteVersion.MatrixF` can be calculated:

$$
\begin{pmatrix}
e^{\frac{-k^2}{20A_{10}Flow}} & 0 \\[2mm]
\frac{A_{10}k^2}{A_{20}Flow^2-A_{10}k^2}\left(e^{\frac{-k^4}{20A_{20}Flow^3}} - e^{\frac{-k^2}{20A_{10}Flow}}\right) & e^{\frac{-k^4}{20A_{20}Flow^3}}
\end{pmatrix}
$$

Even if the world model now is very reduced, the possibilities of the underlying tools can be utilised. For example, simulating the impulse response of the linearised tank system between $0 \leq t \leq 20$ can be accomplished using the visualization capabilities of Mathematica directly:

```
Input = {{ Delta[t] }}
{{ Plot[@LinearVersion.TimeResponse,{t,0,20}] }}.
```

3.3 Tank networks

This example demonstrates the modeling of networks of linear, time invariant concentration processes. Processes are expressed by transfer functions, which are generally dependent of the flow F through the process:

$$
P(s, F) = \begin{cases}
\frac{1}{\frac{V}{F}s+1} & \text{for an ideal mixer} \\
e^{-\frac{V}{F}s} & \text{for an ideal delay element} \\
1 & \text{for no delay}
\end{cases}
$$

These basic processes represent the perfectly stirred tank and the plug flow element, both having constant volume, and an 'identity' process, having no volume. In addition to the processes, the world model now consists of *nodes* that connect proceses together, corresponding to pipe connections. In what follows, the subscripts i refer to the corresponding node, so that the flow through the node i is denoted F_i, and the transfer function is $G_i(s)$. Mathematically, the modeling of flow networks is done in three steps.

1. **Initialization of variables.** Define two artificial variables f_i and $g_i(s)$ for each node in the network. These vectors are initialized as follows:

$$
f_i = \begin{cases}
F_{i,in} & \text{if node } i \text{ is an input node} \\
0 & \text{otherwise}
\end{cases}
$$

$$
g_i(s) = \begin{cases}
1 & \text{if node } i \text{ is an input node} \\
0 & \text{otherwise}
\end{cases}
$$

The role of the *input node* is special, because the transfer functions of other nodes will be calculated with respect to the input node. In the implementation, the final amount of nodes need not be known beforehand, but the initialization is carried out as a new node is introduced.

2. **Addition of a pipe from node i to j.** First, the node j is added to the set S of the *sink nodes*. Next, suppose proportion k_{ij} of the total flow in node i flows through the new pipe having transfer function $P_{ij}(s, k_{ij}F_i)$. The vectors f_j and g_j are updated as follows:

$$f_j \leftarrow f_j + k_{ij}F_i$$
$$g_j(s) \leftarrow g_j(s) + \frac{k_{ij}F_i}{F_j}G_i(s)\,P_{ij}(s, k_{ij}F_i)$$

The above expressions are symbolic—for example, actual values of F_i are not known at the time of applying these expressions. This step of adding pipes may be repeated as many times as necessary, so that even complex connections between nodes can be modeled. Multiple processes between the same nodes cause only notational problems.

3. **Solution of flows and transfer functions.** After all connections are defined, first solve the linear set of flow equations in terms of all nodes i in the network:

$$F_i = f_i$$

Next, the set of equations for transfer functions is solved:

$$G_i(s) = g_i(s)$$

This set is of equations is also linear, *after* the flows have been solved and substituted. At least in principle, solving sets of linear equations is always feasible, and it can be accomplished efficiently even in symbolic form. If a node is not in the set S of sink nodes, its flow and transfer function will remain undetermined, as symbolic quantities, in the resulting expressions.

The **CES** rules to accomplish the above operations are given next. First, the basic processes are defined, **Tank** being a perfect mixer and **Pipe** being a tube with no delay.

```
DEFINE Process,         %% The prototype for all processes
  WITH Response == {{ Simplify[InverseLaplace[@TransferFunction
    *Laplace[@InputFunction,t,s] /. @Values,s,t]] }}
  AND Simulate: {{ Plot[@Response, {t,0,@EndTime}] }}

DEFINE Tank OF Volume IS_A Process,
  WITH TimeConstant == {{ @Volume/@Flow }}
  AND TransferFunction == {{ Simplify[1/(@TimeConstant*s+1)] }}

DEFINE Pipe IS_A Process,
  WITH TransferFunction == {{ 1 }}
```

Next, the **Node** is defined with initially total flow and transfer function set to zero. The total flow and the transfer function are updated every time a new pipe is added. The difference between notations '=' and '==' is that the former evaluates the right hand side once and for all, while the latter is like a *definition*, it is evaluated dynamically, every time something changes on the right hand side.

```
DEFINE Node,
  WITH TotalFlow = {{ 0 }}
  AND TransferFunction = {{ 0 }}
  AND AddPipe:
      TotalFlow = {{ @TotalFlow + @Ratio @StartNode.NAME"Flow" }}
      TransferFunction =
          {{ @TransferFunction + @StartNode.NAME"TransferFunction"
           * @FlowProc.TransferFunction @Ratio @StartNode.NAME"Flow"
           / @NAME"Flow" }}
```

Finally, the `TankSystem` is the framework that binds nodes and processes together. A set variable `SinkNodes` contains names of all those nodes that are sinks for at least one tube. Solving sets of equations is only possible with respect to sink nodes. Property `Flows` calculates the flows in the network, and after that, `TransferFunctions` can be used to solve the transfer functions for each node. When the tank system is declared, the Input node is given. Information of the network input and output is needed only when the whole network transfer function is selected among the transfer functions of the individual nodes.

```
DEFINE TankSystem OF Node Input
  IS_A Process,
  WITH SinkNodes = -
  AND AddPipe:
      WHILE Flow = {{ @Ratio @StartNode.NAME"Flow" }} :
          DO EndNode.AddPipe
          EndNode TO SinkNodes
  AND Flows == {{ Simplify[Solve[Table[{@SinkNodes:TotalFlow}[[i]] ==
          {@SinkNodes:NAME"Flow"}[[i]],{i,@SinkNodes:COUNT}],
          {@SinkNodes:NAME"Flow"}]][[1]] }}
  AND TransferFunctions ==
          {{ Simplify[Solve[Table[{@SinkNodes:TransferFunction}[[i]]
          == {@SinkNodes:NAME"TransferFunction"}[[i]],
          {i,@SinkNodes:COUNT}],{@SinkNodes:NAME"TransferFunction"}]
          /. @Flows][[1]] /. @Input.NAME"TransferFunction" -> 1 }}
```

The elements of the sets can be referred to using colons (':'), that means, having a set variable like `SinkNodes`, this notation refers to the corresponding properties of *all* of the set elements separately. The special set property `COUNT` returns the number of elements in the set.

After the above world model is compiled, it can be used on-line. The following example illustrates its use in a simple modeling task shown in Fig. 4. The processes and nodes are first defined:

```
P1 IS_A Pipe
P2 IS_A Tank OF {{ V }}    %% Volume not yet fixed
P3 IS_A Pipe
P4 IS_A Pipe

N1 IS_A Node
N2 IS_A Node
N3 IS_A Node
```

N4 IS_A Node

These elements are used to create the tank system:

Network IS_A TankSystem OF N1

```
WHILE StartNode = N1  AND  EndNode = N2  AND
      Ratio = {{ 1 }}  AND  FlowProc = P1 :
  Network.AddPipe

WHILE StartNode = N2  AND  EndNode = N3  AND
      Ratio = {{ 1 }}  AND  FlowProc = P2 :
  Network.AddPipe

WHILE StartNode = N3  AND  EndNode = N2  AND
      Ratio = {{ k }}  AND  FlowProc = P3 :
  Network.AddPipe

WHILE StartNode = N3  AND  EndNode = N4  AND
      Ratio = {{ 1-k }}  AND  FlowProc = P4 :
  Network.AddPipe
```

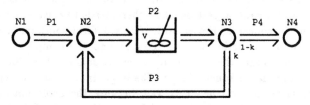

Fig. 4. The network example

The flows of the network are calculated by **Network.Flows**, and the result follows **Mathematica syntax:**

$$N_{2Flow} \rightarrow \frac{N_{1Flow}}{1-k},$$
$$N_{3Flow} \rightarrow \frac{N_{1Flow}}{1-k},$$
$$N_{4Flow} \rightarrow N_{1Flow}$$

The transfer function from node **N1** to node **N4**, for example, can be calculated by {{ **N4TransferFunction /. @Network.TransferFunctions** }} and reads

$$\frac{N_{1Flow}}{N_{1Flow} + V\,s}.$$

Also a state space version of the networks environment has been implemented, so that systems with nonlinearities, time variant process properties or flow distribution can be modeled. It would be easy to simplify modeling environment to tackle with pure *information flow* diagrams, where no actual material flow takes place, and no solution of flows is needed.

A prototype version of the **CES** system is functional in Unix, and the above examples have resulted from actual simulation runs. To make the system easier to use, a user interface is being developed in the **X11** windows environment, with mouse controlled menus. The user interface has to be dynamic, the menus changing according to definitions and operations made by the user, to facilitate the easy access to the data structures.

4 Discussion

Compared to other object-oriented model formalization languages, the main difference in **CES** is its powerful structure processing capability. Often the novel modeling languages are intended mainly for process formalization purposes, just to emphasize some special modeling aspect. The other difference is the modifiability of the system, the definitions being based on the metadescriptions. On the other hand, the adaptability of the system to various tasks is also a consequence of the fact that there are no predefined data structures. In Omola [9], for example, the 'terminals' are predefined, and the data types are fixed—it is doubtful whether this kind of a formalism could be a general purpose modeling language, applicable to varying needs.

The proposed approach is most suitable if the domain area is narrow and well-understood. To gain biggest benefits, there needs to exist exact mathematical definitions to concepts. Better still, if closed form solutions are available. In that case, user interaction can be reduced and the computer can best be utilized—this means that the best applications are small scale problems that are common, for example, in education [7]. The large, industrial scale applications, can probably never be attached this way, because the explosion of complexity in symbolic expressions will always defy even the most efficient computers, not because of technical, but simply theoretical reasons. Some subproblems can, however, be solved symbolically, and the flexibility of the tools must be maintained to preserve their applicability in the special tasks. The currently active application fields of the proposed methodology are the following:

1. **Computer aided education** is facilitated by the system, when automatic generation of dynamic models for the user defined processes makes free experimenting and analysis feasible. The *'what if...'* analyses are possible, and this may motivate further studies in the field.
2. **'Researcher's workbench'** is an environment for tailoring special purpose applications. The system is used as a highly sophisticated calculator, relieving the reserchers burden of redundant data manipulation.
3. **Expert system interface** is being developed, such that the design of multivariable LQG controllers would be automated. The expert system utilizes the developed system to communicate with the process data.

In this paper, only automation of modeling was discussed. Automation of *design* is a still more ambitious target, because design is always based on compromises between specifications that can only be assessed heuristically. Combining expert systems in the environment, or realizing the whole structure of Fig. 1, so that also expert knowledge could be utilized automatically, would be a giant step.

References

1. Cellier, F.E.: *Continuous System Modeling*. Springer-Verlag, New York, 1991.
2. Cellier, F.E., Zeigler, B.P., and Cutler, A.H.: *Object-Oriented Modeling: Tools and Techniques for Capturing Properties of Physical Systems in Computer Code*. Preprints of the IFAC Symposium on Computer Aided Design in Control Systems, Swansea, UK, July 15–17, 1991.
3. Grace, A.C.W.: *SIMULAB, an Integrated Environment for Simulation and Control*. International Conference on Control, Edinburgh, UK, March 25–28, 1991.
4. Hulthage, I.: *Quantitative and Qualitative Models in Artificial Intelligence*. In "Coupling Symbolic and Numerical Computing in Expert Systems II" (Eds. Kowalik, J.S. and Kitsmiller, C.T.), Elsevier Science Publishers B.V. (North-Holland), Amsterdam, 1988.
5. Hyötyniemi, H.: *Tools for Coupling Expert Systems on Control Engineering Software*. In "Lecture Notes on Computer Science 585" (eds. Pichler, F. and Moreno Días R.), Springer-Verlag, Berlin, 1992, pp. 652–667.
6. Hyötyniemi, H.: *Symbolic Calculation in Computer Aided Modeling*. Presented during the Joint Finnish-Russian Symposium on Computer-Aided Control Engineering, Espoo, Finland, October 13–15, 1992 (in print).
7. Hyötyniemi, H.: *Computer-Aided Education of Control Engineers*. International Workshop on Computer Aided Education in Automation and Control Technology (CAE in ACT), Prague, Czechoslovakia, May 20–22, 1992.
8. Mattsson, S.E.: *On Model Structuring Concepts*. Proceedings of the Fourth IFAC Symposium on Computer Aided Design in Control Systems (CADCS'88), Peking, PRC, August 23–25, 1988, pp. 269–274.
9. Mattsson, S.E. and Andersson, M.: *The Ideas behind Omola*. Preprints of the IEEE Symposium on Computer Aided Control System Design (CACSD'92), Napa, California, March 17–19, 1992.
10. Taylor, J.H., Rimvall, M., and Sutherland, H.A.: *Future Developments in Modern Environments for CADCS*. Preprints of the IFAC Symposium on Computer Aided Design in Control Systems, Swansea, UK, July 15–17, 1991.
11. Wolfram, S.: *Mathematica: A System for Doing Mathematics by Computer*. Addison-Wesley Publishing Co., Redwood City, California, 1988.
12. Zeigler, B.P.: *Multifaceted Modeling and Discrete Event Simulation*. Academic Press, London 1984.

2 SPECIFIC METHODS

Formal Methods and Their Future

G. Musgrave[1], S. Finn[2], M. Francis[2], R. Harris[2] and R.B. Hughes[2]

[1] Department of Electrical Engineering and Electronics,
Brunel University, Uxbridge, Middlesex,
UB8 3PH, U.K.
[2] Abstract Hardware Limited, The Science Park - Bldg.2,
Brunel University, Uxbridge, Middlesex,
UB8 3PQ, U.K.

Abstract. The complexity of future systems level problems is driving the need for alternative approaches to examining the problem of architectural synthesis at higher levels of abstraction. In this paper we show how formal reasoning tools may be used to help address this complexity problem and allow the designer to explore the design space with impunity, thanks to the rigour afforded by the mathematical formalism, in the sure knowledge that the final design behaviour will satisfy the specification.

1 Introduction

Design verification has been a very important part of chip design for the past decade. The cost of not verifying the chip by simulation is very high but the cost of simulation for large platforms is proving costly despite enhancement of tools and platforms. More importantly, one has to question the quality of that design verification. Many of today's designs represent a major sequential machine (billions of states) where the simulation waveforms have only exercised a small percentage of the available states. These factors together with an increasing number of safety-critical applications in avionics, nuclear and medical fields have provided the motivation to use formal methods to build engineering design tools to provide a basis of design verification which does not rely on simulation.

Formal methods have already been used [1, 2, 3, 4] in a variety of complex case studies and this is a considerable enhancement over the post design activities for which formal methods initially gained fame such as the formal verification of the floating point unit of the INMOS T800 transputer [5]. The tools described in this paper use a Higher-Order Predicate Logic to represent and manipulate behaviour. There are two major differences between the tool set called LAMBDA and other theorem provers. Firstly, the system uses rules (an approach pioneered in ISABELLE [6]) rather than theorems so as to reason about partial designs as well as completed ones. The proof of design correctness is combined with the initial design phase, thus obviating the need for post-design verification. This speeds up the design process, partly because errors are caught as they occur, but also because the task of verification is incorporated within each design step. The second major difference between the LAMBDA system and other tools is its graphical interface, called DIALOG. This represents the partial design as a schematic, and enables the proof and design process to be viewed by the designer in a form he can use and understand.

In other words, the tools rely on the mathematical rigour of formal methods yet, by utilising a graphical interface, present the designer with a familiar working environment. These tools are not intended to replace existing ones but, to work in conjunction with them so as to extend the designer's capability to achieve reliable, right-first-time designs. The integration of formally based tools into existing tool sets will provide benefits at all stages of a product's life-cycle.

The paper will outline the principles of the core tools and give examples of its use as well as point the way to further developments which enable design-flow graphs to be part of the specification mode and an insight as to how *animation* could help the early synthesis of the specification problem.

2 The LAMBDA (Logic And Mathematics Behind Design Automation) System

Representation of behaviour as predicates has been exploited for formal verification [7, 8]. This representation allows us to compose the implemented behaviour of a complex module from the behaviours of its parts, using logical connectives to model connection and hiding of internal wires. A design may be verified by showing, using formal proof, that its implemented behaviour satisfies the specification. Using these methods, it is now possible to specify and verify the behaviour of digital systems whose behaviour is so complex that exhaustive simulations are not feasible [9].

Published examples of formal verification have started from a completed design. The LAMBDA approach integrates proof and design, allowing full exploitation of the benefits of abstraction and hierarchy. This integration is made possible by formally representing three things: the ultimate goal, what has already been done (the *partial design*) and the *outstanding tasks*. We need to ensure that IF, by making some additions to the present design, we achieve the outstanding tasks THEN we will have achieved the originally specified behaviour. This relationship can be represented as a rule of logic:

$$
\begin{array}{ll}
\text{IF} & \text{current\_design} + \text{further\_work ACHIEVES} \quad \text{task\_n} \\
\text{AND} & \cdots \\
\text{AND} & \text{current\_design} + \text{further\_work ACHIEVES} \quad \text{task\_1} \\
\hline
\text{THEN} & \text{current\_design} + \text{further\_work ACHIEVES} \quad \text{specification}
\end{array}
$$

As the design progresses the rule representing the design state is transformed to keep track of what hass been done and what remains to be achieved. The system ensures that the rule is always logically valid. The design is successfully completed when we have no more tasks:

$$
\overline{\text{completed\_design} + \text{external\_environment ACHIEVES specification}}
$$

To summarise: LAMBDA's representation of the current state of a design is always a valid rule of logic, which is altered with each refinement made by the designer.

These refinements transform the rule, by changing the *partial design* and totally re-forming the outstanding tasks, to give a rule of the same basic form but with a simpler set of outstanding tasks and a more refined partial design. This manipulation of rules faithfully represents the behavioural consequences of manipulations of the *design rules*; outputs must not be connected together, and there should be no zero-delay loops or floating inputs. (N.B. The notation normally used for rules uses the symbol ⊢ meaning logical entailment rather than ACHIEVES and IF, AND and THEN are ommitted as being superfluous.).

The LAMBDA core is composed of three parts:

- a theorem prover
- a library of rules (already proved in the system)
- a help system (a browser)

These are all embodied in a windows environment with a mouse driven interface. With a schematic window it is possible to represent the design as the connection of components that have already been proven as partial dsign steps.

A simple example to illustrate the tools is extensively discussed in [10] and Figure 1 gives a screen dump of the windows, a more interesting example was presented [3] at Euromicro in 1991. Here an infinite impulse response filter, whose output at time t is given by a constant multiple of the input at time $t-1$ plus a constant multiple of the output at time $t-1$. Hence it requires recursion which is notoriously difficult to synthesise using algorithmically based tools.

The paper shows how difficult implementations can be explored and the trade-offs being speed and extra components can be investigated – a vital part of the design process. More importantly the formal proof is integrated with the design realisation, with automated procedures such as scheduling interleaved with interactive synthesis. A more complex example of an IIR filter is detailed in section 4 for completeness.

An important feature of this core system is that it can relate to any existing automation tools through its VHDL output and other interfaces are relatively easily achieved because of its structure. Hence some of the users have been able to relate to their in-house tool sets. These invariably are involved further down through the implementation path but the real challenges are handling the behavioural systems level problems.

3 Systems Level Problems

Undoubtedly writing the initial specification remains a major task, and there can never be an absolute certainty that what is written corresponds to the designer's intentions; clearly some way of checking this would be beneficial. *Animation* is the term adopted for *running* the specification for the purpose of observing it's behaviour. To date this is still the work of researchers, but initial work on real-world examples, culminating in a behavioural-level waveform display of stimuli and responses, has already proved invaluable in helping the designer to 'debug' the initial specifications (see Fig. 2).

Another dimension to the problem of system design is the question of how the architectural synthesis can interface formal reasoning tools to other software components [1, 11, 12] which form an integral part of the design process. This has led to

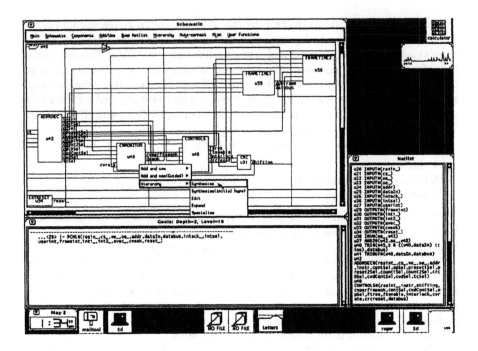

Fig. 1. Typical Window Environment

the development of design-flow graphs (encompassing both data- and control-flow) and thus partitioning. This is part of the ESPRIT Project PATRICIA (No. 5020) and has resulted not only in architechtural synthesis [13] but also giving a useful insight into how the problem of test-pattern generation can be eased. By representing these flow graphs formally inside LAMBDA/DIALOG, re-grouping, expanding, and partitioning can be carried out with impunity thanks to the rigour afforded by the formalism, in the sure knowledge that the behaviour will be unchanged [14].

4 The example - an IIR filter

Our example is an Infinite Impulse Response filter with several feedback paths. The example is of particular interest because it uses recursion, which is notoriously difficult to synthesise using algorithmically based tools. In our example we show the interactive design of the data-flow for the IIR filter, followed by the fully automatic control-flow design.

The filter's output at a time t is given by a constant multiple of the input at a time $t - 3$ added to constant multiples of the outputs at times $t - 3$, $t - 2$ and $t - 1$ respectively. We also wish the output of the filter to be set to zero if a reset signal

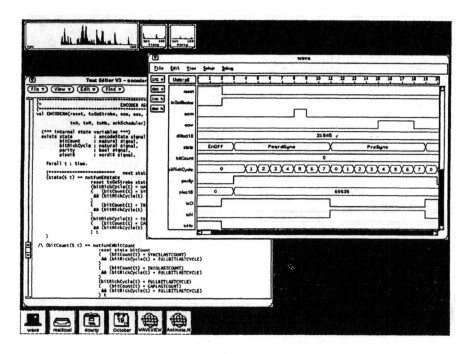

Fig. 2. Waveform debugging of Specification

has been activated during the last three time periods. We can write a specification (abbreviated SPEC#(...)) for this filter in a formal manner using LAMBDA.

```
val SPEC#(reset, input, output) =
    forall t.
    (reset t == true ->> output t == Zero)
/\ (output (t + 3) ==
        if reset (t + 3) || reset (t + 2) || reset (t + 1)
        then Zero
        else a *! output (t + 2) +!
             b *! output (t + 1) +!
             c *! output t       +!
             d *! input t );
```

Here we have used the *! and +! to represent multiplication and addition on some representation of real numbers. The actual representation used is immaterial at this level of abstraction, since we can specify that all components act at on the same representation. The syntax used for the body of the definition is that of ML [15].

The object of the synthesis exercise will be to achieve an implementation which satisfies the specification, i.e. given an input and an output, find an implementation

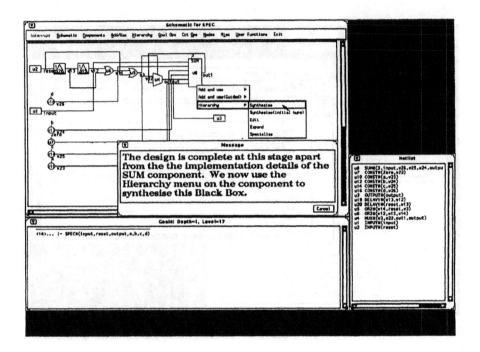

Fig. 3. The Partial design

whose input/output behaviour satisfies the specification.

The implementation proceeds by starting with it entered into the *Goals* window of the DIALOG graphical interface (c.f. Fig. 1). A multiplexor element is used to discharge the *if...then...else* part of the specification. Similarly, delay elements and primitive gates are used to further simplify the specification and further enhance the partial implementation. To illustrate, albeit trivially, aspects of design partitioning we create a *black box* with a latency of 3 to represent the summation expression containing the feedback elements. This is illustrated in Fig. 3 where the implementation is complete apart from the fact that no details of the internal structure of the black box *SUM* component are known. We thus need to synthesise the details of what is in this box.

Using pipelined adders and multipliers which take two and four clock cycles respectively to compute their results, it is clearly not possible to calculate the summation expression in only 3 clock cycles! The external clock must be slowed down by some ratio. That is the purpose of the MICRO and MACRO components (shown in Fig. 4) which show a change in the clock speed. The choice of clock ratio depends on the particular data-flow chosen for the implementation; the minimisation of this clock ratio is a by-product of the overall design process.

The engineer can make several differet decisions regarding the optimum data-

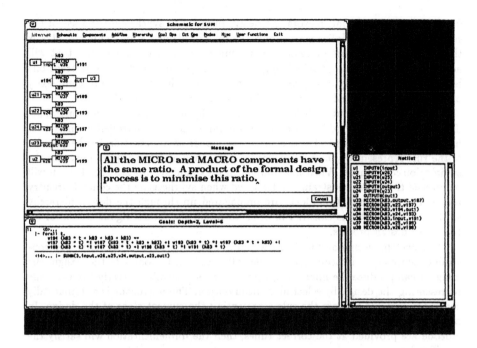

Fig. 4. Changing the external clock speed

flow at this point. It would be possible to use various combinations of adders and multipliers to maximise the amount of hardware so as to achive the output in the minimum time possible, thus reducing the value of the clock ratio, or he could use minimal hardware and reduce silicon area. Of course, the control circuitry required for such a minimal configuration would be significant and might well offset gains in reduction of area. In the example, we choose to use just one adder to implement all the additions and one multiplier to implement all the multiplications. To re-use the same component for different calculations requires that the component be used at different times, i.e. *scheduling* must be carried out to find out when the component may be used. The choice of a single adder and a single multiplier is an implementation decision affecting the data-flow.

Once the data-flow has been decided, the optimum clock ratio required for the design can be calculated completely automatically by carrying out the automatic scheduling [16]. During the scheduling process, guided by minimisation requirements on various variables (in this case only the clock ratio), the final data-flow for the summation component is calculated. This shows how automated procedures (such as scheduling) can be interleaved with interactive synthesis using our approach to formal design reasoning. When the data-flow has been correctly scheduled, it is then possible to create the control-flow part of the design fully automatically. This control-flow involves adding multiplexors and delays in order to synchronise the use of the single adder and multiplier used in our example implementation. This circuitry is the simplest possible set of components, connections and control inputs that will achieve the desired design scheduling. The designer may thus examine trade-offs between the complexity of the controller (maximal in the example) and complexity of the hardware required for the data-flow (minimal in the example).

Synthesis of the control-flow circuitry naturally introduces various constraints. These contraints are conditions on the control inputs for the controller circuitry that has just been synthesised. These constraints appear as an assertion in the premise of the rule representing the current design state; they are thus requirements for the summation component. When this component is expanded out at the top level, it will be noticed that these constraints, introduced when synthesising the control circuitry required in the summation block, are propagated up through the design hierarchy to the top level. This is essential in order to retain the formal rigour esseential for the whole design process.

To leave the design at this stage would not be very engineer-friendly. The constraints represent microcode for the controller and so, to complete the design, the designer can produce the microcode required for the controller directly from the rule representing the design by selecting a menu option. The microcode is automatically generated (see Fig.5) from the rule representing the current state of the design. In our formal approach, the design is now complete. If the signals specified by the microcode are provided at the correct times, then the implementation will satisfy the specification.

5 The Remaining Problem

There are many technical problems that remain, but there is one aspect of this technique which needs tackling now. The basic problems to acceptance of formal methods lie in the fact that there is a large knowledge gap between practising systems engineers and the mathematical knowledge necessary for manipulating a proof of the sort required for the verification and/or synthesis of complex systems. As such, formal methods have tended to only be of interest to logicians and not to practicing designers, whether they are from the software or hardware field. The gap between their knowledge and the level of mathematics required to handle formal methods is far too wide. There is considerable difficulty in attempting to give sufficient mathematical training to the whole of the engineering community so that they have adequate knowledge to be able to apply these methods. Some of the difficulties lie in persuading people that it is a good idea to learn, but the predominant factor lies in the sheer size of the knowledge gap that is present, the mathematics required is just too far removed from the training and experiences of systems designers. Equally, the design knowledge of the engineer is far removed from the mathematics of the logician.

Thus, in order for the systems engineer to be able to use formal methods in practice, some means of simplifying the mathematics are required so as to minimise the amount of specialist training required. The DIALOG approach suggested in this paper helps address this problem. Using these techniques it is possible for the engineer to examine different aspects of the system and to develop different design

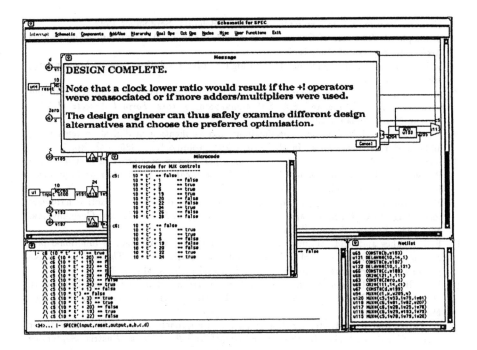

Fig. 5. Complete expanded design showing microcode

solutions, all of which are correct, to ascertain which one is optimal. Such high-level reasoning necessitates the use of formal methods and by bridging the knowledge gap it is possible for the engineer to manipulate the design in ways that are familiar to him, the difficulties of the mathematical proof being hidden. Overcoming the knowledge gap is essential to the greater acceptance of formal methods in industry. That is, the existing designers now need to experiment with such tools in order for the educational process to enhance their capabilities to cope with the complexities of tomorrow.

References

1. R.B. Hughes. Automatic Software Verification and Synthesis. In *Proc. 2nd. Int. Conf. on Software Engineering for Real-Time Systems*, pages 219–223. Institution of Electrical Engineers, September 1989.
2. E. Mayger and M. Fourman. Integration of formal methods with system design. In A. Halaas and P. B. Denyer, editors, *VLSI 91*, Edinburgh, Scotland, August 1991.
3. E. Mayger, M. Francis, R. Harris, G. Musgrave, and M. Fourman. The need for a core method. In *EUROMICRO 91*, September 1991.
4. R. B. Hughes, M. D. Francis, S. P. Finn, and G. Musgrave. Formal tools for tri-state design in busses. In *Proceedings of the 1992 International Workshop on the HOL*

Theorem Prover and its Applications, Leuven, Belgium, September 1992.

5. D. Shepherd. The role of Occam in the design of the T800. *Communicating Process Architecture*, pages 93–103, 1988.

6. L. Paulson. Natural deduction proof as higher-order resolution. *Logic Programming*, 3:237–258, 1987.

7. M. Gordon. Why Higher-Order Logic is a good conclusion for specifying and verifying hardware. In G. Milne and P.A. Subrahmanyam, editors, *Formal Aspects of VLSI Design*. North-Holland, 1986.

8. F.K. Hanna and N. Daeche. Specifications and verification of digital systems using higher-order predicate logic. In *IEE Proceedings*, volume 133, 1986. PtE No. 5.

9. Avra Cohn. A proof of correctness of the viper microprocessor: The first level. In Birtwistle and Subrahmanyam, editors, *VLSI Specification, Verification and Synthesis*. Kluwer Academic Publishers, 1988.

10. M. Fourman and E. Mayger. Formally based system design – interactive hardware scheduling. In G. Musgrave and U. Lauther, editors, *VLSI 89*, Munich, Germany, August 1989. Elsevier Science Publishers.

11. R.B. Hughes and R.M. Zimmer. Automated interactive verification of functional programming languages. In C.M.I. Rattray and R.G. Clark, editors, *The Unified Computation Laboratory*, pages 411–423. Oxford University Press, May 1992.

12. R. B. Hughes. *Automated Interactive Software Verification and Synthesis*. PhD thesis, Department of Electrical Engineering and Electronics, Brunel University, Uxbridge, Middx, U.K., July 1992.

13. A. Antola and F. Distante. DFG: a graph based approach for algorithmic flow driven architechture synthesis. In *Proc. of EUROMICRO 91*, Vienna, Austria, September 1991.

14. R. B. Hughes and G. Musgrave. Design-flow graph partitioning. In *Proceedings of the 1992 International Workshop on the HOL Theorem Prover and its Applications*, Leuven, Belgium, September 1992.

15. R. Milner, M. Tofte, and R. Harper. *The Definition of ML*. The MIT Press, Cambridge, Massachusetts, 1990.

16. M.P. Fourman and Eleanor M. Mayger. Formally based systems design—Interactive hardware scheduling. In *VLSI '89*. North-Holland, 1989.

Formal Description of Bus Interfaces
Using Methods of System Theory

Erwin M. Thurner

Siemens AG, ZFE ST SN 13
Otto-Hahn-Ring 6, D - 8000 München 83

Abstract. While constructing a bus interface, mainly two problems arise: On the one hand, the *protocols* of both sides of the interface have to be described, on the other hand, the *transformation* between these protocols has to be done. To cope with the first problem, a graph-based representation of Timing Diagrams has been developed, the *Trigger-Graph*. For the second problem, the partial protocols of a bus are considered by their meaning and are represented as a *Simple Operation (SIOP)*. The SIOPs are independent of their realisation and can be transformed into each other by formal means.

1   Formal Description of Bus Interfaces

The construction of bus interfaces is up to now a highly informal task. To improve the correctness of the hardware, to reduce its complexity, and to give the user a high-level view of the hardware, formal methods for the description of bus interfaces and protocol are desirable.

To consider this problem more deeply, an example of a simple bus-based computing system is given: Fig. 1 shows a system which consists of two modules, e.g. a CPU module and a memory module. The modules are connected by a system bus. In addition, each module has its own private bus. In this system architecture, the bus interfaces have the task to transform the subset of the system bus protocols into the protocols of their private bus. Do be able to do this, two types of problems have to be solved:

(1) Each partial protocol of every bus has to be described.
(2) A prescription has to be given, which partial protocol of, say, private bus #1 corresponds to a partial protocol of the system bus.

Every bus interface has to deal with both of these problems.

The problem (1) has been solved by representing the Timing Diagrams of the partial protocols of the buses by a special kind of graph, the Trigger-Graph. This graph can be integrated by homomorphic mappings into wellknown methods of system theory, such as Petri-Nets and Finite Automata. The problem (2) is solved by representing the *meaning* of the partial protocols. Because the meaning of protocols is independent of their

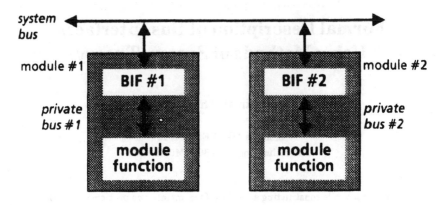

Fig. 1: Simple bus-based computing system

realisation, the protocols of different buses can be transformed into one another using a set of formal rules.

2 Specification of Bus Protocols

As a first step, we define bus protocols in a formal way. From these definitions, the Trigger-Graph can be developed. Furthermore, the basic idea is given how to transform this graph into other methods of system theory.

2.1 Definition of Bus Protocols

Timing Diagrams are often used to represent hardware-protocols, e.g. the set of partial protocols of a parallel bus. To be a protocols π, a Timing Diagram must have some properties:

– It has to be *closed*, i.e. the signals have the same "idle-value" before and after the protocol.

– It is *complete*, i.e. all temporal and causal interdependences (except the first "start-trigger") have their causes and effects within the protocol.

Timing Diagrams that fulfill these definitions, will be called partial protocols π. One of the resulting restrictions is, that protocols π are considered without arbitration phase; protocols with arbitration phase are not closed. The complete protocol (e.g. of a bus) is a set Π of partial protocols.

Now, a partial protocol π can be defined as a vector of bus signals bs:
Every element bs_j of this vector is a quadrupel
$$bs_j := \langle\, signal_j\,,\, Source_j(t)\,,\, Z_j(t)\,,\, Trigger_j\,\rangle$$

$$
\mathbf{m_b} := \begin{pmatrix} \langle\, signal_1\,,\, Source_1(t)\,,\, Z_1(t)\,,\, Trigger_1\,\rangle \\ \langle\, signal_2\,,\, Source_2(t)\,,\, Z_2(t)\,,\, Trigger_2\,\rangle \\ \vdots \\ \langle\, signal_j\,,\, Source_j(t)\,,\, Z_j(t)\,,\, Trigger_j\,\rangle \\ \vdots \\ \langle\, signal_n\,,\, Source_n(t)\,,\, Z_n(t)\,,\, Trigger_n\,\rangle \end{pmatrix}
$$

with the components:

- $signal_j$: name of the bus signal j
- $Source_j(t)$: source of the bus signal j

  The source of a bus signal may change within a partial protocol (e. g. for multiplexed bus signals). That's why it makes sense to describe the source of a bus signal as a set of pairs

  $$Source_j(t) = \{\langle\, source\,,\, \tau\,\rangle\,,\,...\,\}$$

  with:

  $$source \in \{\, CSM\,,\, MT,\, ML,\, ST,\, SL\,\}$$

  τ is the time of the change of *source*.

  CSM are signals from the Central Service Module; MT, ML, ST, SL are from the Master-Talker, Master-Listener, Slave-Talker, and Slave-Listener, respectively. Signals with *source* = MT are often represented by an arrow "→", with *source* = MT by "←".

- $Z_j(t)$: set of transitions ζ of $value_j(t)$

  with $dom\ value := \{\, 0,\, 1,\, A,\, Z,\, X\,\}$

  The values 0 and 1 are for *high* resp. *low*. The value = A (active) is, when the signal gets the value "*0 or 1*"; the value = A is wide spread for signal bundles. Z means "*not logically active*", resp. at tristate signals "not driven". With value = X, the state of this signal is not relevant for the protocol at this time. Transitions from one to another value are defined as tripels

  $$\zeta = \langle\, from,\, to,\, \tau\,\rangle, \text{ with: } from,\, to \in dom\ value.$$

  With this definition, e.g. a rising edge at τ_1 is represented as a transition $\zeta = \langle 0, 1, \tau_1\rangle$.

- $Trigger\,j$: set of triggering pulses for every transition ζ

  with $Trigger_j = \{\langle\, t_\zeta\,,\, signal_{Trig}\,,\, source_{Trig}\,,\, \zeta_{Trig}\,\rangle\,,\,...\,\}$

  with

t_ζ	: time of the transition $\zeta_j(t)$
$signal_{Trig}$	: name of the triggering signal
$source_{Trig}$	: (current) source of the triggering signal
ζ_{Trig}	: trigger transition

  The set *Trigger* may be empty, e.g. for constant signals.

2.2 Definition of the Trigger-Graph

From a protocol π, the Trigger-Graph can be derived easily. The Trigger-Graph of a protocol TG (π) consists of the set of nodes V_{TG} (vertices) and the set of directed edges E_{TG}, with the starting edge $v_{TGs} \in V_{TG}$

$$TG (\pi) := \langle V_{TG}, E_{TG}, v_{TGs} \rangle$$

Every TG-node $v \in V$ is a tripel

$$v := \langle signal_{Trig}, source_{Trig}, \zeta_{Trig} \rangle$$

The TG-nodes correspond with the TG-transitions ζ.

An TG-edge $e_{xy} \in E$ is defined as a tripel

$$e_{xy} := \langle v_x, v_y, \theta \rangle$$

where v_x and v_y are nodes (e_{xy} is directed from v_x to v_y), and θ is the temporal distance between v_x and v_y. The Trigger-Graph describes two types of interdependences between the nodes:
- Purely temporal interdependences, between nodes with the same signal source.
- Causal interdependences, between nodes with different sources.

The TG has these properties:
- Every node $v_i \in V \setminus \{v_s\}$ must be reacheable from v_s. That's why, nodes must not be isolated. Additionally, every node except v_s must have a predecessor.
- Cycles and self-loops are not allowed.
- Nodes without successor are end-nodes V_E.

Now, we can define even the notion of protocol more strictly: A Timing Diagram is a protocol, iff TG (π) exists. The TG can also be used to check the uniqueness of a protocol: Two partial protocols of a bus have the same type, iff they have the same TG.

Another benefit of the TG is, that other graphs can be derived from it in a formal way. In this paper this will be shown for Petri Nets and Finite Automata.

2.3 Example: Read-Protocol of the Multibus I

As a simple real-world example, we consider the Read Protocol of the Multibus I (see [3]). This standard bus has the advantage, that its protocol can be shown easily and clearly. Furthermore, the interdependences are described (informally) very well in its protocol specification [3].

First, the Timing Diagram is considered: A Timing Diagram usually represents the structural model of a communication and gives a class of protocols, whose parameters depend on the partners involved in this protocol. In [3], the partial protocol for a read operation of the Multibus I is represented like this (fig.2):

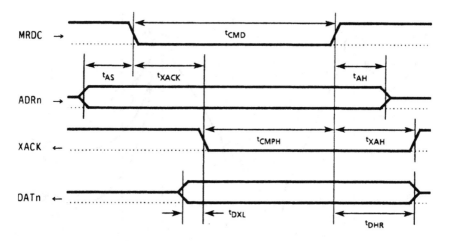

Fig. 2: Memory-Read protocol of the Multibus I (cf. [3], p. I,3-8)

The implicit parameters of this protocol are the times the bus agents need for sending data onto the bus resp. latching data from the bus. These temporal parameters are added as θ_1 and θ_2 to the protocol.

From this protocol, the Trigger-Graph can be derived like defined above. The nodes of the Trigger-Graph are:

$v1 = \langle$ IORC, $\rightarrow, \langle Z,0, \tau_2 \rangle \rangle$, $\qquad v2 = \langle$ IORC, $\rightarrow, \langle 0,Z, \tau_5 \rangle \rangle$,

$v3 = \langle$ ADR[0..23], $\rightarrow, \langle Z,A, \tau_1 \rangle \rangle$, $v4 = \langle$ ADR[0..23], $\rightarrow, \langle A,Z, \tau_6 \rangle \rangle$,

$v5 = \langle$ XACK, $\leftarrow, \langle Z,0, \tau_4 \rangle \rangle$, $\qquad v6 = \langle$ XACK, $\leftarrow, \langle 0,Z, \tau_7 \rangle \rangle$,

$v7 = \langle$ DAT[0..15], $\leftarrow, \langle Z,A, \tau_3 \rangle \rangle$, $v8 = \langle$ DAT[0..15], $\leftarrow, \langle A,Z, \tau_7 \rangle \rangle$

For the edges, these times are given:

$e_{12} = \langle v1,v2, t_{CMD} \rangle$, $e_{15} = \langle v1,v5, t_{XACK} + \theta_1 \rangle$,

$e_{24} = \langle v2,v4, t_{AH} \rangle$, $e_{26} = \langle v2,v6, t_{XAH} \rangle$, $\qquad e_{28} = \langle v2,v8, t_{DHR} \rangle$,

$e_{31} = \langle v3,v1, t_{AS} \rangle$, $e_{52} = \langle v5,v2, t_{CMPH} + \theta_2 \rangle$, $e_{75} = \langle v7,v5, t_{DXL} \rangle$

Where θ_1 is the time the talker needs to put the data onto the bus, and θ_2 is the listener's time to latch the data. θ_1 and θ_2 are parameters which depend on the respective bus agents.

In fig. 2, the Trigger-Graph TG (IO-Read) is shown by graphical means. Note, that the times $\tau_1..\tau_7$ are only node-attributes in the graph.

It can be seen at the first glance, that two nodes have no predecessor: v3 and v7. v3 is selected as the starting node ($v_s = v3$, marked with an arrow). To become a TG, v7 must get a predecessor-node. To provide this, v7 has to "adopt" the predecessor-node v1 of its successor v5 (broken line: $e_{17} = \langle$ v1, v5, t $_{XACK}$ – t $_{DXL}$ + $\theta_1 \rangle$). The edge e_{15} is redundant now, because it is the transitive closure of the edges e_{17} and e_{75}, and therefore it can be left out here (dotted line).

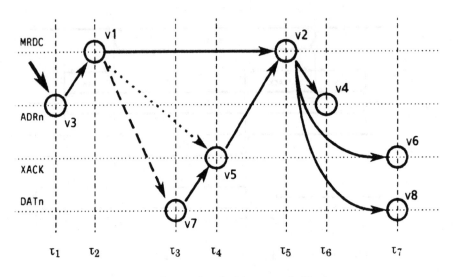

Fig. 3: Trigger-Graph of the protocol of fig. 2

2.4 Deriving Petri-Nets from the Trigger-Graph

In this section, only the basic idea of the transformation procedures is given. For a more detailed description see [7], where this topic has been pointed out.

The transformation ρ from a Trigger-Graph into Timed Petri-Nets is a surjective mapping

$$\rho : TG \rightarrow TPN$$

The *transformation rules* are:

- $\rho\,(v_{TG})$ $:= t_{PN}$
- $\rho\,(e_{TGxy})$ $:= \langle\, f\,(t_x, s), s, f\,(s, t_y)\,\rangle$

The ρ-transformation makes the *TG-nodes* become transitions T of the Petri Net, and the *TG-edges* become PN subnets, which consist of the edge $t \rightarrow s$, the place s and the edge $s \rightarrow t$. The most natural way to map the delays (timing constraints) θ of the TG-edges into a timed Petri Net, is to associate the *places* s within the PN subnets with the respective times θ_{PN} as the edges of the TG are attributed with (see [6]). The transformation rule for a TG-edge with time θ_{TG} is:

$$\rho\,(\langle\, v_x, v_y, \theta_{TG}\,\rangle) = \langle\, f\,(t_x, s), s\,(\theta_{PN}), f\,(s, t_y)\,\rangle$$

3 Simple Operations (SIOPs)

The next problem which has to be solved in order to generate bus interfaces automatically is the transformation of the partial protocols of different buses. To do this, the Simple Operations have been developed. In this chapter, the notion of SIOPs is decribed.

3.1 The Basic Idea

To provide a closer look to the problem, let us consider a bus interface between a bus b1 and a bus b2, with the protocol sets Π_{b1} and Π_{b2}, respectively (fig. 4). The sender $b1_T$ works as a talker on bus b1, the receiver $b2_L$ works as a listener on bus b2. Generally spoken, bus b1 is not the same as b2. Due to this, there need not be a "classical" equivalence-relation between the partial protocols of b1 and b2, because the protocols can be realized in a broad variety of ways. However, there must be a kind of equivalence between at least some of the protocols – elsewise it would be impossible to build bus interfaces between b1 and b2.

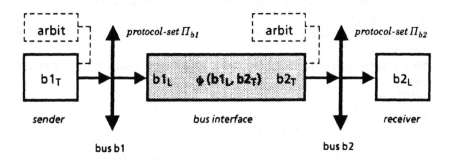

Fig. 4: Bus interface between bus b1 and bus b2

The solution is: we have to represent the meaning of the protocols. By doing this, we come to the notion of Simple Operations, because this is the simpliest uniform way to describe the meaning of the protocols.

3.2 Definition of Simple Operations

We define a particular SIOP ω_i, which corresponds to a particular protocol π_{bi}, as a tripel

$$\omega_i := \langle g_i, e_i, r_i \rangle$$

with
- SIOP-Group g_i

$g_i \in G$:= { Write, Read, ReadWrite, WriteIntern, ReadIntern, ... } ∪ SYSTEM

SYSTEM := { Reset, Powerfail, ... }
- End-State e_i

$e_i \subseteq E$:= { OK, Parity/ECC-Error, Time-Out, Address-Error, Protocol Violation, Attribute Error, TryAgain Later, LateError, ... }
- Real Attribute r_i

$r_i \in R$:= < has-Talker-Address, has-Listener-Broadcast, has-Data, ... >

Note: All the definitions are open for extensions, because particularly on the area of bus protocols some inventions are to be expected.

SIOP-Groups (G). The SIOP-Group is the most significant feature of a particular protocol π_i of a bus. By the SIOP-Group there is a very fast and reasonable distinction between bus protocols.

The *meanings* of the SIOP-Groups are:
- Write: Transfer-operation, data from the talker to the listener.
- Read: Transfer-operation, data from the listener to the talker.
- ReadWrite: Atomic operation, first Read from address A, afterwards Write to address A; using multi-port RAMs, exclusive access to this address must be provided.
- WriteIntern: Writing into registers which are used by the interface internally.
- ReadIntern: Reading from registers which are used by the interface internally.
- SYSTEM:

Reset: reset modules.

Powerfail: power supply interrrupt.

End-States (E). While older bus protocols do not care about the success of there operations, modern bus protocols have precise error message, which are expressed by their End-State.

These End-States are very wide-spread:
- OK: Transfer without errors.
- Parity/ECC-Error: Inconsistency (data error) appeared during the data transfer.
- Time-Out: There was no responding listener within a distinct period of time.
- Address-Error: This address or address space was not defined for this bus.
- Protocol Violation: This protocol was not defined for this bus.
- Attribute Error: Error, typically with data size, length of a block transfer, etc.
- Try Again Later: The addressed listener is just busy.

- LateError: Detection of the error after end of transfer.

Considering End-States, transformation without loss is only possible, if $E_A \subseteq E_B$.

Real Attributes (R). Decomposing the meanings of protocols into SIOPs, there a two classes of attributes: Real Attributes and Numeric Attributes:

a) *Real Attributes:* To two bus protocols, which differ by their Real Attributes, two different SIOPs ω_i and ω_j are assigned.

b) *Numeric Attributes:* To two bus protocols, which only differ by their Numeric Attributes, the same SIOP ω_k is assigned.

Examples for Numeric Attributes are different data sizes in a transfer operation, or the difference between block and single transfer.

We only consider Real Attributes, which can be represented by the boolean tripel

$R = \,< r_1, r_2, r_3 >$

r_1 : *has-Talker-Address*:

Can the sender of a transfer (the talker) be identified? Examples for "true" are interrupts and Message Passing.

r_2 : *has-Listener-Broadcast*:

Is a broadcast protocol (all listeners get the data) supported? For bus-broadcast, the acknowledge problem has to be solved.

r_3 : *has-Data*:

Bus-transfers usually transport some data between talker and listener. But there are also "degenerated transfers", whose data are meaningless (r_3 = false); this bus-operations only triggers some actions on the listener.

3.3 SIOP-SIOP-Transformation

After the definition of SIOPs we have to draw up rules, which SIOPs are allowed to be transformed into others.

The Transformation Rules

$\sigma : \omega \rightarrow \{ \omega_1, \omega_2, ..., \omega_n \}$ with $n \in N_0$

for the given set of SIOP-Groups are:

- Transformations between the same g_i are allowed, except for the SIOPs *WriteIntern* and *ReadIntern*:

    $\sigma (\omega_i) \supseteq \{ \omega_j \}$ with $g_i \notin \{ \text{WriteIntern, ReadIntern} \}$

- Transformations of the SIOPs *WriteIntern* and *ReadIntern* are not allowed:

    $\sigma (g_i = \text{WriteIntern}) = \varnothing,$

    $\sigma (g_i = \text{ReadIntern}) = \varnothing$

SYSTEM-SIOPs and *WriteIntern, ReadIntern* can change the listener maschines:

$\lambda = f (\text{SYSTEM, WriteIntern, ReadIntern})$

- *WriteIntern* and *ReadIntern* can be generated from *Write* and *Read* :

    $\sigma(\omega_i \mid g_i = \text{Read}) \supseteq \{\text{ReadIntern}\}$,

    $\sigma(\omega_i \mid g_i = \text{Write}) \supseteq \{\text{WriteIntern}\}$

- All SYSTEM operations can be transformed into other SIOPs:

    $\sigma(\omega_i \in \text{SYSTEM}) \supseteq \omega_j$

- Transformations without loss are only allowed between SIOPs whose end-states content the end-states of the requesting bus:

    $\sigma(\omega_i) \supseteq \{\omega_j\}$ for $E_i \subseteq E_j$

- Transformations without loss are only possible with the same set of Real Attributes:

    $\sigma(\omega_i) \supseteq \{\omega_j\}$ for $R_i = R_j$

- *Concentrated:*

    $\sigma(\omega_i) \supseteq \{\omega_j\}$

    for $(g_i = g_j \vee g_i \in \text{SYSTEM}) \wedge$

    $g_i \notin \{\text{WriteIntern, ReadIntern}\} \wedge$

    $E_i \subseteq E_j \wedge R_i = R_j$

If the protocol of the target systems has more informations than the protocol of the source system, then the hardware-designer has to choose default values for the missing values.

Reduced Transformations for SIOPs. As mentioned above, the very strict transformation σ has to be extended to fit better for real-world requirements: In most cases, it is better to have some kind of transformation between two buses at all, than to refuse any kind of transformation that is not complete.

Due to this, we define a Reduced Transformation σ_R

$\sigma_R : \omega \rightarrow \{\omega_1, \omega_2, \ldots, \omega_n\}$ with $n \in N_0$

at which on the one hand the transformation is *with loss* of meaning, but on the other hand, many more transformations between SIOPs are allowed than only using σ-transformations. So, the decision if an allowed transformation is actually used, remains the decision of the bus-interface designer.

The Reduced Transformation Rules σ_R are:

- All the SIOPs (except of WriteIntern, ReadIntern) may be σ_R-transformed into SYSTEM-operationes. Data are lost by this transformation.

    $\sigma_R(\omega_i) \supseteq \text{SYSTEM}$

    for $g_i \notin \{\text{WriteIntern, ReadIntern}\}$

- σ_R-transformations are allowed between SIOPs with a smaller set of End-Status.

    $\sigma_R(\omega_i) \supseteq \omega_j$ for $E_i \supset E_j$

- σ_R-transformations are also allowed having a less powerful set of Real Attributes.

    $\sigma_R(\omega_i) \supseteq \omega_j$ for $R_i > R_j$

- *Concentrated:*

    $\sigma_R(\omega_i) \supseteq \{\omega_j\}$

$$for \ (g_i = g_j \lor g_j \in SYSTEM\,) \land$$
$$g_i \notin \{\,WriteIntern, ReadIntern\,\} \land$$
$$((\,E_i \subseteq E_j \land R_i > R_j\,) \lor$$
$$(\,E_i \supset E_j \land R_i \le R_j\,) \lor$$
$$(\,E_i \supset E_j \land R_i > R_j\,)).$$

Sequential SIOPs. In some cases, a protocol-meaning cannot be expressed by a partial operation on this bus, but it must be "emulated" by a *sequence*

$$seq\,(\,\omega_1, \omega_2, \dots, \omega_n\,)\text{with } n = 0, 1, \dots k_{max}$$

This sequence has to be considered by all bus agents as one atomic operation. A very well-known example for the use of this technique is the construction of an "atomic read-write" on buses, whose basic protocol does not content thisoperation: "(1) Write on system address A1, (2) read from memory address A2, (3) write on memory address A2, (4) read from system address A1".

To handle these constructions, we can build a sequential (pseudo) SIOP

$$\omega_S := seq\,(\,\omega_1, \omega_2, \dots, \omega_n\,)$$

with $n = 0, 1, \dots k_{max}$

Like a partial-SIOP, it can be represented by a tripel

$$\omega_S := \,< g_S, e_S, r_S >$$

with

$$g_S \in G$$
$$e_S = ok, \text{für } \cap\, e_i = ok$$
$$= e_i, \text{für } \exists\, e_i \,|\, e_i \ne ok$$
$$r_S \in \{r_1, r_2, \dots, r_n\}$$

Trivially:

$$seq\,(\,\omega\,) := \omega, seq\,(\,\varnothing\,) := \varnothing$$

Transformation-Matrix S. Usually, bus interfaces deal with *sets* of protocols, i. e. one protocol-set Π_A of the bus A has to be transformed into the protocol-set Π_B of the bus B. We can express this by the boolean Transformation-Matrix S (1, 2)

$$S := \begin{pmatrix} \sigma\,(\omega_{11}) \supseteq \omega_{21} \ \dots \ \sigma\,(\omega_{11}) \supseteq \omega_{2n} \\ \sigma\,(\omega_{12}) \supseteq \omega_{21} \ \dots \ \sigma\,(\omega_{12}) \supseteq \omega_{2n} \\ \vdots \\ \sigma\,(\omega_{1i}) \supseteq \omega_{21} \ \dots \ \sigma\,(\omega_{1i}) \supseteq \omega_{2n} \\ \vdots \\ \sigma\,(\omega_{1m}) \supseteq \omega_{21} \ \dots \ \sigma\,(\omega_{1m}) \supseteq \omega_{2n} \end{pmatrix}$$

The expression "$\sigma\,(\omega_{ij}) \supseteq \omega_{ij}$" has three values: *true*, iff σ exists; *reduced*, iff σ_R exists and σ does not; *false*, in all other cases.

The matrix of Sequential SIOPs S_S is defined analogousely. Note that S_S is a superset of S.

3.4 Correlation between SIOPs and Protocols

Up to now, we have the "core" of a bus interface: A possibility to transform some operations of one bus A to another bus B, and vice versa. To build a real bus interface, we have to add two more transformations:

- We need to map a partial protocol Π_{Ai} to a set of partial SIOPs ω_{Ai} (Listener-Transformation λ)

    $\lambda \quad : \{\Pi_{Ai}\} \rightarrow \{\omega_{Ai}\}$

- Furthermore, we need a transformation from a partial SIOP to a partial protocol (Talker-Transformation μ)

    $\mu \quad : \omega_{Bj} \rightarrow \{\Pi_{Bj}\}$

To simplify these transformations, we suppose:

1. The signals of a bus only have the discrete values 0 and 1.
2. The signals change at discrete time-slots.

With this proviso, the Listener-Transformation λ can be treated as a parsing problem: A protocol can be recognized, iff it is unambigous after a lookahead of k bus cycles. To get this, one only has to concentrate all the signals a talker sends to a numerical expression; e. g. if $sig1 = 1$ and $sig2 = 0$, the numerical expression is 10 (decimal: 2). Usually, the bus-states are clustered to classes, e. g. to detect reads from an area of memory addresses.

With the proviso we did for λ, we can express the protocol of a bus by a set of Finite State Machines (FSMs). Each FSM can be considered as being triggered by the corresponding SIOP.

3.5 Application to a Bus Interface

With the definitions made above, we can represent a bus interface as a three-step mapping (see fig. 5):

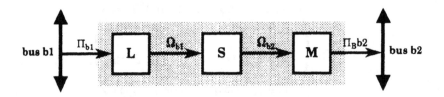

Fig. 5: Bus interface as a three-step mapping

The set of protocols Π_{b1} of bus b1 is λ-transformed by the Matrix L to its corresponding set of SIOPs Ω_{b1} of b1. Matrix S transforms the SIOPs Ω_A of bus b1 to the SIOPs Ω_{b2} of bus b2. At last, M transforms the SIOPs Ω_{b2} of b2 to its corresponding set of protocols Π_{b2}.

4 Conclusion

A method has been sketched for the description of bus interfaces by system-theoretical means. Using this method, a way can be seen to implement bus interfaces automatically.

For this purpose, the Trigger-Graph as an intermediate representation has been introduced. The benefits of this approach are to provide a bridge between hardware designers and system analysts: On the one hand, the Timing Diagram can easily be checked for some properties via the TG, like completeness, lack of temporal contradictions and redundancies in the specified protocol, and the uniqueness of this protocol within the complete protocol of a bus. On the other hand, the derived Petri Net can be examined for some important aspects like correctness and performance by well-known methods of Petri Net theory.

Furthermore, the Simple Operation approach has been introduced. Using this, it becomes possible to transform protocols of different buses. In addition, it becomes possible to express the capability of buses in an uniform way. For real-world problems, the SIOP transformation has been extended to Reduced SIOPs for transformation with loss.

References

1. Ajmone Marsan, M.; Balbo, G.; Conte, G.: Performance Models of Multiprocessor Systems. The MIT Press: Cambridge (Mass.), 1986
2. Brauer, W.: Automatentheorie. Teubner: Stuttgart 1984 (in German)
3. Intel: Multibus I Architecture Reference Book. Santa Clara (CA) 1983
4. Leszak, M.; Eggert, H.: Petri-Netz-Methoden und -Werkzeuge. Informatik-Fachberichte 197. Springer-Verlag: Berlin 1988 (in German)
5. Reisig, W.: Petrinetze. Eine Einführung. Springer-Verlag: Berlin 2 1986 (in German)
6. Sifakis, J.: Use of Petri Nets for Performance Evaluation. Third Int. Workshop on Modeling and Performance Evaluation of Computer Systems. Amsterdam, 1977
7. Thurner, E.M.; Simon, F.: Verfahren zur Umsetzung von Zeitdiagrammen in zeitbehaftete Petri-Netze. In: Kropf, Th.; Kumar, R.; Schmid, D.: GI/ITG-Workshop "Formale Methoden zum Entwurf korrekter Systeme", Bad Herrenalb, Interner Bericht Nr. 10/93 Universität Karlsruhe, Fakultät für Informatik, 1993, S. 61-68 (in German)

CAST Tools for Intelligent Control in Manufacturing Automation

Witold Jacak[1] and Jerzy Rozenblit[2]

[1] Institute of Systems Science
Johannes Kepler University Linz
A-4040 Linz, Austria
[2] Department of Electrical and Computer Engineering
The University of Arizona
Tucson, Arizona 85721, U.S.A.

1 Introduction

In recent years, the use of programmable and flexible systems has enabled partial or complete automation of machining and assembly of products. Such the cellular flexible manufacturing systems (FMS) are data-intensive. They consist of three main components: *a production system, a material handling system,* and *a hierarchical computer assisted control system.*

A cellular manufacturing system is intelligent if it can *self-determine choices in its decisions* based upon the simulation of a needed solution in a virtual world or upon the experience gained in the past from both failures and successful solutions stored in the form of rules in the system's knowledge base. The intelligent manufacturing system discussed here is a *computer integrated cellular system* consisting of fully automated robotic cells. Planning and control within a cell is carried out off-line and on-line by the hierarchical controller which itself is regarded as an integral part of the cell. We call such a cell **Computer Assisted Workcell (CAW).**

Given a technological task, an intelligent Computer Assisted Workcell should be able to determine control algorithms so that: a) a task is realized, b) deadlocks are avoided, c) the flow time is minimal, d) the work-in-process is minimal, e) geometric constraints are satisfied, and f) collisions are avoided.

To synthesize an autonomous, or semi-autonomous, Computer Assisted Workcell we use artificial intelligence and general system theory concepts, methods, and tools such that as *hierarchical decomposition of control problems, hierarchy of models specification, discrete and continuous simulation* from system theory ideas and *action planning methods, model's state-graph searching methods* from artificial intelligence concepts.

The control laws which govern the operation of CAW are structured hierarchically. We distinguish three basic levels: a) *the workstation (execution) level,* b) *the cell (coordination) level,* and c) *the organization level.*

• The organizer accepts and interprets related feedback from the lower levels, defines the strategy of task sequencing to be executed in real-time, and processes large amounts of knowledge with little or no precision. Its functions are: reasoning, decision making, learning, feedback, and long term memory exchange.

- The coordination level defines part routing in logical and geometric aspects and coordinates the activities of workstations and robots. The robots coordinate the activities of the equipment in the workstation. This level is concerned with the formulation of the actual control task to be executed by the lowest level.
- The execution level consists of device controllers. It executes the programs generated by the coordinator.

Consider the CAW with a hierarchical structure shown in Figure 1. It is composed of three interactive levels of organization, coordination and execution, modeled with the aid the theory of intelligent system [3]. All planning and decision making actions are performed within the higher levels. In general, performance of such systems is improved through self-planning with different planning methods and through self-modification with learning algorithms and schemes interpreted as interactive procedures for the determination of the best possible cell action.

There are two major problems in synthesis of such complex control systems: The first depends on the *coordination and integration* at all levels in the system, from that of the cell, where a number of machines must cooperate, to that of the whole manufacturing workshop, where all cells must be coordinated. The second problem is that of *automatic programming* of the elements of the system. Thus we distinguish two major levels of the control problem: the *logical (operational)* and the *geometric control.*

The intelligent control of CAW is synthesized and executed in two phases, namely:
- planning and off-line simulation phase and
- on-line simulation based monitoring and control phase.

In the first phase, a hierarchical simulation model of a robotic workcell called **virtual cell** is created. Workcell components such as NC-machine tools, robots conveyors, etc., are modelled as elementary DEVS systems [13]. This type of simulation is used for verification and testing of different variants of task realizations (processes) obtained from the route planner. To simulate variants of a process, the system must have the knowledge of how individual robots actions are carried out. For detailed modeling of cell components continuous simulation and motion planning methods are used. The geometrical interpretations of cell-actions are obtained from the motion planner and are tested in a geometric cell-simulator. This allows us to select the optimal task realizations which establish logical control of the system.

In the second phase, a real-time discrete event simulator of CAW is used to generate a sequence of future events of the virtual cell in a given time-window. These events are compared with current states of the real cell and are used to predict motion commands of robots and to monitor the process flow. Since a CAW has to make many subjective decisions based on deterministic programming methods, knowledge bases and knowledge based decision support should be available as an advisory layer. Such knowledge bases are part of the so-called virtual workcell.

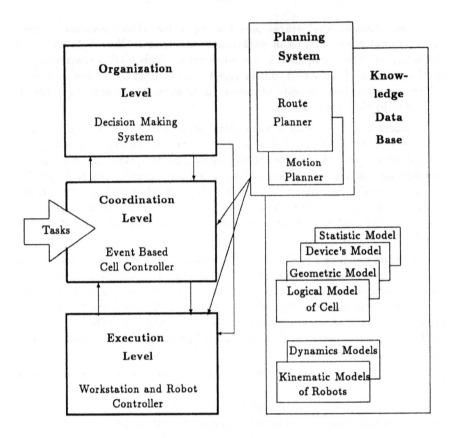

Fig. 1. Hierarchical structure of CAW control system

2 Virtual Workcell and Technological Task

A real cell is a fixed, physical group of machines D (or stores M), and robots R. A *virtual cell* is a formal representation (computer model) of a workcell. To synthesize CAW's control, we first specify the family of technological tasks realized in the cell.

2.1 Technological Task

The *technological task* realized by the robotic cell is represented by a triple:

$$Task = (O, \prec, \alpha) \tag{1}$$

where: O is a finite set of *technological operations* (machine, test, etc.) required to process the parts, $\prec \subset O \times O$ is the partial order (*precedence relation*) on the set O, and $\alpha \subset O \times (D \cup M)$ is a relation of device or store assignment.

The partial order represents an operational precedence i.e.; $q \prec o$ means that the operation q is to be completed before the operation o can begin. $(o, d) \in \alpha$ means that the operation o can be performed on the workstation d, and if $(o, m) \in \alpha$, then m is the production store from the set M where the parts can be stored after the operation o has been completed.

The technological task described above can be realized in a virtual cell. The virtual robotic cell has a hierarchical structure which comprises models. The models represent the knowledge about a real production environment.

2.2 Geometric Model of Workcell

The lower level of virtual cell representation describes the geometry of a virtual cell. Formally, the geometry of the cell is defined as follows [15]:

$$Cell_{Geometry} = (G, H) \tag{2}$$

The first component of the cell geometry description

$$G = \{\mathcal{G}_d = (E_d, V_d) | d \in D \cup M\} \tag{3}$$

represents the set of geometric models of the cell's objects. E_d is the coordinate frame of object (device) d and V_d is the polyhedral approximation of the d-th object geometry in E_i.

$$H = \{\mathcal{H}_d : E_d \rightarrow E_0 | d \in D \cup M\} \tag{4}$$

represents the cell layout as the set of transformations between an object's coordinate frames E_d and the base coordinate frame E_0 [18].

Consequently, a geometric model has two components: (1) workcell objects models and (2) workcell layout model.

Workcell Object Model: The geometry model of each object is created by using solid modeling [9]. Solid modeling incorporates the design and analysis of virtual objects created from primitives of solids stored in a geometric database. The complex virtual objects of a workcell V_d (such as technological devices, robots, auxiliary devices or static obstacles) are composed of solid primitives such as cuboid, pyramids, regular polyhedrons, prisms, and cylinders.

Workcell Layout The workcell's objects can be placed in a robot's workscene at any position and in any orientation. The virtual objects (devices, stores) are loaded from a library (model base) into the Cartesian base frame. They can be located anywhere in the cell using translation and rotation operations in the base coordinate frame [14].

2.3 Logical Model of Workcell

Based on a geometric model of the workcell, a logical structure of the cell must be created. The group of machines is divided into subgroups, called *machining centers* serviced by separate robots.

Parts are transferred between machines by the robots from set R, which service the cell. A robot $r \in R$ can service only those machines that are within its service space $Serv\_Sp(r) \subset E_0$ (E_0 - Cartesian base frame). The set of devices which lie in the r-th robot service space is denoted by $Group(r) \subset D \cup M$. More specifically, a device belongs to group serviced by robot r (i.e., $d \in Group(r)$) if all positions of its buffer lie in the service space of robot r. Consequently, the logical model of a workcell is represented by:

$$Cell_{Logic} = (D \cup M, \ R, \ \{Group(r) \mid r \in R\}) \tag{5}$$

Based on the sets $Group(r)$ and the description of the task, we can define the relation β which describes the transfer of parts after each technological operation of the task.

$$\beta \subset (O \times O) \times R \tag{6}$$

where: $((o_i, o_j), r) \in \beta) \Leftrightarrow (\{d_i, d_j\} \subset Group(r) \vee \{m_i, d_j\} \subset Group(r))$
and $\alpha(o_i) = \{d_i, m_i\}$ and $\alpha(o_j) = \{d_j, m_j\}$. In a special case β is a partial function; i.e.: $\beta : O \times O \to R$.
The virtual cell concept is the basis for designing the CAW planners.

3 Logical and Geometric Control Planners

The realization of a technological task depends on the sequencing of its operations. The execution sequence can be represented as a list of machines, called a *production route* (or *logical control of process*), along which parts flow during the manufacturing process. Each route determines a different topology of robot motion trajectories, different deadlock avoidance conditions, and a different job flow time. Therefore, route planning is a critical issue in all manufacturing problems.

3.1 Technological Route Planner

Searching for the most efficient route requires that we define route comparison criteria. One such criterion is that the sequence of operations and robot/machine actions related to it minimize the mean flow time of parts [6]. The flow time of every job is the sum of *processing* time, *waiting* time, and the time of interoperational moves called *transfer* times.

The route planner should find an ordered execution sequence of L operations called a *sequential machining process*:

$$p = (o_1, o_2, ..., o_L) \tag{7}$$

with the following constraints:

(i) if for two operations o_i and o_j from $Task$, $o_i \prec o_j$, then $i < j$

(ii) for each $i = 1, ..., L - 1$ there exists a robot which can transfer a part from machine d_i (or store m_i) assigned to operation o_i to machine d_{i+1} assigned to operation o_{i+1}; i.e., $(\exists r \in R)(((o_i, o_{i+1}), r) \in \beta)$

A process can be realized by different sequences of technological devices (called *resources*) required by successive operations from the list p when they are being executed. This set, denoted by $\mathbf{P}$, is called production routes. The route planning algorithm should take into account the conditions for deadlock avoidance. In order to formulate a quality criterion of process planning, we use the procedure of deadlock avoidance presented in [8], [7].

The quality criterion used to evaluate the technological route being planned is the probability of waiting for a resource when deadlock avoidance conditions are in force. The problem solved by the route planner is to find an ordered sequence of operations from the technological task which is feasible and which minimizes this quality criterion. This is a permutation problem with potentially more than one solution. To solve the planning problem under consideration, the route planner uses a backtracking graph search algorithm [14].

Each route from the output of the task planner determines a different logical sequence of robot and machine actions and a different input data for the geometric control of cell actions.

3.2 Motion Planner

The production route p for a machining task determines the parameters of the robot's movements and manipulations (such as initial and final positions) to carry out this task. The set of all robot's motions between devices and stores needed to carry out a given process is called a *geometric route* or *geometric control*.

First, based on the sequence of operations *Process* and its production route p, the positions table (Frames_Table) for all motions of each robot is created. The Frames_Table determines the initial and final geometric positions and orientations of the robot's effector for each robot movement. The robot's motion trajectory planning process is performed in two stages: a) planning of the geometric track of motion, and b) planning of the motion dynamics along a computed track.

The Collision-free Path Planner In order to apply fast methods of geometric path planning, the robot's kinematics model should facilitate a direct analysis of the robot's location with respect to objects (virtual devices) in its environment. Moreover, such a model should facilitate 3D graphic simulation of robot movements. One possible description of the manipulator's kinematics is a discrete dynamic system [16],[17],[18].

The discrete kinematics model of robots with n degrees of freedom has the following form:

$$Robot = (C, U, \delta) \tag{8}$$

where: C denote the set of robot configurations (manipulator states) in the Cartesian base frame; $U = \{-1, 0, +1\}^n$ is the discrete input signal set, and $\delta : C \times U \to C$ is the one-step state transition function of a dynamic system of the form $c(k + 1) = \delta(c(k), u(k))$ and is defined recursively [16],[18].

The complete formal explanation of the discrete model of robot kinematics is presented in [16],[18].

Motion Trajectory Planning The trajectory planner receives the geometric tracks as inputs and determines a time history of position, velocity, acceleration, and input torques which can be then fed to the trajectory tracker. In the trajectory planner, the robot is represented by the model of path parameterized manipulator dynamics [19]:

$$m(s)\dot{\mu} + n(s)\mu^2 + r(s)\mu + g(s) = f \tag{9}$$

$$\dot{s} = \mu \tag{10}$$

where pseudo-velocity μ is the time derivative of the parameter s.
To optimize the time trajectory of motion the following cost function is chosen:

$$I = \int_0^{s_{final}} (\frac{1}{\mu(s)} \lambda_1 + \lambda_2 \sum_{i=1}^{n} ||f_i||)ds \text{ and } \lambda_1 + \lambda_2 = 1. \tag{11}$$

The trajectory planning problem is then reduced to cost minimization, subject to the dynamic model specification and the constraints of torque f. This is a classical dynamical optimization problem for a nonlinear system and a effective bidirectional search method is used [20], [21].

The Motion Planner generates the set of optimal trajectories of parts transfer movements. This set is establishes the geometric control laws of a CAW.

4 Discrete Event, Real-Time Cell Controller

The process planning is based on the description of task operations and their precedence relation. As a result of this stage, the fundamental plan of cell-actions, i.e., an ordered sequence of technological operations, is created.

In the second phase, the real time event-based simulator of CAW is synthesized. The simulator generates a sequence of future events of the virtual cell in a given time-window. These events are compared with current states of the real cell and are used to predict motion commands of robots and to monitor the process flow. The simulation is event oriented and is based on the DEVS system concept introduced by Zeigler [13]. The structure of CAW is shown in Figure 2.

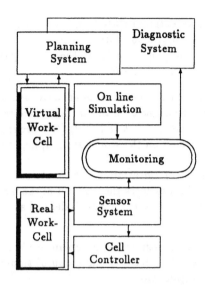

Fig. 2. Structure of Computer Assisted Workcell

4.1 Synchronized, Event-Based Model of Workstation

Each workstation $d \in D$ is modelled by two coupled atomic discrete event systems (DEVS) called active and passive models of a device, i.e., $DEVS_d = (Dev_d^A, Dev_d^P)$, where Dev_d^A is the active model and Dev_d^P is the passive model.

The active atomic DEVS performs the simulation process and a passive atomic DEVS represents the register of real states of the workstation. It acts as a synchronizer between the real and the simulated processes. This model is a modified version of the specification proposed by Zeigler and has following structure:

The active model of workstation:

$$Dev_d^A = (X_V(d), S_V(d), \delta_{int}^d, \delta_{ext}^d, \text{ta}^d)$$

$X_V(d)$ is a set, the external input virtual event types, $S_V(d)$ is a set, the sequential virtual states, δ_{int}^d is a set of functions, the internal transition specification, δ_{ext}^d is a function, the external transition specification, ta^d is a function, the time advance function, with the following constraints:

(a) The total virtual event-set of the system specified by Dev_d^A is
$$E_V(d) = \{(s,t)|s \in S_V(d), 0 \le t \le \text{ta}^d(s)\};$$
(b) δ_{int} is a set of parameterized internal state-transition functions:
$$\delta_{int} = \{\delta_{int}^u | u \in U\} \text{ and}$$
$$\delta_{int}^u : S_V \to S_V;$$

(c) δ_{ext} is an external state-transition function:
$$\delta_{ext} : E_V \times X_V \rightarrow S_V;$$
(d) ta is a mapping from S_V to the non–negative reals with infinity:
$$\text{ta} : S_V \rightarrow R,$$

We now describe each component of an active model. Each workstation $d \in D$ can have a buffer. The capacity of this buffer is denoted by $C(d)$. If a workstation has no buffer, then $C(d) = 1$. Let NC(d) (NC program register) denote the set of operations performed on the workstation d, i.e., NC$(d) = \{o \in O|(o,d) \in \alpha\}$.

The virtual state set of workstation: The state set of d is defined by $S_V(d) = SD_d \times SB_d$, where SD_d denotes the state set of the machine and SB_d denotes the state set of its buffer. The state set SD_d is defined as $SD_d = S_{ready} \cup S_{busy} \cup S_{done}$ where:

- $S_{ready} = \{ready\}$ signifies that machine is free
- $S_{busy} = \{(busy, q)|q \in$ NC$(d)\}$ and $(busy, q)$ signifies that **machine is busy processing q-operation**
- $S_{done} = \{(done, q)|q \in$ NC$(d)\}$ and $(done, q)$ signifies that **machine has completed q-operation and is not free**

The state set of a workstation's buffer SB_d is specified as

$$SB_d = (O \times \{0, 1, \#\})^{C(d)}.$$

Let $C(d) = K$ and $b_d = (b_i|i = 1, ..., K) \in SB_d$, then

$$b_i = \begin{cases} (0,0) & \Leftrightarrow i - \text{th position of the buffer is free} \\ (0,\#) & \Leftrightarrow i - \text{th position of the buffer is reserved for a part being currently} \\ & \text{processed} \\ (q,0) & \Leftrightarrow i - \text{th position of the buffer is occupied by a part before operation } q \\ (q,1) & \Leftrightarrow i - \text{th position of the buffer is occupied by a part after operation } q \end{cases}$$

We assume that the i-th position denotes a location at which a part is placed in the buffer.

Model of Workstation's Controller: Given the virtual state set, we define the internal state-transition functions δ_{int} to model the workstation. They are parameterized by an external parameter u which is loaded into the workstation from a higher level of control called the workcell management system. The parameter $u \in U$ is the operation's choice function. It represents the priority strategy of workstation, i.e.,

$$u : 2^{\text{NC}(d)} \rightarrow \{o\} \subset \text{NC}(d)$$

and $u(\emptyset) = \emptyset$, $u(\{o\}) = \{o\}$.

The choice function u defines a priority rule under which the operations are chosen from the device's buffer.

Let $s(d) = (s_d, (b_1, b_2, ..., b_K)) = (s_d, b) \in S_V(d)$ and

$$Wait(d) = \{q|((\exists b_j)((o, 0) = b_j\}$$

be the set of operations waiting to be processed.

Then the internal transition function

$$\delta_{int}^u : S_V(d) \to S_V(d)$$

is specified as follows:

$$\delta_{int}^u(s(d)) = \begin{cases} ((done, q), (b_1, b_2, ..., b_K)) & \Leftrightarrow s_d = (busy, q) \\ ((busy, q), (b_1, ., b_{i-1}, (0, \#), b_{i+1}, ., b_K)) & \Leftrightarrow s_d = ready \land \\ & u(Wait(d)) = \{q\} \land (q, 0) = b_i \\ (ready, (b_1, b_2, ..., b_K)) & \Leftrightarrow s_d = ready \land Wait(d) = \emptyset \\ (ready, (b_1, ., b_{i-1}, (q, 1), b_{i+1}, ., b_K)) & \Leftrightarrow s_d = (done, q) \land b_i = (0, \#) \end{cases}$$

Corollary 1. *It is clear that if* $s_d = (done, q) \lor s_d = (busy, q)$ *than* $(\exists! i)(b_i = (0, \#)$ *because the workstation can not process two parts simultaneously.*

Model of Workstation-Robot Interaction: The interaction between the robots and workstation is modelled by the external state transition function.

The set of external virtual events for the workstation's model Dev_d^A is defined as follows:

$$X(d) = \{e_d^1(q, i), e_d^2(q, i), e^0 | q \in NC(d) \land i = 1, ..., K\}$$

where:

$e_d^1(q, i) = $ PLACE $(q, 0)$ ON *i-th position* - signifies that a part before q operation is placed on i-th position of d-workstation's buffer

$e_d^2(q, i) = $ PICKUP $(q, 1)$ AT *i-th position* - signifies that a part after q operation is removed from i-th position of d-workstation's buffer,

$e^0 = $ DO NOTHING (empty event)

The external transition function δ_{ext}^d for each workstation d is defined below.

Let $s(d) = (s_d, (b_i | i = 1, .., K)) \in S_V(d)$ then

$\delta_{ext}((s(d), t), e^0) = s(d)$

$\delta_{ext}((s(d), t), e_d^1(q, i)) = (s_d, (b_1, ., b_{i-1}, (q, 0), b_{i+1}, ., b_K)) \Leftrightarrow b_i = (0, 0)$

$\delta_{ext}((s(d), t), e_d^2(q, i)) = (s_d, (b_1, ., b_{i-1}, (0, 0), b_{i+1}, ., b_K)) \Leftrightarrow b_i = (q, 1)$

$\delta_{ext}((., .), .) = failure$ for all other states

Time Model The time advance function for Dev_d^A determines the time needed to process a part on the d-th workstation. It is defined as follows:

Let $s(d) = (s_d, (b_i | i = 1, .., K))$. Then

$$ta^d(s(d)) = \begin{cases} \tau_{process}(q) & \Leftrightarrow s_d = (busy, q) \\ \tau_{load} + \tau_{setup}(q) & \Leftrightarrow s_d = ready \land u(Wait(d)) = \{q\} \\ \tau_{unload} & \Leftrightarrow s_d = (done, q) \\ \infty & otherwise \end{cases}$$

$\tau_{process}(q)$ denotes the tooling/assembly time of operation q for the workstation d. $\tau_{load} + \tau_{setup}(q)$, τ_{unload} denote the loading, setup, and unloading times for d, respectively.

At time moment t, let the workstation be in the active state s which has begun at time moment t_o. After this time, the workstation transfers its state from s into $\delta^u_{int}(s)$, which will be active in the next time interval $[t_o + \mathrm{ta}(s), t_o + \mathrm{ta}(s) + \mathrm{ta}(\delta^u_{int}(s))]$.

Corollary 2. *It is easy to observe that the external events cannot change the advance time for each active state. Let $x = (e, t)$ be an external event different from e^0 which occurs at time moment $t \in [t_o, t_o + \mathrm{ta}(s)]$. The workstation changes its state to state $s' = \delta_{ext}((s, t), e)$ but the advance time of the new state s' is equal to $\mathrm{ta}(s') = t_o + \mathrm{ta}(s) - t$.*

The passive model of workstation: The passive model is represented by a finite state machine (FSM) :

$$Dev^P_d = (X_R(d), Q_R(d), \varphi^d_{ext}, \lambda^d)$$

where:

$X_R(d)$ is a set, the external input real event types
$Q_R(d)$ is a set, the sequential real states
φ^d_{ext} is a function, the external real-state transition specification
λ^d is a function, the updating function

with the following constraints:

(a) φ^d_{ext} is a real-state transition function (one-step-transition function):
$$\varphi^d_{ext} : Q_R(d) \times X_R(d) \to Q_R(d);$$
(b) λ^d is a updating function (output function):
$$\lambda : UP \times Q_R(d) \times S_V(d) \to S_V(d);$$
where: $UP = \{0, 1\}$ and 0 denotes that the updating process is stopped and 1 denotes that updating should take place.

The interaction between external sensors and the workstation is modelled by the state transition function φ_{ext}. The external sensor system generates real events, which can be used to synchronize the simulated technological process with the real technological process.

The set of real events for the workstation's model Dev^P_d can be defined as follows:

$$X_R(d) = \{re^1_d(q, i), re^2_d(q, i), re^3_d(q, i), re^4_d(q, i) | q \in \mathrm{NC}(d) \wedge i = 1, ..., K\}$$

where:

$re^1_d(q, i)$ - signifies that a part before q operation is placed on i-th position of d-workstation's buffer

$re^2_d(q, i)$ - signifies that a part after q operation is removed from i-th position of d-workstation's buffer,

$re^3_d(q, i)$ - signifies that a part before q operation is loaded into machine and processing is began,

$re_d^4(q, i)$ - signifies that machine has completed the q operation, machine is unloaded and part is placed on i-th position of d-workstation's buffer.

We assume that the state set $Q_R(d)$ of the passive model is the same as that of the state set of the active one, i.e., $Q_R(d) = S_V(d)$.

Then, the state transition function φ_{ext}^d for the workstation d can be defined in a similar manner as the external transition function δ_{ext}. This transition function reflects real state changes of the workstation. The output function λ produces the real state in the active model of workstation when updating is required, i.e.,

$$\lambda(q, s, 1) = q \in S_V(d) \quad \& \quad \lambda(q, s, 0) = s \in S_V(d).$$

The production store model: A production store model can be defined in a similar manner. The state set of a store m is specified as a vector of states of each store position, i.e.,

$$S_V(m) = (O \cup \{0\})^{C(m)}$$

where: if $s_m = (s_i | i = 1, ..., C(m)) \in S_V(m)$, then

$$s_i = \begin{cases} 0 \Leftrightarrow i - \text{th position of store is free} \\ q \Leftrightarrow i - \text{th position of store is occupied by a part after operation } q \end{cases}$$

The set of external events for m is similar to the set of external events of a workstation.

The internal transition δ_{int}^m is an identity function which can be omitted. The external transition function δ_{ext}^m for each store m can be defined in the same manner as the function δ_{ext}^d. The time advance function for Dev_m is equal to ∞.

The model of robot: Each robot $r \in R$ is modelled by two coupled systems, again called active and passive models.

$$REVS_r = (Rob_r^A, Reg_r^P)$$

where Rob_r^A is the active discrete event model and Reg_r^P is the passive model. The active model is used for simulation while the passive one represents the register of the robot's real states. It acts as a synchronizer between the real and simulated processes.

The virtual workstations and robots (DEVS models) together with the real devices and robots represent the execution level of CAW control system.

The model of cell controller: The activation of each machine is caused by an external event generated by the model of a robot. The events generated by the cell's robots depend on the states of the workstations d_i and a given fundamental plan $p \in Processes$. Thus, we define a system (called an *acceptor*) which observes the states of each workstation. The acceptor is defined as the following discrete event system [13]:

$$A = (X_A, Processes, Y_A, \lambda_A) \tag{12}$$

where:

$$X_A = \{(s,t)|s \in S_{cell} \wedge t \in Time\}$$
$$S_{cell} = S(d_1) \times S(d_2) \times ... \times S(d_D) \times S(m_1) \times ... \times S(m_M)$$

and *Time* is the time base.

The role of the acceptor is to select the events which will invoke the robots to service the workstations. The output of the acceptor can be an empty set or it can contain indexes of only those operations which can execute without deadlock. Not all the operations can be performed simultaneously. Therefore, to select the operations which can be executed concurrently, we introduce a discrete event system called the *cell controller*. First, this system eliminates all unrealizable operations from the set operations given as the acceptor's output function. Thus, a set of *executable operations* is created.

Then, from such a set, operations with the highest priority are selected. The priorities are established by the *workcell organization layer* as shown in Figure 3.

Each function δ_{int} is parameterized by an external parameter g which is loaded into the cell controller from the workcell organizer. The parameter $g \in Strategies$ is the operation's choice function, and represents the priority strategy of CAW, i.e.,

$$g : 2^{OP} \rightarrow \Re$$

The choice function g defines a type of priority rule under which the operations will be chosen for processing.

5 Real Time Control and Monitoring

The external events generated sequentially by the cell controller and robots activate the workcell's devices and coordinate the transfer's actions.

The discrete event model of workcell generates a sequence of future events of the virtual cell in a given time-window which we called a *trace*. A trace of the behaviour of a device d is a finite sequence of events (state changes) in which the process has been engaged up to some moment in time. A trace in a time window is denoted as a sequence of pairs (state,time) , i.e.,

$$tr_{[t_0,t_0+T]} = < (s_1,t_1),(s_2,t_2),(s_3,t_3),..(s_n,t_n) >$$

where $t_1 \geq t_0 \wedge t_n \leq t_0 + T$.

Organization Level

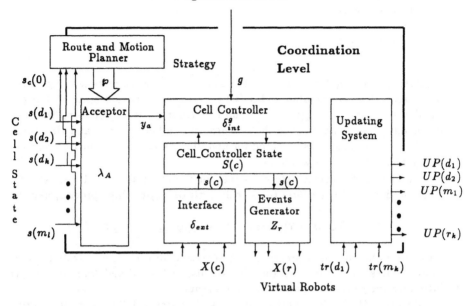

Fig. 3. The coordination level of CAW

The events from a trace are compared with current states of the real cell and are used to predict motion commands of robots and to monitor the process flow.

The simulation model is modified if the states of the real cell change and the current real states are introduced to the model.

5.1 Monitoring

Let $tr_T(d)$ be a trace of virtual events of device d on time-interval $T = [t_o, t_o + T]$ where t_o is the moment of the last updating:

$$tr_T(d) = < (s_o, t_o), (s_1, t_1)...., (s_k, t_k) >$$

where $t_k \leq t_o + T$ and s_i is the virtual state of DEVS model of d-device.

The monitoring process is performed as follows:

Let $q(t)$ be the current real state of d-device, modelled by passive Dev_d^P automaton and $t \geq t_o$.

If
$$q(t) \neq s_j \text{ for } t \in [t_j - \tau_0, t_j + \tau_0]$$

then
call diagnosis
else
call updating

Updating: Let $q_d(t)$ be the current real state of d-device in time t, registered by passive Dev_d^P automaton and $s_d(t') \in S_V(d)$ be a virtual state of d-device simulator in time t'. If $q(t) = s(t')$ and $t' \neq t$ but $t \in [t' - \tau_0, t' + \tau_0]$ where τ_0 is tolerance time-window. then the synchronization between real and virtual cell should be performed. Easy method for updating process is to synchronize every device and robot of the workcell.

The device d generate external signal for updating module of workcell controller and controller perform the so called **global updating** process, namely:

$$(\forall x \in D \cup M \cup R)(up(x) = 1 \Rightarrow \lambda_x(s_x, q_x, 1) = q_x \in S(x))$$

Such the global updating process is not necessary for each device. Only the part of all devices should be updating. To specify such *local updating* process we introduce the causality relation between events.

Let predicate $Occ(e)$ denotes that the event e has occurred. The causality relation is defined as follows:

$$e \rightsquigarrow e' \Leftrightarrow (\neg\, Occ(e) \Rightarrow \neg\, Occ(e'))$$

and expresses that event e is one of causes of event e'.

The causality relation is reflexive, asymmetrical and transitive relation.

Let $Trace_{[t',t]} = \bigcup \{tr_{[t',t]}(x) | x \in D \cup M \cup R\}$ be the union of all devices and robots traces.

Based on above definition we can construct the set of devices and robots for which the updating the virtual process is needed.

$$UpDate(s(t')) = \{x \in D \cup M \cup R | (\exists e \in tr_{[t',t]}(x))((s(t'), t') \rightsquigarrow e)\}$$

Now, we define **local updating** process as follows:

$$(\forall x \in UpDate(s(t')))(up(x) = 1 \Rightarrow \lambda_x(s_x, q_x, 1) = q_x \in S_V(x))$$

Corollary 3. *It is easy to proof that so defined local updating process is equivalent to global one, i.e., synchronization of devices and robots from the set UpDate is equivalent to the synchronization of all devices and robots of CAW.*

Moreover

Corollary 4. *If $t' > t$ then $UpDate = \{d\}$ and only for the d-device the synchronization is necessary to performed.*

Pre-Diagnosis: Let $q_d(t)$ be the current real state of d-device in time t, registered by passive Dev_d^P automaton and $s_d(t') \in S_V(d)$ be a virtual state of d-device simulator in time t'. Let $q(t) = s(t')$ and $t' \neq t$ and $t < t' - \tau_0$ or $t > t' + \tau_0$ where τ_0 is tolerance time-window.

In this case the diagnosis of real cell should be performed. To reduce the complexity of such process we use these same causality relation $\rightsquigarrow$ in order to eliminate the devices which do not need the diagnostics process. By $CON(s(t'))$ we denote the set of events which are direct causes of event $(s(t'), t')$, i.e.

$$CON(s(t')) = \{e \in Trace_{[t',t]} | e \rightsquigarrow (s(t'), t') \wedge (\neg \exists e')(e \rightsquigarrow e' \wedge e' \rightsquigarrow (s(t'), t'))\}$$

From the set $CON(s(t'))$ we eliminate these events which ware previously monitored, i.e.

$$Con(s(t')) = CON(s(t')) - Monitored\_Events$$

Based on the set $Con(s(t'))$ we can find the devices or robots for which diagnosis is needed as follows:

$$Diag(s(t')) = \{x \in D \cup M \cup R | (\exists e \in tr_{[t',t]}(x))(e \in Con(s(t')))\}$$

Additionally the taxonomy of failure type can be performed.

6 Summary

A comprehensive framework for design of an intelligent cell-controller requires integration of several layers of support methods and tools. We have proposed an architecture that facilitates an automatic generation of different plans of sequencing operations, synthesis of action plan for robots servicing the devices, synthesis of the workcell's simulation model, and verification of control variants based on simulation of the overall cell's architecture. The real-time discrete event simulator is used next to generate a sequence of future events of the virtual cell in a given time-window. These events are compared with current states of the real cell and are used to predict motion commands of robots and to monitor the process flow. The architecture, called Computer Assisted Workcell, offers support methods and tools at the following layers of control: organization, coordination, and execution.

References

1. E.D. Sacerdot, *Planning in a hierarchy of abstraction spaces*, Artificial Intelligence, vol. 15, no. 2, 1981
2. D. McDermott, *A temporal logic for reasoning about processes and plans*, Cognitive Science, vol. 6, 1982
3. G. N. Saridis, "Intelligent robotic control", *IEEE Trans. Autom. Contr.* vol. AC-28,5,1983
4. G.N. Saridis, "Analytical formulation of the principle of increasing precision with decreasing intelligence for intelligent machines", *Automatica*, 1989

5. A. Meystel, "Intelligent control in robotics", *J. Robotic Syst.*, vol.5, 5, 1988

6. J.A. Buzacott "Modelling manufacturing systems", *Robotics and Comp. Integr. Manufac.*, vol. 2, 1, 1985

7. E.G. Coffman, M. Elphick, A. Shoshani. System Deadlock. *Computing Surveys*, 3(2),67-78, 1971

8. Z.Banaszak, E.Roszkowska. Deadlock Avoidance in Concurrent Processes. *Foundations of Control*, 18(1-2),3-17, 1988

9. P.G.Ranky, C.Y.Ho *Robot Modeling. Control and Applications with Software*, Springer Verlag. 1985

10. A. Kusiak *Intelligent Manufacturing Systems*. Prentice Hall. 1990

11. J.E. Lenz. *Flexible Manufacturing*. Marcel Dekker, Inc., 1989

12. A.C. Sanderson, L.S. Homem De Mello and H. Zhang. Assembly Sequence Planning. *AI Magazine*, 11(1), Spring 1990

13. B.P. Zeigler. *Multifacetted Modeling and Discrete Event Simulation*, Academic Press, 1984

14. W. Jacak, ICARS Simulation based CAPP/CAM System for Intelligent Control Design in Manufacturing Automation, in *The International Handbook on Robot Simulation System*, ed. D.W. Wloka, John Wiley, 1993

15. W. Jacak and J.W. Rozenblit. Automatic Simulation of a Robot Program for a Sequential Manufacturing Process, *Robotica* 10/3, 45-56, 1992.

16. W. Jacak. Strategies for Searching Collision-Free Manipulator Motions: Automata Theory Approach. *Robotica*, 7, 129-138, 1989.

17. W. Jacak. A Discrete Kinematic Model of Robot in the Cartesian Space. *IEEE Trans. on Robotics and Automation*, 5(4), 435-446, 1989

18. W. Jacak. Discrete Kinematic Modeling Techniques in Cartesian Space for Robotic System. in: *Advances in Control and Dynamics Systems*, ed. C.T. Leondes, Academic Press, 1991

19. K.Shin, N.McKay, A Dynamic Programming Approach to Trajectory Planning of Robotic Manipulators. *IEEE Trans. on Automatic Control*, 31(6), 491-500, 1986

20. K.Shin, N.McKay, Minimum Time Control of Robotic Manipulator with Geometric Path Constrains. *IEEE Trans. on Automatic Control*, 30(6), 531-541, 1985

21. W.Jacak, I.Dulęba, P.Rogaliński. A Graph- Searching Approach to Trajectory Planning of Robot's Manipulator, *Robotica*, 10/6, 38-47, 1992

An Algebraic Transformation of the Minimum Automaton Identification Problem

Ireneusz Sierocki

Institute of Technical Cybernetics
Technical University of Wrocław
50-370 Wrocław, Poland

Abstract. The paper proposes a problem transformation method for solving the minimum automaton identification problem. An algebraic characterization of a set of all simplest hypotheses explaining a given set of input-experiments is performed. It is shown that the minimum identification problem is polynomially transformable into a problem of determining a simplest congruence of the so-called basic hypothesis. It is proved that the method produces a weakly exclusive set of simplest hypotheses.
Keywords: finite automaton, identification, simplest hypothesis, congruence,

1 Introduction

An automaton identification problem (AIP) is formulated as follows: given a set E of input-output experiments performed on a finite automaton $S_?$, which is regarded as a black box with only input and output terminals accessible; one should determine a finite automaton S such that S is behaviourally equivalent with $S_?$. (An automaton compatible with E is called also a hypothesis explaining E.) The AIP is solvable; i.e., there exists a deterministic algorithm, which always produces a correct hypothesis, if there is known in an advance an upper bound n on the size of the state-set of $S_?$ and E contains all input-output pairs of the length at most $2n - 2$. Because these conditions are too strong to be realistic in a typical experimental framework then the AIP is reformulated to a problem of finding a minimum state finite automaton compatible with E. This problem is called the *minimum automaton identification problem* ($MAIP$). It is worth noting that a process of a minimum automaton identification is based on the metahypothesis about the simplicity of explanation, i.e., it is believed that the simplest hypothesis seems to be the most likely to be a correct one.

The methods solving the AIP can be classified into two categories : the enumerative and constructive methods [1], [4]. An enumerative method relies on searching a class of all finite automata, ordered with respect to a number of states, until it meets a first finite automaton compatible with E. However, a constructive method produces a solution to the AIP as a result of performing the set-theoretical and algebraic operations on E. The constructive algorithms are described in [2],[5],[10],[11],[13],[15],[16],[20].

It has been proved [3], [7],that the $MAIP$ is computationally hard, i.e., an amount of time needed to solve it is exponential in the size of E. In order to reduce a problem solving complexity, some problem manipulation methods are needed. The most known strategy is *divide and conquer*, called also a *problem decomposition method*, which consists in dividing a problem into intermediate problems, solving them in parallel or sequentially, and combining the results. If the intermediate problems are simple they are solved directly, otherwise they are solved by applying the divide-and-conquer strategy recursively. It should be pointed out that the problem decomposition method requires an appropriate, algebraic structure of a problem being solved. It is interesting to note that the algebraic properties, needed to solve a problem by applying divide and conquer, strictly depend on a representation of a problem in question. As mentioned in [21], it may happen that we have two isomorphic representations of a problem in question and only one of them allows us to solve a problem by the problem decomposition method. This means that, in general, a decomposition structure of a problem is not preserved under an isomorphic relation, which can be treated as a special case of a problem transformation relation.

In this paper we are going to apply a strategy, which we call *transform and divide* or a *problem transformation method*. Roughly speaking, a transform and conquer is method for solving a problem by replacing it by another problem, solving it and translating a solution. A formal and general description of this approach is contained in [18]. It is assumed that a problem transformation method is methodologically and computationally justified if a target-problem is simpler or better recognized than a source-problem. Additionally, we are interested in such the problem transformation, which allows us to solve a target-problem by divide and conquer. In other words, one can say the most effective strategy is a strategy, which is a combination of transform-and-conquer and divide-and-conquer.

This paper proposes a problem transformation method for solving the $MAIP$. First, an algebraic characterization of a set of simplest hypotheses explaining E is performed. It is shown that the $MAIP$ is polynomially transformable into a problem of determining a simplest congruence of the so-called basic hypothesis. A complete description of the method is presented. It is proved that the method produces a weakly exclusive set of simplest hypotheses.

2 Basic Notions

In this section we state several notational conventions, make some preliminary definitions and some results which are needed later on.

Definition 2.1 A *finite, incompletely specified automaton (sequential system)* S is a 5-tuple $S = (U, X, Y, F, G, x^o)$, where: U is a finite input-alphabet, X is a finite state-set, Y is a finite output-alphabet, $F : X \times U \longrightarrow X$ is a an one-step transition partial function, $X \longrightarrow Y$ is an output partial function, $x^o \in X$ is an initial state. A cardinality of a state-set X is called a *dimension (complexity)* of

S and is denoted by $dim(S)$. If F and G are functions are then S is said to be a *finite automaton.*

Let Z be a finite set. Z^* denotes a set of all finite sequences (strings) over Z. An element $\overline{z} \in Z^*$ will be written $\overline{z} = z(1)...z(n)$. λ denotes the empty sequence. Then $Z^+ = Z^* - \{\lambda\}$. The length of $\overline{z}$ is denoted by $le(\overline{z})$. The concatenation of the sequences $\overline{z}$, $\overline{w}$ is denoted by $\overline{zw}$. The elements of U^* and Y^* are called the input and output strings, respectively. Let a finite, incompletely specified automaton $S = (U, X, Y, F, G, x^o)$ be given. Then one can define, in the inductive way, the partial functions: $F^* : X \times U^* \longrightarrow X$, $G^* : X \times U^* \longrightarrow Y$ and $G^{**} : X \times U^* \longrightarrow Y^+$, which are called a transition partial function, an ouput-generating partial function and an input-output partial function, respectively.

Definition 2.2 A *behaviour of S*, denoted $B(S)$, is a partial function $B(S) : U^* \longrightarrow Y^+$ defined as follows: $B(S)(\overline{u}) = G^{**}(x^o, \overline{u})$.

Definition 2.3 Let two finite, incompletely specified automata $S_i = (U, X_i, Y, F_i, G_i, x_i^o)$, for $i = 1, 2$, be given.

(i) S_1 is *behaviourally equivalent* with S_2 iff $B(S_1) = B(S_2)$.

(ii) S_1 is *isomorphic* with S_2 iff there exists a bijection $H : X_1 \longrightarrow X_2$ such that $H(x_1^o) = x_2^o$ and the following conditions are satisfied:

(ii1) for every $x \in X_1$ and $u \in U$, $F_1(x, u)$ is defined iff $F_2(H(x), u)$ is defined, and $H(F_1(x, u)) = F_2(H(x), u)$,

(ii2) for every $x \in X_1$, $G_1(x)$ is defined iff $G_2(H(x))$ is defined, and $G_1(x) = G_2(H(x))$.

Definition 2.4 A finite automaton S is said to be *minimal* iff $dim(S) \leq dim(T)$, for every finite automaton T such that S and T are behaviourally equivalent.

Proposition 2.5 (i) If two finite, incompletely specified automata are isomorphic then they are behaviourally equivalent. (ii) If two minimal, finite automata are behaviourally equivalent then they are isomorphic.

Let $S_? = (U, X_?, Y, F_?, G_?, x_?^o)$ denote a finite automaton, which is given for an investigator (an experimenter) as a black box with only input and ouput terminals accessible. It means that an one-step transition function and an output function are not available for an investigator. By an experiment performed on $S_?$, being at an initial time in a state $x_?^o$, it is meant a pair $e = (\overline{u}, \overline{y})$, where $\overline{u}$ is an input string applying to the input terminal and $\overline{y}$ is an observed output string, i.e., $e \in B(S_?)$. In general, one can distiguish two modes of experimentation: passive and active experimentation. More specifically, an experiment $e = (\overline{u}, \overline{y})$ is said to *passive* if an investigator has no choice about $\overline{u}$, otherwise e is said to *active.*

Let a finite set E of passive experiments be given. (E is called also a multiple, passive experiment). A class of all finite, incompletely specified automata can be divided into the following subclasses:

- a finite, incompletely specified automaton S is said to a *partial hypothesis explaining* E, written $S \in PHYP(E)$, iff $E \subset B(S)$, i.e., S is compatible (consistent) with E,
- a finite automaton S is said to be a *hypothesis explaining* E, written $S \in HYP(E)$, iff $S \in PHYP(E)$,
- a finite, incompletely specified automaton S is said to a *simplest, partial hypothesis explaining* E, written $S \in PHYP^{sim}(E)$, iff $S \in PHYP(E)$ and, for every $T \in PHYP(E)$, $dim(S) \leq dim(T)$,
- a finite automaton S said to a *simplest hypothesis explaining* E, written $S \in HYP^{sim}(E)$, iff $S \in HYP(E)$ and, for every $T \in HYP(E)$, $dim(S) \leq dim(T)$.

With these notions, the *minimum automaton identification problem (MAIP)* can be stated as a problem of determiming a simplest hypothesis explaining E.

To show that this problem is transformable into a problem of determining a simplest congruence of the so-called basic hypothesis explaining E, we must recall some basic definitions and facts from the algebraic theory of an automaton identification problem [17].

Definition 2.6 Let $S = (U, X, Y, F, G, x^o)$ be a finite, incompletely specified automaton. An equivalence relation R on a set X is said to a *congruence relation of S*, written $R \in CON(S)$, iff the following conditions are satisfied:

(i) for every $x_1, x_2 \in X$ and $u \in U$, if $(x_1, x_2) \in R$ and $F(x_1, u)$, $F(x_2, u)$ are defined then $(F(x_1, u), F(x_2, u)) \in R$,

(ii) for every $x_1, x_2 \in X$, if $(x_1, x_2) \in R$ and $G(x_1)$, $G(x_2)$ are defined then $G(x_1) = G(x_2)$.

Definition 2.7 $R \in CON(S)$ is said to a *simplest congruence relation of S*, written $R \in CON(S)^{sim}$ iff for every $P \in CON(S)$, $card(X/R) \leq card(X/P)$.

Definition 2.8 Let $R \in CON(S)$. A *quotient of S modulo R* is a finite, incompletely specified automaton $S/R = (U, X/R, Y, F/R, G/R, [x^o]_R)$ determined as follows:

(i) $F/R([x]_R, u) = [F(z, u)]_R$ iff $(z, x) \in R$ and $F(z, u)$ is defined,

(ii) $G/R([x]_R) = G(z)$ iff $(z, x) \in R$ and $G(z)$ is defined.

Definition 2.9 Let $E = \{e_i = (\overline{u_i}, \overline{y_i}) : i = 1, ..., k\}$ be a multiple, passive experiment. A *basic hypothesis explaining* E is a finite, incompletely specified automaton $S_E = (U, X_E, Y, F_E, G_E, x_E^o)$ constructed according to the following rules:

(i) $X_E = \bigcup_{i=1}^{k} Pre(\overline{u_i})$, where $Pre(\overline{u_i})$ is a set of all prefixes of $\overline{u_i}$; $x_E^o = \lambda$,

(ii) for every $x_E \in X_E$ and $u \in U$, if $x_E u \in X_E$ then $F_E(x_E, u)$ is defined and $F_E(x_E, u) = x_E u$,

(iii) let $x_E = u(1)_i...u(l)_i$ be a prefix of $\overline{u_i}$, then $G_E(x_E) = y(l)_i$, where $y(l)_i$ is the l-th element of $\overline{y_i}$; $G_E(x_E^o) = G_?(x_?^o)$.

It can be easily observed that a basic hypothesis can be constructed in a linear time as a function of a size of E.

Definition 2.10 A partial hypothesis S explaining E is *structure-complete* , written $S \in PHYP^{sc}(E)$ iff every state-transition from S is used in a generation of E.

Definition 2.11 Let $S = (U, X, Y, F, G, x^o)$ be a partial hypothesis explaining E. A function $H_S : X_E \longrightarrow X$ defined as follows: $H_S(x_E) = F^*(x^o, x_E)$ is said to be a *string-to-state assigment of S_E into S*. An equivalence relation R_S induced by H_S is called an *equivalence relation associated with S*.

In [17] it has been performed an algebraic characterization of a class of all partial, structure-complete hypotheses explaining E. It has been shown that every partial, structure-complete hypothesis explaining E can be obtained (up to an isomorphism) from a basic hypothesis by constructing an appropriate quotient of S_E. More specifically, a main result of the quoted paper asserts that:

Theorem 2.12 A partial hypothesis S explaining E is structure-complete if and only if an equivalence relation R_S associated with S is a congruence relation of a basic hypothesis S_E explaining E and a quotient of S_E modulo R_S is isomorphic with S.

3 Algebraic Characterization of the Minimum Automaton Identification Problem

A purpose of this section is to provide an algebraic description of a class of all simplest hypotheses explaining E. A first step in this direction is to derive a necessary and sufficient condition for a partial hypothesis explaining E to be simplest one within the class of all structure-complete hypotheses explaining E.

Theorem 3.1 A structure-complete, partial hypothesis S explaining E is simplest iff an equivalence relation R_S associated with S is a simplest congruence relation of a basic hypothesis explaining E.

Proof

($\Longrightarrow$) Let $S \in PHYP^{sc}(E)$. Assume that $R_S \notin CON^{sim}(S_E)$. This means that there exists $R \in CON(S_E)$ such that $card(X_E/R) < card(X_E/R_S)$. Because $S_E/R \in PHYP^{sc}(E)$ and $dim(S_E/R) < dim(S_E/R_S) = dim(S)$ then $S \notin PHYP^{sim}(E) \cap PHYP^{sc}(E)$.

($\Longleftarrow$) Let $S \in PHYP^{sc}(E)$. Assume that $S \notin PHYP^{sim}(E)$. This means that there exists $T \in PHYP^{sim}(E) \cap PHYP^{sc}(E)$ such that $dim(T) < dim(S)$. By Theorem 2.12, $dim(S) = dim(S_E/R_S)$ and $dim(T) = dim(S_E/R_T)$. So, we have shown that $R_S \notin CON^{sim}(S_E)$. $\square$

Note that, in general, a quotient of S_E modulo a simplest congruence relation is an incompletely specified automaton. Then it raises a question of how to obtain a completely specified automaton being a simplest hypothesis explaining E. To this end, we must employ some auxiliary definitions.

Definition 3.2 A partial hypothesis S explaining E is *state-complete* iff every state from S is used in a generation of E.

It is clear that if S is structure-complete then it is state-complete.

Definition 3.3 Let $S = (U, X, Y, F, G, x^o)$ be a partial hypothesis explaining E. An *active part of* S is a finite, incompletely specified automaton $S_A = (U, X_A, Y, F_A, G_A, x_A^o)$ defined as follows:

(i) $X_A = \{x \in X : (\exists x_E \in X_E)(F^*(x^o, x_E) = x\}$, $x_A^o = x^o$,

(ii) $F_A(x, u) = F(x, u)$ iff there exists $x_E \in X_E$ such that $x_E u \in X_E$ and $F^*(x^o, x_E) = x$, otherwise $F_A(x, u)$ is undefined,

(iii) $G_A(x) = G(x)$ iff there exists $x_E \in X_E$ such that $F^*(x^o, x_E) = x$, otherwise $G_A(x)$ is undefined.

We have the following obvious consequences of the previous definitions:

Proposition 3.4 (i) S_A is a partial hypothesis explaining E. (ii) S_A is structure-complete. (iii) The congruence relations associated with S and S_A are equal. (iv) S is state-complete iff $dim(S) = dim(S_A)$.

Let $S = (U, X, Y, F, G, x^o)$ be a finite, incompletely specified automaton such that G is a function. (Note that from S being a structure-complete, partial hypothesis explaining E it follows that G is a function.) Define a set $V_S \subset X \times U$ as follows: $(x, u) \in V_S$ iff $F(x, u)$ is undefined. By $COM(S)$ it will be denoted a set of all functions from V_S into X.

Definition 3.5 A *completion of* S *w.r.t.* $C \in COM(S)$ is finite automaton $C(S) = (U, X, Y, C(F), G, x^o)$ constructed as follows: if $F(x, u)$ is defined then $C(F)(x, u) = F(x, u)$, otherwise $C(F)(x, u) = C(x, u)$.

From this definition it follows straightforwardly the following fact:

Proposition 3.6 (i) $dim(S) = dim(C(S))$. (ii) If $S \in PHYP(E)$ then $C(S) \in HYP(E)$.

Now, we have everything needed to formulate and prove a main theoretical result of this paper.

Theorem 3.7 A hypothesis S explaining E is simplest if and only if S is state-complete and an active part S_A of S is a simplest, structure-complete, partial hypothesis explaining E.

Proof

($\Longrightarrow$) First, assume that S is not state-complete. By Proposition 3.4 (iv), $dim(S_A) < dim(S)$. Let $C(S_A)$ be an arbitrary completion of S. From Proposition 3.6, it follows that $dim(S_A) = dim(C(S_A))$ and $C(S_A) \in HYP(E)$. Hence, $S \notin HYP^{sim}(E)$. Next, assume that $S_A \notin PHYP^{sim}(E) \cap PHYP^{sc}(E)$, i.e., there exists $T \in PHYP^{*c}(E)$ such that $dim(T) < dim(S_A)$. Let $C(T)$ be an arbitrary completion of T. Because $dim(T) = dim(C(T))$, $dim(S_A) \leq dim(S)$ and $C(T) \in HYP(E)$ then $S \notin HYP^{*im}(E)$.

($\Longleftarrow$) Assume that $S \notin HYP^{*im}(E)$, i.e., there exists $T \in HYP^{*im}(E)$ such that $dim(T) < dim(S)$. Because $T \in HYP^{*im}(E)$ then $dim(T_A) = dim(T)$, where T_A is an active part of T. On the other hand, from the fact that S is state-complete it follows that $dim(S) = dim(S_A)$. So, we have proved that $S_A \notin PHYP^{*im}(E) \cap PHYP^{*c}(E)$. $\square$

The theorem, we have just proved, and Theorem 2.12 allows us to give the following constructive characterization of a simplest hypothesis explaining E.

Theorem 3.8 A hypothesis S explaining E is simplest iff there exists a simplest congruence relation R of a basic hypothesis explaining E and there exists $C \in COM(S_E/R)$ such $C(S_E/R)$ is isomorphic with S.

This theorem suggests the following structure of the problem transformation method for solving the minimum automaton identification problem.

Input: A multiple, passive experiment E performed on an automaton $S_?$ being identified.

Output: A simplest hypothesis explaining E.

1. Construct a basic hypothesis S_E explaining E.
2. Find a simplest congruence R of S_E.
3. Construct a quotient S_E/R of S modulo R, if S_E/R is a finite automaton then S_E/R is a simplest hypothesis explaining E, otherwise continue.
4. Choice a function C from $COM(S_E/R)$ and construct $C(S_E/R)$.

Following [6], we will say that a finite class of hypotheses explaining E is *weakly exclusive* if every two elements of the class are distinguishable, i.e., they are not behaviourally equivalent. It is worth noting that this property is a sufficient condition for the automaton identification problem to be solved by the so-called multiple, crucis experiment [13]. Observe that the proposed method is nondeterministic, i.e., it can produce a set of all simplest hypotheses explaining E. Hence, it is interesting to ask whether the method generates a weakly exclusive set of simplest hypotheses explaining E. To answer this question, we need to prove a sequence of additional propositions.

Proposition 3.9 If two partial hypotheses explaining E are isomorphic then the congruence relations associated with them are equal.
Proof
Let $S_i = (U, X_i, Y, F_i, G_i, x_i^o) \in PHYP(E)$, for $i = 1, 2$. Assume that S_1 is isomorphic with S_2 via a bijection $H : X_1 \longrightarrow X_2$. Let R_1 and R_2 denote the congruence relations associated with S_1 and S_2, respectively. Because H is an isomorphism and $E \subset B(S_1)$, $E \subset B(S_2)$ then we have the following equality: $F_2^*(x_2^o, x_E) = H(F_1^*(x_1^o, x_E))$, for all $x_E \in X_{\Sigma}$. Let $x_E^1, x_E^2 \in X_E$ and $(x_E^1, x_E^2) \in R_1$, i.e., $F_1^*(x_1^o, x_E^1) = F_1^*(x_1^o, x_E^2)$, see Definition 2.11. From this assumption and the previous equality, it follows that $F_2^*(x_2^o, x_E^1) = F_2^*(x_2^o, x_E^2)$, i.e., $(x_E^1, x_E^2) \in R_2$. Thus we have shown that $R_1 \subseteq R_2$. Applying the same argument for H^{-1}, one can easily verify that $R_2 \subseteq R_1$. $\square$

From the proposition we have just proved and Theorem 2.12, we obtain the following:

Proposition 3.10 Two structure-complete, partial hypotheses explaining E are isomorphic if and only if the congruence relations associated with them are equal.

Proposition 3.11 (i) Every two different completions of a structure-complete, partial hypotheses explaining E are not isomorphic. (ii) If two structure-complete, partial hypotheses S_1 and S_2 explaining E are not isomorphic then any completion of S_1 is not isomorphic with any completion of S_2.
Proof
(i) Let $S = (U, X, Y, F, G, x^o) \in PHYP^{sc}(E)$ and, $C, D \in COM(S)$ and $C \neq D$. For a proof by contradiction, assume that $C(S)$ is isomorphic with $D(S)$ via a bijection $H : X \longrightarrow X$. Because $H(x^o) = x^o$ and $S \in PHYP^{sc}(E)$ then H must be the identity function. Take an arbitrary, undefined transition $F(x, u)$ and assume that $C(F)(x, u) = C(x, u) = x_1$, $D(F)(x, u) = D(x, u) = x_2$. Because H is the identity function then $x_1 = x_2$. This means that $C = D$.
(ii) Assume that there exist $C \in COM(S_1)$ and $D \in COM(S_2)$ such that $C(S_1)$ is isomorphic with $D(S_2)$. By Proposition 3.9, $R_{C(S_1)} = R_{D(S_2)}$, where $R_{C(S_1)}$ and $R_{D(S_2)}$ are the congruence relations associated with $C(S_1)$ and $D(D_2)$, respectively. Because $S_1, S_2 \in PHYP^{sc}(E)$ then S_1 and S_2 are the active parts of $C(S_1)$ and $D(S_2)$, respectively. Proposition 3.4 (iii) allows us to assert that $R_{S_1} = R_{C(S_1)}$ and $R_{S_2} = R_{D(S_2)}$, where R_{S_1} and R_{S_2} are congruence relations associated with S_1 and S_2, respectively. Hence, we have showed that $R_{S_1} = R_{S_2}$. Finally, from Proposition 3.9, it follows that S_1 is isomorphic with S_2. $\square$

Proposition 3.12 Two simplest hypotheses explaining E are isomorphic if and only if they behaviourally equivalent.
Proof
Because of Proposition 2.5 it is enough to prove that a simplest hypothesis explaining E must be a minimal, finite automaton, see Definition 2.4. For a proof by a contradiction, assume that $S \in HYP^{sim}(E)$ and S is not a minimal, finite automaton. This means that there exists a finite automaton T such that $B(S) = B(T)$ and $dim(T) < dim(S)$. Because $E \subset B(S)$ and $B(S) = B(T)$ then $T \in HYP(E)$. So, we have shown that $S \notin HYP^{sim}(E)$. $\square$

Now, we are ready to prove that the method is not redundant, i.e., a set of simplest hypotheses produced by the method is weakly exclusive.

Theorem 3.13 Every two simplest hypotheses generated by the problem transformation method are not behaviourally equivalent.
Proof
One can distinguish two cases of generation of two simplest hypotheses explaining E.

The case (A): let R be a simplest congruence of S_E such that S_E/R is not a finite automato. Take into account two different completions $C(S_E/R)$ and $D(S_E/R)$ of S_E/R. By Proposition 3.11, they are not isomorphic. In turn, Proposition 3.12 allows us to conclude that they are not behaviourally equivalent.

The case (B): Let R_1 and R_2 be two different, simplest congruences of S_E. By Proposition 3.10, S_E/R_1 and S_E/R_2 are not isomorphic. Let $C \in COM(S_E/R_1)$ and $D \in COM(S_E/R_2)$. From Proposition 3.11 it follows that $C(S_E/R_1)$ and $D(S_E/R_2)$ are not isomorphic. Finally, by Proposition 3.12, $C(S_E/R_1)$ and $D(S_E/R_2)$ are not behaviourally equivalent. $\square$

For a good understanding of our formal approach it might be helpful to go through the following:

Example. Let $S_?$ be a finite, automaton being identified such that $U = Y = \{0, 1\}$. Assume that $e = (101001, 000110)$ is a simple input-output experiment performed on $S_?$ and that $G_?(x_?^o) = 0$. In accordance with Definition 2.9, a basic hypothesis explaining $E = \{e\}$ has the following form:

F_E	0	1	G_E
λ	-	1	0
1	10	-	0
10	-	101	0
101	1010	-	0
1010	10100	-	1
10100	-	101001	1
101001	-	-	0

One can convince oneself that there are four simplest congruence relations of S_E, namely R_1, R_2, R_3, R_4 where:
$X_E/R_1 = \{a = \{\lambda, 1\}, b = \{10, 101, 101001\}, c = \{1010, 10100\}\}$,
$X_E/R_2 = \{d = \{\lambda, 101\}, e = \{1, 10, 101001\}, f = \{1010, 10100\}\}$,
$X_E/R_3 = \{g = \{\lambda, 101, 101001\}, h = \{1, 10\}, i = \{1010, 10100\}\}$,
$X_E/R_4 = \{j = \{\lambda, 1, 101001\}, k = \{10, 101\}, l = \{1010, 10100\}\}$.

Using a quotient construction, see Definition 2.8, one obtains a weakly exclusive set $\{S_i = S_E/R_i : i = 1, ..., 4\}$ of all simplest hypotheses explaining E, where:

F_1	0	1	G_1		F_2	0	1	G_2		F_3	0	1	G_3		F_4	0	1	G_4
a	b	a	0		d	f	e	0		g	i	h	0		j	k	j	0
b	c	b	0		e	e	d	0		h	h	g	0		k	l	k	0
c	c	b	1		f	f	e	1		i	i	g	1		l	l	j	1

4 Final Remarks

We have shown that the minimum automaton identification problem is polynomially transformable into a problem of a determining a certain congruence relation of an incompletely specified automaton. It can be easily computed that a search space for the problem of simplest congruence relation is smaller than a search space for the problem of minimum automaton identification. Hence, a targer-problem is simpler, with respect to the enumerative method, than a source-problem. A target-problem is also better recognized than a source-problem because many automata problems are transformable into a problem of determining a congruence relation of a finite automaton or an incompletely specified automaton. The typical examples of this class of problems are: the problem of structural-decomposition of an automaton [8], [9], the problem of reduction of an automaton [8],[9] the problem of minimization of an incompletely specified automaton [14], the problem of a realization of a behaviour [8], the problem of a state-assignment [9]. In turn, it has been proved [19] that a target-problem is parallelly and serially decomposable. This means that the minimum automaton identification problem can be solved by transform, divide-and-conquer strategy. Thus, we have shown that our method of a problem transformation is methodologically and computationally justified.

Finally, it worth noting that this paper can treated as a theoretical and computational contribution to a field called Computer Aided Systems Theory (CAST) [12]. More specifically, our method can be incorporated into a part of CAST, namely into the Finite State Machine Method Bank System.

References

[1] Angluin D., Smith C.H., "Inductive inference: Theory and Methods". *Computing Surveys*, 15, pp. 237-269, 1983.

[2] Angluin D., "Learning regular sets from queries and counterexamples". *Information and Computation*, 75, pp. 87-106, 1987.

[3] Angluin D., "On the complexity of minimum inference of regular sets". *Information and Control*, 39, pp. 337-350, 1978.

[4] Biermann A.W., "Fundamental mechanisms in machine learning and inductive inference". *Lecture Notes in Computer Science*, 322, pp. 133-171, 1985.

[5] Gerardy R., "New methods for the identification of finite state systems". *Intern. J. General Systems*, 15, pp. 97-112, 1989.

[6] Gill A., *Introduction to the Theory of Finite-State Machines*, McGraw-Hill, New York, 1962.

[7] Gold E.M., "Complexity of automaton identification from given data". *Information and Control*, 37, pp. 302-320, 1978.

[8] Kalman R.E., Falb P.I. and Arbib M.A., *Topics in Mathematical System Theory*, McGraw-Hill, New York, 1969

[9] Kohavi Z., *Switching and Finite Automata Theory*, McGraw-Hill, New York, 1969.

[10] Lunea P., Richetin M., Cayla C., "Sequential learning of automata from input-ouput behaviour". *Robotica*, 1, pp. 151-159, 1984.

[11] Miclet L., "Regular inference with a tail-clustering method". *IEEE Trans. Syst. Man. Cybern.*, 10, pp. 737-743, 1980.

[12] Pichler F., Schwärtzel H., *CAST: Methods in Modelling*, Springer-Verlag, Berlin, 1992

[13] Pao T.W., Carr III J.W., "A solution of the syntactical induction-inference problem for regular languages". *Computer Languages*, 3, pp. 53-64, 1978.

[14] Reusch B., Merznich W., "Minimal coverings for incompletely specified sequential machines". *Acta Informatica*, 22, pp. 663-678, 1986.

[15] Richetin M., Naranjo M., " Inference of automata by dialectic learning". *Robotica*, 3, pp. 159-163, 1985.

[16] Rivest R.L., Schapire R.E., "Inference of finite automata using homing sequences". *Proc. of ACM Symposium on Theory of Computing, Seattle, Washington, May, 1989*, pp. 1-10.

[17] Sierocki I., "A description of a set of hypotheses for system identification problem". *Intern. J. General Systems*, 15, pp. 301-319, 1989.

[18] Sierocki I., "A transformation of the problems of minimal satisfaction of constraints". *Lecture Notes in Computer Science*, 585, pp. 218-224, 1992.

[19] Sierocki I., "A problem transformation and decomposition method for solving the minimum automaton identification problem", *submitted for publication*

[20] Veelenturf L.P.Y., "Inference of sequential machines from sample computations". *IEEE Trans.Comput.*, 27, pp. 167-170, 1978.

[21] Korf R.E., "Macro-operators: A weak method of learning", *Artificial Intellignce*, 26, pp.35-77, 1985.

On Possibilistic Automata

Cliff Joslyn * **

Systems Science, SUNY at Binghamton, Binghamton, New York

Abstract. General automata are considered with respect to normalization over semirings. Possibilistic automata are defined as normal pessimistic fuzzy automata. Possibilistic automata are analogous to stochastic automata where stochastic $(+/\times)$ semirings are replaced by possibilistic $(\vee/\wedge)$ semirings; but where stochastic automata must be normal, fuzzy automata may be (resulting in possibilistic automata) or may not be (resulting in fuzzy automata proper). While possibilistic automata are direct generalizations of nondeterministic automata, stochastic automata are not. Some properties of possibilistic automata are considered, and the identity of possibilistic automata strongly consistent with stochastic automata is shown.

1 Introduction

Fuzzy automata were introduced and examined in the late 1960's in a series of papers by Santos [20], Santos and Wee [22], and Wee and Fu [25]. Their most commonly considered form is the **pessimistic max-min fuzzy automata**, a system $\tilde{\mathcal{A}} := \langle Q, A, \mathbf{F}, \mu^0 \rangle$, where $Q := \{q_i\}$ is a finite set of **states**, $1 \le i \le N = |Q|$; $A := \{a_k\}$ is a finite **input alphabet**; $\mathbf{F} := \{F^k\}$ is a set of fuzzy binary relations on Q, denoted $F^k \tilde{\subset} Q^2$, one for each input symbol a_k; and the **initial state** $\mu^0 \tilde{\subset} Q$ is a fuzzy subset of Q. The **transition function** is then $\mu^n := F(n) \circ \mu^{n-1}$ at time $n \in \{1, 2, \ldots\}$, where $\circ$ is max-min composition and $F(n) := F(a_k(n)) = F^k \in \mathbf{F}$ is the binary relation for the symbol $a_k(n)$ input to $\tilde{\mathcal{A}}$ at time n.

Some properties of fuzzy automata and languages were investigated [3, 18], and they were incorporated into a general theory of automata in the context of category theory [1]. Fuzzy automata and languages have received considerably less attention since the 1980's, but some progress continues [16, 17].

Gaines and Kohout [6] expanded the idea of fuzzy automata to include "possible automata", and in so doing described the basic concepts of **possibility theory** that would later be developed by Lotfi Zadeh [26] and others [5, 10]. Researchers have advanced possibility theory as a form of information theory which is distinct from, but related to, both probability theory and fuzzy theory. Therefore, an explicit definition of **possibilistic automata** is now required.

---

* Current mailing address: 327 Spring St. # 2, Portland ME, 04102, USA, cjoslyn@bingsuns.cc.binghamton.edu, joslyn@kong.gsfc.nasa.gov.

** Supported under NASA Grant # NGT 50757.

2 Automata

2.1 General Automata

A general automaton can be defined [6, 21] as a process over an ordered, commutative semiring $\mathcal{V} := \langle V, \oplus, \otimes, 0, 1, \leq \rangle$, where $V := \{v\}$; $\oplus$ and $\otimes$ are general operators over V; $0, 1 \in V$ are the identities for $\oplus$ and $\otimes$ respectively; usually $V \subseteq \mathbb{R}$ with $\{0, 1\} \subseteq V \subseteq [0, 1]$; and $\leq$ is a (usually complete) order on V. For convenience, denote $\mathcal{V} = \langle V, \oplus, \otimes \rangle$.

Definition 1 General Finite Automaton. A system $\mathcal{A} := \langle Q, A, \mathcal{V}, \sigma, \phi^0 \rangle$ is a general automaton where:

- $\sigma: Q \times A \times Q \mapsto V$ is the **transition function**;
- $\phi^0: Q \mapsto V$ is the **initial state**; and
- $\phi^n: Q \mapsto V$ is the **state function** for time n with

$$\phi^n(q_i) := \bigoplus_{q_j \in Q} \sigma(q_i, a_k(n), q_j) \otimes \phi^{n-1}(q_j) \ . \tag{1}$$

$\tilde{\mathcal{A}}$ results when: $\mathcal{V} = \langle [0, 1], \vee, \wedge \rangle$; $\vee$ and $\wedge$ are the maximum and minimum operators; $\mu^n = \phi^n$; and $F(n) = \sigma(\cdot, a_k(n), \cdot)$ is taken to be an $N \times N$ **transition matrix**. Other classes of fuzzy automata include **max-product fuzzy automata** for $\mathcal{V} = \langle [0, 1], \vee, \times \rangle$ and **optimistic min-max fuzzy automata** for $\mathcal{V} = \langle [0, 1], \wedge, \vee \rangle$ and the ordering $\leq$ on $[0, 1]$ reversed to $\geq$.

2.2 Normal General Automata

Stochastic automata result for $\mathcal{V} = \langle [0, 1], +, \times \rangle$. Letting i and j be general indices over Q, and denoting $p_i^n := \phi^n(q_i)$ and $p_{ij}^k = p(q_i|a_k, q_j) := \sigma(q_i, a_k, q_j)$ as the state and transition probabilities respectively, then the well-known stochastic transition function is

$$p_i^n = \sum_{j=1}^N p_{ij}^k p_j^{n-1} \ . \tag{2}$$

However, $v, v' \in [0, 1] \not\mapsto v + v' \in [0, 1]$. In general $p_i^n \in [0, 2] \neq V$. To guarantee the closure of state transitions the following **stochastic normalization conditions** are required:

$$\sum_{i=1}^N p_i^0 = 1, \qquad \mathop{\forall}_{a_k \in A} \mathop{\forall}_{j=1}^N \sum_{i=1}^N p_{ij}^k = 1 \ . \tag{3}$$

On generalization of (3) to $\mathcal{V}$ the following **general normalization conditions** are obtained.

Definition 2 State Normalization. A state function $\phi: Q \mapsto V$ is normal when

$$\bigoplus_{q_i \in Q} \phi(q_i) = 1 \ . \tag{4}$$

Definition 3 Automaton Normalization. An automaton $\mathcal{A}$ is normal when ϕ^0 is normal and

$$\mathop{\forall}_{a_k \in A} \mathop{\forall}_{q_j \in Q} \mathop{\bigoplus}_{q_i \in Q} \sigma(q_i, a_k, q_j) = 1 \ . \tag{5}$$

If $\mathcal{A}$ is normal, then state normalization holds in general [6].

Theorem 4. *If $\mathcal{A}$ is normal, then $\forall n \geq 0, \phi^n$ is normal.*

Proof. Proof by induction. (1) Since $\mathcal{A}$ is normal, ϕ^0 is normal. (2) For $n \geq 1$, assume ϕ^{n-1} is normal. Fix a_k and n, and denote

$$\phi_i' := \phi^n(q_i), \quad \phi_j := \phi^{n-1}(q_j), \quad \sigma_{ij} := \sigma(q_i, a_k, q_j), \quad vv' := v \otimes v' \ . \tag{6}$$

Then in virtue of the commutivity of $\oplus$ and $\otimes$, the distributivity of $\otimes$ over $\oplus$, the normality of $\mathcal{A}$, the identity of 1 for $\otimes$, and the normality of ϕ (respectively), we have

$$\bigoplus_i \phi_i' = \bigoplus_i \bigoplus_j \sigma_{ij} \phi_j = \bigoplus_j \bigoplus_i \phi_j \sigma_{ij}$$
$$= \bigoplus_j \phi_j \bigoplus_i \sigma_{ij} = \bigoplus_j \phi_j 1 = \bigoplus_j \phi_j = 1 \ . \tag{7}$$

$\square$

When, and only when, normalization holds, then the final classes of classical automata can be recovered for discrete V: **nondeterministic automata** for $V = \langle \{0,1\}, \vee, \wedge \rangle$ and **deterministic automata** for $V = \langle \{0,1\}, +, \times \rangle$ [6].

3 Possibility Theory

Possibility theory is a new form of mathematical information theory. It is related to, but distinct from, both probability and fuzzy set theory, and arises independently in both Dempster-Shafer evidence theory and fuzzy set theory [5, 7, 14].

3.1 Mathematical Possibility Theory

A set function $m: 2^Q \mapsto [0, 1]$ is an **evidence function** (otherwise known as a **basic probability assignment**) when $m(\emptyset) = 0$ and $\sum_{G \subseteq Q} m(G) = 1$. The **random set** generated by an evidence function is

$$S := \{ \langle G_j, m_j \rangle : m_j > 0 \} \ , \tag{8}$$

where $\langle \cdot \rangle$ is a vector, $G_j \subseteq Q, m_j := m(G_j)$, and $1 \leq j \leq |S| \leq 2^n - 1$. S has **focal set** $\mathcal{F} := \{ G_j : m_j > 0 \}$.

The dual **belief** and **plausibility** measures are

$$\forall G \subseteq Q, \quad \mathrm{Bel}(G) := \sum_{G_j \subseteq G} m_j \quad \mathrm{Pl}(G) := \sum_{G_j \cap G \neq \emptyset} m_j \tag{9}$$

respectively. Both are **fuzzy measures** [24]. The **plausibility assignment** (otherwise known as the **contour function, falling shadow,** or **one-point coverage function**) of S is

$$\mathbf{Pl} := \langle \mathrm{Pl}(\{q_i\}) \rangle = \langle \mathrm{Pl}_i \rangle, \qquad \mathrm{Pl}_i := \sum_{G_j \ni q_i} m_j . \tag{10}$$

When $\forall G_j \in \mathcal{F}, |G_j| = 1$, then S is **specific**, and $\forall G \subseteq Q, \mathrm{Pr}(G) := \mathrm{Bel}(G) = \mathrm{Pl}(G)$ is a **probability measure** with **probability distribution** $p = \langle p_i \rangle := \mathbf{Pl}$, additive normalization $\sum_i p_i = 1$, and operator $\mathrm{Pr}(G) = \sum_{q_i \in G} p_i$.

S is **consonant** ($\mathcal{F}$ is a **nest**) when (without loss of generality for ordering, and letting $G_0 = \emptyset$) $G_{j-1} \subseteq G_j$. Now $\Pi(G) := \mathrm{Pl}(G)$ is a **possibility measure** with dual **necessity measure** $\eta(G) := \mathrm{Bel}(G)$. As Pr is additive, so Π is **maximal** in the sense that

$$\forall G_1, G_2 \in \mathcal{F}, \quad \Pi(G_1 \cup G_2) = \Pi(G_1) \vee \Pi(G_2) . \tag{11}$$

Denoting $G_i := \{q_1, q_2, \ldots, q_i\}$, and assuming that $\mathcal{F}$ is **complete** ($\forall q_i \in Q, \exists G_i \in \mathcal{F}$), then $\pi = \langle \pi_i \rangle := \mathbf{Pl}$ is a **possibility distribution** with maximal normalization and operator

$$\bigvee_{i=1}^{n} \pi_i = \pi_1 = 1, \qquad \Pi(G) = \bigvee_{q_i \in G} \pi_i . \tag{12}$$

3.2 Possibility and Fuzzy Theory

It is clear that π is a normal fuzzy set, and indeed, Zadeh defined possibility strictly in terms of fuzzy sets in his original introduction of possibility theory [26]. But the relation between possibility and fuzziness is actually a bit more involved.

Let the cardinality of a fuzzy set $\phi \,\tilde{\subset}\, Q$ with membership function $\mu_\phi : Q \mapsto [0,1]$ be $|\phi| := \sum_i \mu_\phi(q_i)$. Kampè de Fériet has shown [12] that for countable Q, and for some random set S and fuzzy set ϕ taken as a vector, that:

- $|\phi| \geq 1$ iff ϕ can be taken as a plausibility assignment $\mathbf{Pl}$;
- similarly, $|\phi| \leq 1$ iff ϕ can be taken as a "belief assignment" $\mathbf{Bel} := \langle \mathrm{Bel}(\{q_i\}) \rangle$; and finally
- $|\phi| = 1$ iff ϕ can be taken as a probability distribution p.

In the first two cases, only a mapping back to an equivalence class of random sets is guaranteed, while in the last case each additive fuzzy set maps to a unique specific (probabilistic) random set. Finally, if $\bigvee_i \mu_\phi(q_i) = 1$ then the first case holds, but also $\mathbf{Pl} = \pi$ is a possibility distribution mapping back to an equivalence class of consonant (possibilistic) random sets [7].

3.3 Interpretations of Possibility

While possibility theory has almost invariably been related directly to fuzzy set theory, probability theory has been regarded as independent from it. This confusion may be a result of the fact that possibility theory is a very weak representation of uncertainty, whereas probability makes very strong requirements.

Furthermore, the wedding of possibility theory to fuzzy sets has relegated possibility to be interpreted strictly in accordance with fuzzy semantics. Since the founding of fuzzy theory by Zadeh in the 1960's, fuzziness has been interpreted almost exclusively as a *psychological* form of uncertainty, expressed in natural language (or "linguistic variables"), and measured by the subjective evaluations of human subjects.

A number of researchers are developing possibility theory into a complete, alternative information theory. Unique measures of information, analogs of stochastic entropy, are being developed [13], along with a possibilistic semantics of *natural* systems which is independent of *both* fuzzy sets and probability [9, 10, 11].

4 Possibilistic Automata

4.1 Definition

Let $\Pi(Q) := \{\pi\}$ be the set of all possibility distributions on Q so that $\pi\colon Q \mapsto [0,1]$ and

$$\forall \pi \in \Pi(Q), \quad \bigvee_{q_i \in Q} \pi(q_i) = 1 \tag{13}$$

is the normalization requirement.

Definition 5 Possibilistic Automaton. A system $\mathcal{A}^{\pi} := \langle Q, A, \pi, \pi^0 \rangle$ is a possibilistic automaton if $\langle Q, A, \langle [0,1], \vee, \wedge \rangle, \pi, \pi^0 \rangle$ is a normal general finite automaton, $\pi \in \Pi(Q|A \times Q)$ is the **transition function**, and $\pi^0 \in \Pi(Q)$ is the **initial state**.

$\pi(q_i|a_k, q_j)$ is the **conditional possibility**[3] of transiting to state q_i from state q_j given an input symbol a_k. The normalization condition (5) becomes

$$\forall a_k, q_j, \quad \bigvee_{q_i \in Q} \pi(q_i|a_k, q_j) = 1 , \tag{14}$$

which states that no matter what the current state, there must be at least one state to which it is completely possible to transit.

The state function is

$$\pi^n = \langle \pi_i^n \rangle = \boldsymbol{\pi}(n) \circ \pi^{n-1} , \tag{15}$$

---

[3] An exact mathematical definition or interpretation of conditional possibility is still somewhat controversial [8, 19].

where: $\pi_i^n := \pi^n(q_i)$ is the possibility of being in state q_i at time n; and $\pi(n) = [\pi(n)_{ij}] := \pi(\cdot|a_k(n), \cdot)$ is the $N \times N$ **transition matrix** at time n. π^n is a column vector; i indexes the rows of π^n and $\pi(n)$; and j indexes the columns of $\pi(n)$. In general

$$\pi_i^n = \bigvee_{j=1}^{N} \pi(n)_{ij} \wedge \pi_j^{n-1} \ . \tag{16}$$

$\mathcal{A}^\pi$ is a special case of $\widetilde{\mathcal{A}}$, under the assignments $F(n) = \pi(n)$ and $\phi^n = \pi^n$. Santos identified $\mathcal{A}^\pi$ as a **restricted fuzzy automaton** [20].

4.2 An Example

For a simple example, let $\pi^0 = \langle 0.3, 1.0, 0.6 \rangle^T$ and let π be fixed at

$$\pi = \begin{bmatrix} 0.1 & 0.0 & 0.4 \\ 1.0 & 0.5 & 1.0 \\ 0.2 & 1.0 & 0.7 \end{bmatrix} \ . \tag{17}$$

π^0 is normal, as is π, with a 1 in each column. The state vector changes as follows:

n	0	1	2	3	4
π^n	0.3	0.4	0.4	0.4	0.4
	1.0	0.6	1.0	0.7	1.0
	0.6	1.0	0.7	1.0	0.7

Since $\pi^2 = \pi^4$, the cycle will repeat. This is completely in keeping with the periodic behavior of fuzzy matrix composition [23]. By Thm. 4, each π^n is normal, although the normalizing element rotates among the π_i^n. Further properties are considered in Sec. 5.

4.3 Discussion

It is significant to note that not all fuzzy relations which are normal in the sense of fuzzy sets are normal in the sense of possibilistic transition matrices (while the converse *is* true). Unlike (14), fuzzy relation normalization [14] only requires that

$$\forall a_k, \ \bigvee_{q_i, q_j \in Q} \pi(q_i|a_k, q_j) = 1 \ ; \tag{18}$$

that is, that there is *some* unitary element in each $\pi(n)$, not that there is some unitary element in *each column* of each $\pi(n)$.

When $\oplus = \vee$, then we call $\mathcal{V}$ a **possibilistic semiring**. Contrary to stochastic automata, $v, v' \in [0, 1] \rightarrow v \oplus v' \in [0, 1]$. Therefore for fuzzy automata with possibilistic semirings, $\phi^n(q_i) \in [0, 1]$ *whether $\mathcal{A}$ is normal or not*. By Thm. 4, π^n will be normal in general. But while this is also true for stochastic automata, it is not the case for all fuzzy automata.

In Sec. 2.2 it was shown that both deterministic and stochastic automata are based on the same algebraic structure which we will call a **stochastic semiring** $\langle \oplus, \otimes \rangle = \langle +, \times \rangle$. In fact, stochastic automata are direct generalizations of deterministic automata when $V = \{0, 1\}$ is replaced by its superset $V = [0, 1]$. Similarly, fuzzy automata are direct generalizations of nondeterministic automata for $V = [0, 1]$ with possibilistic semirings. Finally, possibilistic automata result when normalization is imposed on fuzzy automata, while normalization is automatic for both stochastic and deterministic automata. Since for nondeterministic automata

$$\forall i, j \quad \pi_i^n, \pi(n)_{i,j} \in \{0, 1\} , \tag{19}$$

therefore all nondeterministic automata are also possibilistic automata, and are identified as **crisp possibilistic automata**.

Thus proper (non-crisp) possibilistic automata are also direct generalizations of nondeterministic automata. And it is significant to note that stochastic automata are *not* generalizations of nondeterministic automata, despite years of commitment to them as the sole embodiment of machines with quantified uncertainty. These relations among these classes of automata are summarized in Fig. 1.

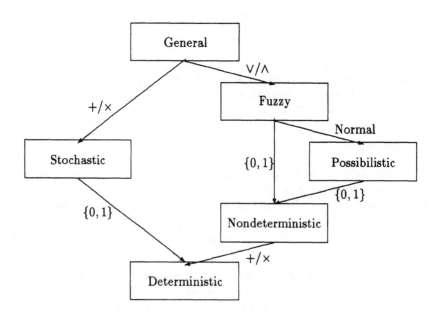

Fig. 1. Classes of automata.

5 Properties

The actions of possibilistic automata are dependent on a maximally normalized initial state vector π^0 and conditional transition matrices $\pi(n)$, and thus on the presence of 1's in π. In investigating their properties, the following lemma will be useful.

Lemma 6. *Let* $R := [r_{ij}], S := [s_{ij}], T := [t_{ij}]$ *be* $N \times N$ *matrices with* $0 \leq r_{ij}, s_{ij}, t_{ij} \leq 1$, *and let* $R \circ S = T$. *If* $\exists r_{i \cdot j \cdot} = 1$, *then* $\forall j, t_{i \cdot j} \geq s_{j \cdot j}$.

Proof. Assume $r_{i \cdot j \cdot} = 1$. Then $\forall j$,

$$t_{i \cdot j} = \bigvee_{k=1}^{N} r_{i \cdot k} \wedge s_{kj}$$

$$= \left(\bigvee_{\substack{1 \leq k \leq N \\ k \neq j \cdot}} r_{i \cdot k} \wedge s_{kj} \right) \vee (r_{i \cdot j \cdot} \wedge s_{j \cdot j}) = \left(\bigvee_{\substack{1 \leq k \leq N \\ k \neq j \cdot}} r_{i \cdot k} \wedge s_{kj} \right) \vee s_{j \cdot j}$$

$$\geq s_{j \cdot j} . \tag{20}$$

□

In the sequel, assume a possibilistic automaton $\mathcal{A}^{\pi}$ with fixed input symbol $a_n(k)$ and single transition matrix $\pi := [\pi_{ij}]$. Given a state vector π, then the next-state vector is $\pi' := \pi \circ \pi$. Where appropriate, the nth step state vector will be denoted π^n.

Whereas each column of π is guaranteed to have at least one unitary value, this is not necessarily the case for the rows. For each row i, let $C(i) := \{c_k^i\}$ be the set of indices (if any) for which

$$\pi_{i,c_k^i} = 1, \quad 1 \leq k \leq |C(i)| \leq N , \tag{21}$$

and let $c(i)$ be the unique such index in each row i, if it exists. The column positions of the unitary elements in a row determine which elements of the state vector will provide lower bounds for the corresponding element of the next-state vector. Where there are multiple unitary elements in a row, then the bound will be the maximum of all corresponding state values.

Theorem 7. $\forall i$, *if* $C(i) \neq \emptyset$, *then*

$$\pi_i' \geq \bigvee_{c_k^i \in C(i)} \pi_{c_k^i} . \tag{22}$$

Proof. $\forall i, \forall c_k^i \in C(i), \pi_{i,c_k^i} = 1$. Under the assignments $R_{N \times N} = \pi, S_{N \times 1} = \pi, T_{N \times 1} = \pi'$, then by Lem. 6, $\forall i, \forall c_k^i \in C(i), \pi_i' \geq \pi_{c_k^i}$, and so the conclusion follows. □

Corollary 8. $\forall i$, if $c(i)$ exists then $\pi'_i \geq \pi_{c(i)}$.

Proof. Follows directly from Thm. 7 under the assumption that $C(i) = \{c(i)\}$. $\square$

If each row of π has a unique unitary element, then $c(i)$ exists and is an injective function creating a cyclic group permuting the elements of Q through at most N steps. Similarly, the bounds of the state vector will cyclically permute through a maximum of N elements of π.

Theorem 9. If $\forall i, c(i)$ exists, then $\forall i, \exists e(i), 1 \leq e(i) \leq N$ such that $\pi_i^{e(i)} \geq \pi_i^0$.

Proof. Let $\lceil x \rceil$ be any value such that $\lceil x \rceil \in [x, 1]$, and let $\pi^0 = \langle \pi_1, \pi_2, \ldots, \pi_N \rangle^T$. By Cor. 8, the action of the automaton in the first step affects the transformation

$$\pi^0 \mapsto \pi^1 = \left\langle \lceil \pi_{c(1)} \rceil, \lceil \pi_{c(2)} \rceil, \ldots, \lceil \pi_{c(N)} \rceil \right\rangle^T . \tag{23}$$

The next step of the automaton affects the further transformation

$$\pi^1 \mapsto \pi^2 = \left\langle \lceil \pi_{c^2(1)} \rceil, \lceil \pi_{c^2(2)} \rceil, \ldots, \lceil \pi_{c^2(N)} \rceil \right\rangle^T . \tag{24}$$

In general,

$$\pi^n = \left\langle \lceil \pi_{c^n(1)} \rceil, \lceil \pi_{c^n(2)} \rceil, \ldots, \lceil \pi_{c^n(N)} \rceil \right\rangle^T . \tag{25}$$

Since $\forall i, \exists e(i) \leq N, c^{e(i)}(i) = i$, therefore

$$\forall i, \quad \pi_i^{e(i)} = \lceil \pi_{c^{e(i)}(i)} \rceil = \lceil \pi_i \rceil \geq \pi_i . \tag{26}$$

$\square$

The placement of unitary values on the diagonal of π guarantee a monotonic increase of state vector values.

Corollary 10. If $\exists i, c(i) = i$, then $\pi'_i \geq \pi_i$.

Proof. Follows immediately from Cor. 8. $\square$

Finally, if π has unitary values on all diagonal elements, then a result of Pedrycz [18] is recovered.

Corollary 11. If $\forall i, c(i) = i$, then $\pi' \geq \pi$.

Proof. Follows immediately from the application of Cor. 10 to all columns. $\square$

6   Consistency with Stochastic Concepts of Possibility

While the mathematical syntax of possibility theory has become relatively well established, the semantics of possibility values has not been developed in its own right. Instead, possibility values have only been interpreted in the context of either fuzzy sets or probability. As mentioned in Sec. 3.3, possibility values are typically interpreted as membership grades of a fuzzy set. Possibility values are then determined from a given fuzzy set, which is itself usually determined by some subjective valuation procedure.

6.1 Strong Probability–Possibility Consistency

The other major set of methods developed to date derive possibility values by converting a given probability or frequency distribution. Such conversion methods have been guided by a principle of **probability-possibility consistency**, briefly stated by Delgado and Moral as "the intuitive idea according to which as an event is more probable, then it is more possible [2]," and summarized most generally by the formula

$$\forall G \subseteq Q, \quad \Pr(G) \leq \Pi(G) \ . \tag{27}$$

In Zadeh's initial introduction of possibility theory [26] he also proposed a quantitative measure of this consistency

$$\gamma(\pi, p) := \sum_i \pi_i p_i \in [0, 1] \ , \tag{28}$$

such that $\gamma(\pi, p) = 0$ indicates complete inconsistency and $\gamma(\pi, p) = 1$ complete consistency. This measure has been generalized by Delgado and Moral [2].

Given a probability distribution $p = \langle p_i \rangle, \sum_i p_i = 1$, then the most traditional conversion formula is "maximum normalization" according to the formula

$$\pi_i = \frac{p_i}{\bigvee_{i=1}^N p_i} \ . \tag{29}$$

Other methods have also been suggested [4, 15].

However, recently Joslyn [10] has proposed a reconsideration of the relation between probability and possibility based on *semantic* criteria, and in particular the natural meaning of probability and possibility. This view incorporates both the modal sense of possibility and necessity, which accommodate only crisp possibility values ($\pi_i \in \{0, 1\}$); and a "stochastic" interpretation of possibility, in which an event with *any* positive probability must be *completely* possible.

The following **strong compatibility principle** [10] results,

$$\forall G \subseteq Q, \quad \Pr(G) > 0 \leftrightarrow \Pi(G) = 1 \tag{30}$$

which implies

$$\forall G \subseteq Q, \quad \Pr(G) = 0 \leftrightarrow \Pi(G) < 1 \ . \tag{31}$$

According to this principle, positive probability and complete possibility are synonymous, and thus the domains of applicability of probability and possibility should be considered as essentially distinct. Furthermore, when (30) holds, then $\gamma(\pi, p) = 1$; (27) holds; and the distributions are also strongly compatible [10]:

$$\forall q_i, \quad p_i > 0 \leftrightarrow \pi_i = 1, \quad p_i = 0 \leftrightarrow \pi_i < 1 \ . \tag{32}$$

6.2 Consistency and Automaton Action

Strong compatibility of (30) is consistent with the operation of possibilistic automata. In particular, the operation of a stochastic automaton which is then converted to a possibilistic automaton according to (30) is equivalent to one which is converted first, and then operated by possibilistic formula.

Assume matrices R and S with $r_{ij}, s_{ij} \in [0, 1]$ (no normalization of any kind assumed). Let $\overline{R} := [\overline{r_{ij}}]$, where

$$\overline{r_{ij}} := \begin{cases} 1, r_{ij} > 0 \\ 0, r_{ij} = 0 \end{cases}, \tag{33}$$

(and similarly for $\overline{S}$), and let RS denote matrix composition under a stochastic semiring (i.e., standard matrix multiplication).

Theorem 12. $\overline{RS} = \overline{R} \circ \overline{S}$.

Proof. Let $T = [t_{ij}] := RS$, so that $\overline{RS} = \overline{T} = [\overline{t_{ij}}]$. Also let $U = [u_{ij}] := \overline{R} \circ \overline{S}$. We need to prove that $\forall i, j, \overline{t_{ij}} = u_{ij}$. First, we have $t_{ij} = \sum_{k=1}^{n} r_{ik} s_{kj}$, so that

$$\overline{t_{ij}} = \begin{cases} 0, \underset{1 \le k \le m}{\forall} r_{ik} = 0 \text{ or } s_{kj} = 0 \\ 1, \text{Otherwise} \end{cases}. \tag{34}$$

Then

$$\begin{aligned} u_{ij} = \bigvee_{k=1}^{m} (\overline{r_{ik}} \wedge \overline{s_{ik}}) &= \begin{cases} 0, \underset{1 \le k \le m}{\forall} \overline{r_{ik}} = 0 \text{ or } \overline{s_{ik}} = 0 \\ 1, \text{Otherwise} \end{cases} \\ &= \begin{cases} 0, \underset{1 \le k \le m}{\forall} r_{ik} = 0 \text{ or } s_{ik} = 0 \\ 1, \text{Otherwise} \end{cases} \\ &= \overline{t_{ij}}. \end{aligned} \tag{35}$$

□

References

1. Arbib, Michael A and Manes, Ernst: (1977) "A Category Theoretic Approach to Systems in a Fuzzy World", in: *Systems: Approahces, Theories, and Applications*, ed. WE Hartnett, pp. 1-26, D. Reidel
2. Delgado, M and Moral, S: (1987) "On the Concept of Possibility-Probability Consistency", *Fuzzy Sets and Systems*, v. 21, pp. 311-318
3. Dubois, Didier and Prade, H: (1980) *Fuzzy Sets and Systems: Theory and Applications*, Academic Press, New York
4. Dubois, Didier and Prade, Henri: (1986) "Fuzzy Sets and Statistical Data", *European J. of Operations Research*, v. 25, pp. 345-356
5. Dubois, Didier and Prade, Henri: (1988) *Possibility Theory*, Plenum Press, New York

6. Gaines, Brian R and Kohout, Ladislav J: (1976) "Logic of Automata", *Int. J. General Systems*, v. **2**:4, pp. 191-208

7. Goodman, IR and Nguyen, HT: (1986) *Uncertainty Models for Knowledge-Based Systems*, North-Holland, Amsterdam

8. Hisdal, E: (1978) "Conditional Possibilities: Independence and Noninteraction", *Fuzzy Sets and Systems*, v. **1**:4, pp. 283-297

9. Joslyn, Cliff: (1992) "Possibilistic Measurement and Set Statistics", in: *Proc. 1992 NAFIPS Conference*, v. **2**, pp. 458-467

10. Joslyn, Cliff: (1993) "Possibilistic Semantics and Measurement Methods in Complex Systems", in: *Proc. of the 2nd International Symposium on Uncertainty Modeling and Analysis*, ed. Bilal Ayyub, pp. 208-215, IEEE Computer Socety Press

11. Joslyn, Cliff: (1993) "Some New Results on Possibilistic Measurement", to appear in: *Proc. 1993 NAFIPS Conference*

12. Kampè de Fériet, J: (1982) "Interpretation of Membership Functions of Fuzzy Sets in Terms of Plausibility and Belief", in: *Fuzzy Information and Decision Processes*, ed. MM Gupta and E Sanchez, pp. 93-98, North-Holland, Amsterdam

13. Klir, George: (1992) "Probabilistic vs. Possibilistic Conceptualization of Uncertainty", in: *Analysis and Management of Uncertainty*, ed. BM Ayyub et al., pp. 13-25, Elsevier

14. Klir, George and Folger, Tina: (1987) *Fuzzy Sets, Uncertainty, and Information*, Prentice Hall

15. Klir, George and Parviz, Behvad: (1992) "Probability-Possibility Transformations: A Comparison", *Int. J. General Systems*, v. **21**:1

16. Močkor, Jiři: (1991) "A Category of Fuzzy Automata", *Int. J. General Systems*, v. **20**, pp. 73-82

17. Peeva, Kety: (1991) "Equivalence, Reduction and Minimization of Finite Automata over Semirings", *Theoretical Computer Science*, v. **88**:2, pp. 269-285

18. Pedrycz, W: (1981) "An Approach to the Analysis of Fuzzy Systems", *Int. J. of Control*, v. **34**, pp. 403-421

19. Ramer, Arthur: (1989) "Conditional Possibility Measures", *Cybernetics and Systems*, v. **20**, pp. 233-247

20. Santos, E: (1968) "Maximin Automata", *Information and Control*, v. **13**, pp. 363-377

21. Santos, E: (1977) "Fuzzy and Probabilistic Programs", in: *Fuzzy Automata and Decision Processes*, ed. MM Gupta, pp. 133-147

22. Santos, E and Wee, WG: (1968) "General Formulation of Sequential Machines", *Information Control*, v. **12**, pp. 5-10

23. Thomason, MG: (1973) "Finite Fuzzy Automata, Regular Fuzzy Languages, and Pattern Recognition", *J. Pattern Recognition*, v. **5**, pp. 383-390

24. Wang, Zhenyuan and Klir, George J: (1991) *Fuzzy Measure Theory*, Plenum Press, New York

25. Wee, WG and Fu, KS: (1969) "Formulation of Fuzzy Automata and Application as Model of a Learning Systems", *IEEE Trans. on Systems Science and Cybernetics*, v. **5**, pp. 215-223

26. Zadeh, Lotfi A: (1978) "Fuzzy Sets as the Basis for a Theory of Possibility", *Fuzzy Sets and Systems*, v. **1**, pp. 3-28

On Automatic Adjustment of the Sampling Period

S. Dormido, I. López, F. Morilla, Mª. A. Canto

Dpto. de Informática y Automática, Facultad de Ciencias, UNED
Avda Senda del Rey s/n, 28040 Madrid (Spain)

Abstract. The sampling period can be considered as an additional parameter in the design of a controller. In this paper we present an automatic adjustment procedure of the sampling period that tries to maintain constant the number of samples during the rise time If the sampling period is modified obviously the parameters of the discrete model will change. However from the knowledge of the initial sampling period, the final sampling period and the initial values of the parameters it is possible to deduce a good approximation for the final parameters once the sampling period has been adjusted. In this sense a new approximate method, very simple to implement, is proposed. The results obtained in simulation confirm the possibilities of the method.

1 Introduction

The choice of the sampling period in a sampled system is an important problem. A good choice depends on many factors, such as, the properties of the signal, the purpose of the system, etc.In this paper we study the problem of finding automatic adjustment methods of the sampling period that allow us to maintain the specifications of the process.

The content of the paper is organised as follows: In section 2 some aspects of the problem are presented; they show that the choice of the sampling period is an important factor in the design of a sampled system since it is considered a critical parameter. For this reason it is very interesting to develop methods that permit to adjust the sampling period in an automatic way. The principles of the adjustment mechanism are studied in section 3. The strategy is based on a design method where the sampling period T is selected in relation to the rise time -the time which elapses while the step response rises from 10% to 90% of its final value-.

To apply this procedure it is necessary to use an estimation method that determines the rise time. The direct simulation of the step response has been employed, its principal advantage being that it is very easy to implement since it is based only on counting and comparisons. If the sampling period is modified in this way then the parameter estimates of the sampled system will be modified. A way of accelerating the tuning procedure to the new parameters is by transforming directly the previous estimates of the parameters when the sampling period is adjusted. These new values are considered the initial estimates in the identification method used to determine the parameters of the system with the new sampling period. An approximate procedure that develops these ideas is presented in section 4. In section 5 the properties of the scheme, with automatic adjustment of the sampling period, in different situations.are illustrated through simulation. Finally conclusions are given in section 6.

2 Some aspects of the selection of the sampling period

The different trade offs in the areas of signal processing and control may lead to very different rules to select the sampling period.

2.1 Choice of the sampling period for signal processing

In a problem of signal-processing , the objective is simply to record a signal digitally and to reconstruct it from its samples. In the ideal case the sampling theorem of Shannon provides a very simple rule when the two following conditions are fulfilled:
a) A great delay can be tolerated in the reconstruction process of the signal .
b) The spectrum in frequency of the signal to sample has a limited band.That is to say the Fourier transform of the signal is zero for $|w| > w_0$

If the delay in the reconstruccion of the signal is acceptable, Shannon's theorem says that a sampling frequency of $2w_0$ is sufficient. Nevertheless if the delay in the reconstruction is limited sampling frequencies considerably higher are needed. It is then necessary to use a causal order reconstruction like a zero or first order hold. In [1] the choice of the sampling period for applications in signal processing is analyzed in detail.

2.2 Selection of the sampling period for control systems

Frequently continuous control systems are replaced by its discrete version simply because the implementation is cheaper and more reliable. A direct form is to convert the design of the continuous control system in a sampled system by using a short sampling period and by carrying out some sort of approximation through discretisation of the continuous controller.

One way of determining the sampling period in this case is to use arguments of a continuous type. The sampling system can be approximated by the order zero hold, followed by the continuous system. For short sampling periods, the transfer function of the order zero hold can be approximated as:

$$\frac{1 - e^{-sT}}{sT} \approx \frac{1 - 1 + sT - \frac{(sT)^2}{2} + \dots}{sT} = 1 - \frac{sT}{2} + \dots \tag{1}$$

where T represents the sampling period. The two first terms correspond to the series development of $\exp(-0.5Th)$, for small T, the order zero hold can be approximated by a delay of half of a sampling period. This delay establishes an additional decrease in the phase margin in relation to the continuous system. A way of specifying the sampling period is to establish a range for this decrease of the phase margin (.among 5° and 10°). This observation gives the following empirical rule [2]:

$$5^{\circ} < e^{-j\frac{w_cT}{2}} < 15^{\circ} \Rightarrow \frac{p}{36} < \frac{w_cT}{2} < \frac{p}{12} \Rightarrow w_cT = 0.15 - 0.5 \tag{2}$$

here w_c it is the crossover frequency in radians per second of the continuous system ($|G(jw_c)| = 1$). This rule gives very short sampling periods. The frequency of Nyquist ($w_N = \pi / T$) will be of the order of 5 to 20 times greater than the crossover frequency w_c:

$$w_N = \frac{p}{T} \Rightarrow T = \frac{p}{w_N} \quad \text{substituting in (2)} \Rightarrow \frac{p}{0.5} < \frac{w_N}{w_c} < \frac{p}{0.15} \tag{3}$$

However it is possible to use discrete design methods that permit greater sampling periods. As a rule it is important for the sampling period to be related to the desired behavior of the system operating in closed loop. For example, to apply in a practical way the pole placement design method it is necessary to understand the way in which the properties of the system in closed loop are influenced by the design parameters - for example, the poles in closed loop and the sampling period.

If the desired discrete time system is obtained by sampling a second order, the parameters p1 and p2 in the characteristic equation $z^2 + p_1 z + p_2 = 0$ are:

$$p_1 = -2e^{-\zeta wT}\cos(wT\sqrt{1 - \zeta^2})$$
$$p_2 = -2e^{-\zeta wT}$$

where w is the natural frequency and ζ is the relative damping. The parameter ζ influences the relative damping of the response and w the response speed of the system. If the parameters ζ and w are fixed , the sampling period can still be selected. The choice of T will determine the speed with which the disturbances are detected and also the form of the response curve. It is useful to characterize the sampling period through a dimension free variable that also possesses a good physical interpretation. Åström and Wittenmark have proposed that for nonoscillatory systems the rise time represents a natural factor of normalization In this way N_r is introduced as the number of sampling periods per rise time:

$$N_r = \frac{T_r}{T} \tag{4}$$

where T_r is the rise time. The sampling period can be chosen in such a way that :

$$N_r \approx 4 - 10$$

where N_r is the number of samples per rise time of the closed loop system [2]. This means that the sampling period should be choosen in relation to the desired behaviour of the closed loop system. The sampling period can also be related to the damped frequency of the closed loop system, for which reason it is convenient to introduce the parameter N defined by:

$$N = \frac{2\pi}{wT\sqrt{1 - \zeta^2}} \tag{5}$$

This parameter is the ratio of the damped period and the sampling period. A reasonable rule of thumb is to choose N around 20. This gives wT = 0.44 for ζ = 0.7. Nr $\approx$ 2 - 4 corresponds to N $\approx$ 10 - 20 for the damping ζ = 0.7. From these considerations Åström and Wittenmark suggest the following rule to select the sampling period:

$$wT \approx 0.2 - 0.6 \tag{6}$$

where w is the desired natural frequency of the closed loop system. The election depends on the nature of the disturbances that work on the system. These rules are also valid for pole-placement design based on input-output models.

For a more general control problem , such as the Linear Quadratic Problem (LQP), the choice of the sampling period depends on the way in which the specifications for the control problem are formulated. For the LQP the specifications are supposed to be given in terms of the cost function in continuous time. Then the continuous time LQP controller will minimize the cost function. The cost function increases quadratically with T for small sampling periods, which means that in practice there exists a limit which is not worth reducing any further. It is possible to obtain an approximation of the increase in the cost function due to an increase in the sampling period [3].

For control laws of deadbeat response or minimum variance, the variance of control signal increases when the sampling period decreases. To choose the sampling period for such systems, it is useful to plot the input variance versus the output variance with the sampling period as a parameter. More information about the election of the sampling period for minimum variance controller is found in [2,4,5].

This short review shows that different heuristic rules can be given to choose the sampling period.One of them consists in relating T to the rise time. The adjustment method of the sampling period proposed below is based on this rule.

3 Adjustment of the sampling period

The problem of the choice of the sampling period described previously shows that T is a critical parameter to determine the behaviour of the system. For this reason it is clearly desirable to adjust T in an automatic way.The principles to adjust the sampling period are analyzed in this section. A regulator with adjustment capacity of the sampling period can be considered as composed of four subsystems (Fig. 1):

a) Process, b) Perturbation Generator, c) Parameter Estimator, d) Calculation Block

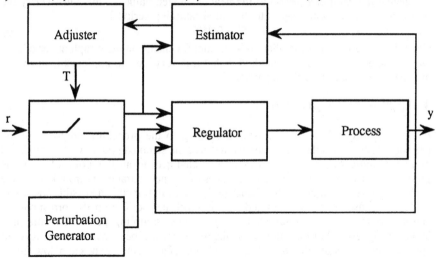

Fig. 1. Structure of the automatic adjustment of the sampling period

The system works in the following way: The "perturbation generator" provides test signals that allow us to estimate the parameters of the process. Then the rise time is estimated, and depending on its value, the sampling period is adjusted or not according to a given criterion. The strategy is based on a design method where the sampling period is selected in relation to the rise time. To use this method it is necessary to use some estimation procedure allowing to determine the rise time. In principle this could be made by transforming the system to its continuous time form and calculating the parameters in this domain. In this case for a first order system the rise time is equal to the time constant of the system. For a second order system with damping ζ and natural frequency w_0, the rise time T_r is given by:

$$T_r = \frac{e^{\frac{\varphi}{tg\varphi}}}{w_0} \tag{6}$$

where $\zeta = \cos\phi$

This procedure, however, consumes too much calculation time and it is not easy to implement. The proposed technique is based on a method for direct determination of the rise time. With this purpose a difference equation is used , describing the relationship between the input and the output of the system:

$$y(t) + a_1 y(t-1) + a_2 y(t-2) +..+ a_n y(t-n) = b_1 u(t-1) +..+ b_m u(t-m) \tag{7}$$

The parameters of the model (7) are determined by using a recursive least square algorithm. In this way the rise time can easily be determined through a direct measure (simulation) of the step response of the system. The new value of the sampling period T' is calculated using the rule of thumb (4) for a given value of Tr, that is to say:

$$T' = \frac{T_r}{N_r} \tag{8}$$

To change the sampling period to the new calculated value T', this must verify the following condition with respect to the initial sampling period T:

$$|T' - T| \geq 0.25T \tag{9}$$

Condition (9) guarantees that there is a meaningful change in the sampling period:
One of the advantages of this method is that it is very easy to implement, since it is only based on counting and comparisons.

4 Approximate procedure

If the sampling period T is adjusted according to the procedure proposed in the previous section, then the parameter estimates of the sampled system will be modified in an automatic way once the value of T is changed. From this point of view the problem to solve is to speed up the estimates of the new parameters from knowledge of the previous estimates and the values of T and T'. In principle a possible way to do this is by transforming the sampled system (with period T) into its corresponding continuous model, that is sampled using the new sampling period T'. This plan however is very laborious and consequently it would be convenient to have an approximate procedure that need less calculations.The method exposed below fulfils the previous requirements and is very simple to carry out.
Let the continuous system be described by:

$$\begin{aligned} \dot{x} &= Ax + Bu \\ y &= Cx \end{aligned} \tag{10}$$

For periodic sampling with period T the discrete state model of the continuous system (10) with a piecewise constant input u is given by:

$$\begin{aligned} x(kT+T) &= Fx(kT) + Gu(kT) \\ y(kT) &= Hx(kT) \end{aligned} \tag{11}$$

where

$$F = e^{AT}, \qquad G = \int_0^T e^{At}d\tau.B, \qquad H = C$$

The matrices **F** and **G** can be calculated in different ways, for example by using the following series expansion:

$$F = \sum_{i=0}^{\infty} \frac{A^i T^i}{i!}, \quad F^0 = I; \quad G = T \sum_{j=0}^{\infty} \frac{A^j T^j}{(j+1)!} \tag{12}$$

One possibility of having a simple procedure consists in approximating (13) to the first two terms of the infinite series, that is to say:

$$F = I + AT; \quad G = T(I + \frac{AT}{2})B \tag{13}$$

The inverse transformation is:

$$A = \frac{1}{T}(F - I); \quad B = \frac{2}{T}(I + F)^{-1}G \tag{14}$$

When the sampling period is changed from T to T ', through simple calculations it can be seen that the equations (14) become:

$$F' = hF + (1 - h)I; \quad G' = h(hF + (2 - h)I)(F + I)^{-1}G \tag{15}$$

Where h = T' / T. The discrete transfer function W(q) for the state space model (11) is:

$$W(q) = H(qI - F)^{-1}G \tag{16}$$

When the sampling period is T' the discrete transfer function is:

$$W'(q) = H(qI - F')^{-1}G' \tag{17}$$

If (15) is substituted in (17), W'(q) can be expressed by:

$$W'(q) = H(qI - hF - (1 - h)I)^{-1}(h(hF + (2 - h)I)(F + I)^{-1})G \tag{18}$$

In this way the discrete transfer function (W'(q)), when the sampling period is modified, can be given in terms of the parameters of the original transfer function (W(q)) and h, that is the ratio between T' and T. For a second order system whose representation in the discrete time state space is:

$$\begin{bmatrix} x_1(k+1) \\ x_2(k+1) \end{bmatrix} = \begin{bmatrix} -a_1 & 1 \\ -a_2 & 0 \end{bmatrix} \begin{bmatrix} x_1(k) \\ x_2(k) \end{bmatrix} + \begin{bmatrix} b_1 \\ b_2 \end{bmatrix} u(k)$$

$$y(k) = \begin{bmatrix} 1 & 0 \end{bmatrix} \begin{bmatrix} x_1(k) \\ x_2(k) \end{bmatrix} \tag{19}$$

if the sampling period is changed from T to T', then the discrete transfer funcion:

$$W(q) = \frac{b_1 q + b_2}{q^2 + a_1 q + a_2} \tag{20}$$

is transformed into:

$$W'(q) = \frac{b'_1 q + b'_2}{q^2 + a'_1 q + a'_2} \tag{21}$$

where

$$a'_1 = ha_1 - 2(1 - h)$$
$$a'_2 = (1 - h)^2 - h(1 - h)a_1 + h^2 a_2$$
$$b'_1 = \frac{h(2 - h)b_1 + h^2(a_2 - a_1)b_1 - 2h(1 - h)b_2}{1 - a_1 + a_2} \tag{22}$$
$$b'_2 = (h - 1)b'_1 + \frac{h(h(1 - h)2a_2 b_1 + h(2 - h)(1 - a_1)b_2 + h^2 a_2)b_2)}{1 - a_1 + a_2}$$

In this way, when the sampling period is changed, the parameters of the second order process are modified according to (22). These values are taken as initial estimates in the parameter estimator and permit to accelerate its convergence.

5 Simulations

In this section results obtained through simulation will be presented in order to show in different situations the properties of the adjustment of the sampling period described. In all the simulations the process that must be identified or controlled is supposed to have the transfer function:

$$G(s) = \frac{w_0^2}{s^2 + 2\zeta w_0 s + w_0^2} \tag{23}$$

where ζ is the damping coefficient and w_0 the natural frequency. The corresponding input output difference equation is:

$$y(t) + a_1 y(t-1) + a_2 y(t-2) = b_1 u(t-1) + b_2 u(t-2) + e(t) \tag{24}$$

where the time unit has been chosen as the sampling period T and e(t) represents a disturbance sequence of unspecified character. With the notation:

$$\varphi(t) = \begin{bmatrix} -y(t-1) & -y(t-2) & u(t-1) & u(t-2) \end{bmatrix}^T$$
$$\theta = \begin{bmatrix} a_1 & a_2 & b_1 & b_2 \end{bmatrix}^T \tag{25}$$

model (24) can be rewritten as:

$$y(t) = \theta^T \varphi(t) + e(t) \tag{26}$$

5.1 Simulation for a parameter estimation case

In this case the parameters of model (24) are unknown. The recursive least squares algorithm has been used for identification of the parameter vector θ in (26). It is described by the following equations:

$$\hat{\theta}(t) = \hat{\theta}(t-1) + K(t) e(t)$$
$$e(t) = \varphi^T(t) \hat{\theta}(t-1)$$
$$P(t) = \frac{1}{\lambda} \left[P(t-1) - \frac{P(t-1) j(t) \varphi^T(t) P(t-1)}{1 + \varphi^T(t) P(t-1) \varphi(t)} \right] \tag{27}$$
$$K(t) = P(t) \varphi(t)$$

where λ is the forgetting factor

We assume that the purpose is to identify the parameters of the process (23). However the right sampling period T cannot be selected , since the values of the parameters are not exactly known. Such situation may also arise when we change from an indetified process to another with the same model structure but with different parameter values.The automatic adjustment procedure of the sampling period can be summarized in the following steps:

Step 1 Initialize the sampling period T using apriori knowledge about the process.

Step 2 Estimate the parameter vector θ in (24) recursively using (27).

Step 3 Replace the coefficients in equation (24) by the estimates obtained in Step. 2 and estimate the rise time T_r calculating it directly from the step response of the process model (20).

Step 4 Calculate the new value of the sampling period T' by (8) and to verify if the condition (9) is fulfilled

Step 5 If the sampling period is changed, then calculate new initial values for parameter estimates with the approximate procedure (22). Otherwise go to Step 2.

Repeat Steps 3-5 every N sampling periods.

Example System (23) with parameters $\zeta=0.7$ and $w=1$ is considered. The rise time for this system is 2.14 sec and according to (8) a suitable sampling period is T = 0.5 sec. The behaviour of the estimator when the sampling period is adjusted is shown in two different situations: a) Initial sampling period smaller than the value recommended by the criterion (0.3 sec). b) Initial sampling period greater than the value recommended by the criterion (0.9 sec). The step response of the sampled system with T = 0.3 is shown in Fig. 2.

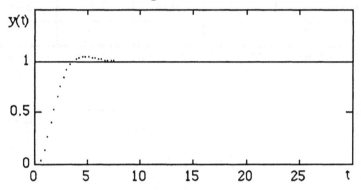

Fig. 2 Step response of the system (24) - $\zeta=0.7$ and $w=1$, with T=0.3 -

In both cases a right estimate of the rise time and of the adjustment of the sampling period to the value recommended is achieved(see Table 1)

Case	Rise Time T_r		Sampling Period T	
	correct	estimated	initial	adjusted
1	2.14	2.1	0.3	0.525
2	2.14	2.1	0.9	0.450

Table 1. The rise time estimates and the adjustment of the sampling period

In Tables 2 and 3 the parameter estimates obtained for cases a) and b) are shown. The estimates appearing under the term "approximated" have been obtained from the knowledge of h and by applying (26) to the initial values; the estimates appearing under the term "correct" are the exact value of the parameter estimates when the sampling period is modified to the new value. It is observed that the approximate values that are obtained with the procedure proposed for the parameter estimates are very close to the right values, in spite of the fact that the change in the sampling period is superior to 50% in both cases.

	a_1	a_2	b_1	b_2
initial	1.58	-0.66	0.039	0.034
approximated	1.27	-0.50	0.110	0.100
correct	1.28	-0.47	0.108	0.080

Table 2 Parameter estimates when the sampling period changes from 0.3 to 0.5

	a_1	a_2	b_1	b_2
initial	0.85	-0.28	0.26	0.170
approximated	1.42	-0.53	0.086	0.064
correct	1.38	-0.53	0.081	0.066

Table 3 Parameter estimates when the sampling period changes from 0.9 to 0.45

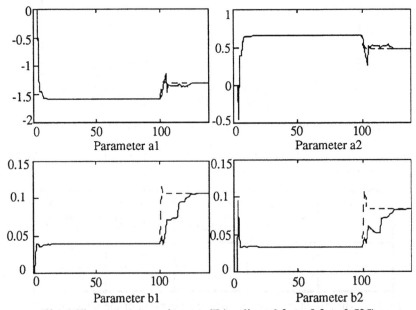

Fig. 3 The parameter estimates. (T is adjusted from 0.3 to 0.525)

The behaviour of the estimator adjusting the sampling period is shown for the case when the initial value T is smaller (0.3 sec) than the correct value. The estimates of parameters a_1, a_2, b_1, and b_2 are shown in Fig. 3, where the estimates of the parameters have been superimposed in the two following cases:
a) without update of the estimates when the sampling period is adjusted (continuous line). b) with update of the estimates by using the approximate procedure studied in

seccion 4 (discontinuous line). A much more rapid convergence to the right values is observed in case b)

5.2 Simulation for a control case

In the previous section the automatic adjustment algorithm of the sampling period was applied to the problem of parameter identification. It is also possible to apply the same algorithm to the problem of control.

As an example, we have considered the case of the process with the feedback regulator (see Fig. 1), which can be described by the transfer function (23). The system is initialized with a sampling period $T=0.5$

Fig.4 shows what happens when the process parameters are abruptly changed from $\zeta=0.7$, w=1 to $\zeta=0.5$, w=2 in the sampling instant n=100, while the sampling period T maintains the same value.

The overshoot and the response time are far from the right values. This example shows that the choice of the sampling period influences the high order process dynamics. If the parameters of the process change, T has to be modified.

The automatic adjustment algorithm of the sampling period (from $T=0.5$ to $T=0.2$) starts in the sampling instant n=200. The second part of Fig. 4 shows that the desired results are obtained.

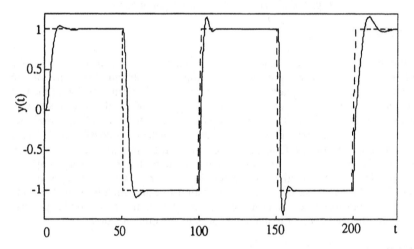

Fig. 4 Simulation with the automatic adjustment of the sampling period

In Fig. 5 the parameter estimate of the discrete model ($T=0.5$) are shown. In n=100 there is a sudden change in the process parameters (from $\zeta=0.7$, w=1 to $\zeta=0.5$, w=2). In n=200 the sampling period is changed (from $T=0.5$ to $T=0.2$). The estimates of the parameters have been superimposed in the the two following cases:

a) without update of the estimates when the sampling period is adjusted (continuous line)

b) with update of the estimates by using the approximate procedure studied in seccion 4 (discontinuous line).

Again a much more rapid convergence to the right values is observed in case b) .

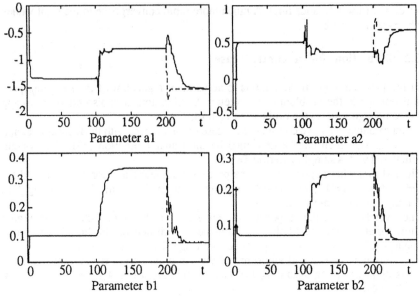

Fig. 5 Estimates of the process parameters

6 Conclusions

The sampling period can be considered an additional parameter in the design of a controller. In this paper we present an automatic adjustment procedure of the sampling period that tries to maintain constant the number of samples during the rise time. If the sampling period is modified the parameters of the discrete model will obviously change. However, from the knowledge of the initial sampling period, the final sampling period and the initial values of the parameters it is possible to deduce a good approximation for the final parameters once the sampling period has been adjusted.In this sense we propose a new approximate method that is very simple to implement. The results obtained in simulation confirm the possibilities of the method. These ideas are susceptible of application to other situations by using the same strategy. For example, the criterion of a constant number of samples per rise time could be changed in order to maintain constant any other relation.

References

1. L.W. Gardenhire: Selection of Sample Rates. ISA Journal, April, 59-64 (1964) .
2. K.J Åström, .B.W. Wittenmark: Computer Controlled System:Theory and Design. Prentice Hall , 2nd Edition (1990).
3. S.M. Melzer, B.C. Kuo: Sampling Period Sensitivity of the Optimal Sampled Data Linear Regulator. Automatica, 7, 367-370 (1971).
4. J.F. MacGregor: Optimal choice of the sampling interval for discrete process control.Technometrics, 18, 2, 151-160 (1976).
5. T. Söderström, B. Lennartsson: On linear optimal control with infrequent output sampling. In: J.E. Marshall et al (eds): Third IMA Conference on Control Theory, New York, Academic Press (1981).

FSM Shift Register Realization for Improved Testability

Thomas Mueller-Wipperfuerth, Josef Scharinger, Franz Pichler

Johannes Kepler University Linz
Austria

Abstract. An approach for the FSM state assignment problem based on an enhanced algorithm for shift register realizations is presented. A FSM is transformed into a generalized shift register structure, where memory elements are put together to one or more shift registers. The proposed procedure considers a final scan path architecture of FSM memory cells already during state assignment, reducing the hardware overhead for testability purposes by exploiting special optimization potentials. Theoretically founded criteria are used to cancel the computation at an early stage if no fruitful shiftregister realization is possible. Experimental results for two-level and multi-level implementations are given for MCNC benchmark machines.

1 Introduction

In the field of controller synthesis many conflicting goals are addressed: decomposed structures achieve flexibility in floorplanning but bring about additional area for wiring, an optimization of total area contradicts extra hardware for testing purposes, etc. The problem of assigning binary codes to symbolic states is mainly driven by area optimization efforts [5, 17]. However, we investigate a state assignment approach for testability purposes. Conventional state assignment procedures do not consider scan paths of controller memory cells. We target at a state assignment which allows single memory elements to be put together to shift registers to be used within a scan path.

Testing sequential circuits is hard because memory states are neither directly controllable nor directly observable. In case that no additional test points are inserted input sequences for synchronization, homing, and diagnosing have to be found to detect faults and to propagate them to the systems output lines. A sufficient fault coverage is mostly not achievable by a mere application of I/O experiments if the system under test has not been designed for testability.

Applying the partition approach for testability purposes separates circuit functionality to improve test characteristics. Known techniques are general decomposition [1] algebraic decomposition [9, 11], or state assignment procedures achieving decoupled memory elements [13], etc.

Another test strategy is introducing additional hardware to obtain a high degree of observability and controllability of states. A popular method for enhancing the testability relies on scan path techniques, where memory elements are lined up to shift registers. In test mode arbitrary test pattern may be shifted into memory and the systems state after one clock period is shifted out. The test generation problem to be solved is finding test patterns for pure combinational logic.

Fig. 1 illustrates the general structure of a shift register realization of a FSM addressed in this paper [14]. Here the hardware overhead of scan path architectures is reduced by

- a decreased number of output lines of combinational logic implementing state transition functionality and
- a smaller wiring area as not all memory cells have to be loaded from the combinational logic.

Even if a complete shift register realization is not possible, the shift registers of a partial realization facilitate the construction of a scan path. Depending on number and length of shift registers complete and partial realizations may be defined:

- *type1*: one shift register SR of length L;
- *type2*: a collection of n different shift registers SR_j, $L_j > 1$, $1 < j \leq n$;
- *type3*: a collection of u shift registers SR_j, $L_j > 1$, $1 < j \leq u$ and v additional single memory elements.

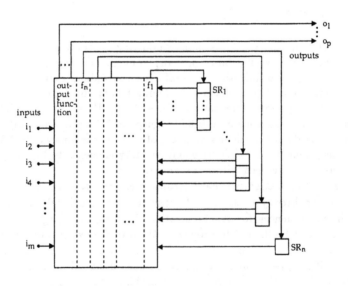

Fig. 1. General Shift Register Structure

Shift register realizations are in many cases not achievable because of the existence of special kinds of states and state transitions: Reset states (reachable from any FSM state with one single state transition) and FSM states which are not changed by some input prevent good shift register structures. We extract those transitions and do the shift register realization for the remaining (essential) FSM. If a good realization is obtained the extracted functionality is reinserted by additional hardware. Here we make use of special properties of elements of the target library elements, especially of reset- and shift-enable functionality. We evaluate costs for a better testability for two-level and multi-level implementations in terms of area and total number of memory elements. Logic minimization of the obtained shift register realizations is performed using ESPRESSO [3] and MIS [4]. We do not take into account that shift register structures reduce the number of connections from combinational logic to memory elements. A final chip implementation is necessary to consider and to evaluate these wiring aspects; but this is subject of future work. The remainder of this paper is organized as follows: In Section 2 we discuss previous and related work. Section 3 gives some preliminaries on the topic, in Section 4 key aspects of the underlying theory are presented. The realization algorithm is sketched in Section 5, followed by a discussion of extensions of the approach in Section 6. Finally some results for two-level and multilevel implementations of shift register realizations of MCNC benchmark FSMs and of machines synthesized by Callas [6] are presented in Section 7.

2 Previous and Related Work

Shift register realizations of finite state machines have been intensively studied in the sixties and seventies [7, 10, 16]. We focus on the original work of [2] and use his theoretical results for the application on benchmark FSMs [19]. We present not only an implementation of the original algorithm for shift register realizations but introduce extensions which broadens the algorithms applicability.

Our approach differs from others in using theoretical founded criteria to evaluate whether a machine is suited for a shift register realization prior to the actual state assignment procedure. We determine the maximal obtainable shift register length and the necessary number of single memory elements which cannot be coupled to shift registers and avoid computational intensive but unsuccessful state assignment efforts for shift register realizations. The algorithm for shift register realization was developed, tested, and applied using CAST.FSM*, a software implementation of a framework for the application of various FSM methods, introduced in [15]. CAST.FSM* was successfully applied for the purpose of this paper. It reduced the implementation work for the shift register realization algorithm because of its predefined building blocks for set-, function-, and FSM-manipulation and supported extending the applicability of the algorithm and carrying out benchmark experiments.

3 Preliminaries

A finite state machine M is a quintupel $(I, O, S, \delta, \lambda)$, where I, O, S are finite sets of inputs, outputs, and states, respectively, where δ is the state transition function which maps $S \times I$ into S, and where λ is the output function mapping $S \times I$ into O.

We define subsets nr_δ and ns_δ:

$$ns_\delta := \{ \, s \in S \mid \forall (i \in I) : \delta(s, i) \text{ undefined } \},$$

$$nr_\delta := \{ \, s \in S \mid \forall (s' \in S) \, \not\exists (i \in I) : \delta(s', i) = s \, \}.$$

Elements of nr_δ and ns_δ are not reachable states and states for which no successor state is defined by δ, respectively. The state transition function δ is said to be *(1,3)-complete* iff $nr_\delta = \emptyset$ and $ns_\delta = \emptyset$.

A partition π on S is a set of disjoint non-empty subsets $b_i \subset S$, whose set-union is S, written $\pi = \{b_1, b_2, \ldots\}$. The b_i 's are called *blocks* of π. We will denote the number of blocks of π by $|\pi|$ and the size of the largest block by $\|\pi\|$. If $|\pi| = 1$ we call π the *1-partition* ($\underline{1}$) with all set elements in the same block; π is denoted the *0-partition* ($\underline{0}$) if $|\pi| = |S|$ where each set element is in a separate block. Given a set P of partitions $P = \{\pi_1, \pi_2, \ldots, \pi_l\}$ we define the *partition product*

$$PP(P) := \pi_1 * \pi_2 * \ldots * \pi_l.$$

An *ordered partition* τ on S is a sequence of disjoint subsets b_i whose set union is S, written $\tau = (b_1 b_2 \ldots)$. The b_i 's are called blocks of τ, too, but each b_i has an unique index $1 \leq i \leq |\tau|$, by which it may be identified. The multiplication of two ordered partitions creates an ordered partition with a possibly different number of blocks and allows in this case an arbitrary block ordering. The partition product PP applied to a set of ordered partitions results in an (arbitrarily ordered) ordered partition.

Any set P of partitions π_i ($1 \leq i \leq n$) on S whose partition product equals $\underline{0}$ can be used for state encoding purposes. If $|\pi_i| = 2$ for all π_i, a binary state encoding is induced for the elements of S. In the following sections we will derive sequences of partitions which will be used for binary state encodings resulting in shift register realizations of the underlying machine M.

4 Shift Register Realization of FSMs

4.1 Shift Register Partitions

For the construction of coding partitions for shift register realizations (SR-realizations) two operations m^+ and m^- on ordered partitions are essential: Let $\mathcal{P}(S)$ be the power set of S and $\mathcal{T}$ be the set of ordered partitions on S. We define a block transition function δ':

$$\delta' : \mathcal{P}(S) \to \mathcal{P}(S) \, , \; \delta'(b) := \{ \, s' : \delta(s, i) = s', \exists (s \in b), \exists (i \in I) \, \},$$

and a block predecessor function δ'^{-1}:

$$\delta'^{-1} : \mathcal{P}(S) \to \mathcal{P}(S) , \quad \delta'^{-1}(b) := \{ s : \delta(s, i) = s', \exists(s' \in b), \exists(i \in I) \}.$$

m^+ and m^- will be defined by means of a function $m : T \to T$:

$$m((b_1 \ldots b_n)) := \begin{cases} (b_1 \ldots b_n), & when \quad b_i \cap b_j = \emptyset, 1 \leq i, j \leq n, i \neq j \\ join(b_1 \ldots b_n), & otherwise \end{cases}$$

m is the identity operation if the argument to m is a sequence of pairwise disjoint blocks; if two or more argument blocks have non empty intersections m joins them into one block, such that the result of m is again a sequence of disjoint blocks. In the latter case the number of blocks is reduced and hence a new ordering of blocks evolves.

Now m^+ and m^- are easily defined:

$$m^+ : T \to T, m^+(\tau) = m^+((b_1 b_2 \ldots)) = m((\delta'(b_1)\delta'(b_2)\ldots)),$$

$$m^- : T \to T, m^-(\tau) = m^-((b_1 b_2 \ldots)) = m((\delta'^{-1}(b_1)\delta'^{-1}(b_2)\ldots)).$$

In [12] *comapping* partitions were introduced, which can be easily rewritten now by means of m^+ and m^-: An ordered partition τ_1 comaps into an ordered partition τ_2, iff $\tau_2 = m^+(\tau_1)$ and $\tau_1 = m^-(\tau_2)$. A sequence of k ordered partitions is denoted a comapping chain of length k iff τ_i comaps into τ_{i+1} for all $1 \leq i \leq k$. In this paper a comapping chain of length k is denoted DPS, *(Shift Register) Defining Partition Sequence* of length k.

Let D be a set of DPSs. We define the product of all partition products of the elements of D:

$$\Theta(D) := PP(DPS_1) * PP(DPS_2) * \ldots * PP(DPS_n), DPS_i \in D.$$

D induces a shift register realization if $\Theta(D)$ equals $\underline{0}$ (see [2] for proofs and details). If $|D| = 1$ a *type 1* SR is realized, otherwise a *type 2* realization is induced by D. If $\Theta(D) = \underline{0}$ cannot be achieved because suitable DPSs do not exist or cannot be found, the use of additional single memory elements accomplishes valid state encodings resulting in a *type 3* shift register structure. A DPS where $|\tau_i| = 2, 1 \leq i \leq n$ is called a *binary DPS*.

4.2 V-Graphs

V-Graphs provide criteria to decide whether further calculation efforts will result in a good shift register structure and give initial sets of DPSs from which coding partitions may be derived. A V-Graph is constructed by means of m^+ and m^- as shown in Fig. 2. According to [2] each row of length k forms a DPS of length k. It was also shown in [2] that each row of the V-Graph is minimal in the sense, that no DPS of length k exists with a smaller partition product Θ of its partitions.

- The first criterion for successful realizations is the minimum number DM_2 of discrete binary memory elements (memory elements which cannot be chained together to shift registers). DM_2 determines whether a good shift register realization is feasible and is defined by

$$DM_{val} \geq \log_{val}(\|\tau_{2,1} * \tau_{2,2}\|).$$

A proof is given in [2].

- If the $\underline{1}$-Partition terminates the V-Graph, the maximum possible shift register length LL_δ can be extracted form the Graph. It is defined by the number of ordered partitions of the row directly below $\underline{1}$. If the V-Graph is degenerated to $\underline{1}$ directly above $\underline{0}$ the maximal shift register length LL_δ equals 1, which implies that no nontrivial shift register realization is possible and the procedure stops.

If the $\underline{1}$-Partition does not terminate the V-Graph, a value for LL_δ can not be derived from the V-Graph. Examples show that in this case SR-realizations of high quality are very likely to exist. Since no upper bound is predictable the presented algorithm tries repeatedly to lengthen $DPSs$ (see [18]).

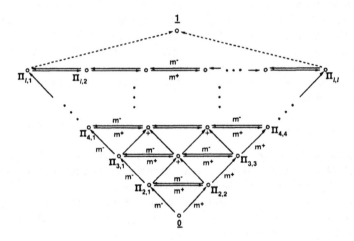

Fig. 2. Typical Structure of a V-Graph

5 SR-Realization Algorithm

1. FSM-Analysis

The SR-realization starts with checking the state transition function for (1,3)-completeness and identifying nr_δ and ns_δ. For now, only (1,3)-complete state transition functions are considered; the treatment of non empty sets ns_δ and nr_δ is discusses in section 6.

A further analysis of the state transition function achieves a lower bound for the number of necessary shift registers. State transitions which map a state s into itself ($\delta(s, i) = s$, for some $i \in I$) will enforce unchanged contents of the shift registers during FSM operation. For binary encodings only two codes ($00 \ldots 0$ and $11 \ldots 1$) allow this special behavior. If more than two states are mapped into themselves the number of necessary shift registers increases unless an additional shift/non-shift mechanism as proposed in [8] is applied. If there is a reset state defined for M, i.e., if there is a state which is reachable from every state, the shift register realization of M will not be successful. An example is the MCNC benchmark machine *scf* for which no shift register state assignment can be found. However, if the reset-functionality is separated, the essential behavior of *scf* has an acceptable shift register realization. The presented algorithm realizes *scf* (without reset) with two shift registers of length 3 and two discrete memory elements.

2. Computation of Basis Partitions

After constructing the V-Graph LL_δ and DM_2 are already known at this early stage of the algorithm. The SR-calculation may be canceled if DM_2 prevents a good SR-Realization or if LL_δ is too small.

Rows of the V-Graph are the building blocks of the SR-Realization process; they form basic DPSs. If there exists a row of binary partitions, whose partition product equals $\underline{0}$ no further calculation is necessary and a complete binary SR-realization is found. The partitions of this DPS may directly be used for state encoding resulting in a *type 1* SR-realization.

Table 1 shows a description of the V-Graph of *planet*. Each row characterizes one basic DPS: k denotes the number of partitions, *val* describes the number of blocks, and *quality* states what this DPS would contribute to a SR-realization. Quality (lower is better) equals the size of the largest block of the partition product ($\|PP(.)\|$).

Planet has a binary basis DPS of length 8, but its quality is only 35. An 8 bit shift register may be defined, but a block of 35 states remains to be split up. One single bit can achieve a quality of 24 (48/2) and thus it makes no sense to take this 8 bit shift register into consideration. The DPS with best quality (2) requires multi-valued (25-valued) memory elements and hence this DPS is also not directly suited for state encoding purposes.

3. Combination of Basis Partitions

This step of the algorithm combines partition blocks of multi-valued DPSs to a set of binary DPSs. In step 4 promising candidates will be selected form these DPSs. [16] presented an approach where an intertwined combination and selection is done and an optimal solution is worked out. However, as partitions of the basic DPSs in row 2 of *planet's* V-Graph contain 25 blocks, there are 2^{25} possibilities to combine these blocks into two blocks. For *scf* 2^{88} combinations are possible, which are hardly computable within a finite amount of time.

Table 1. Description of basis DPSs of *planet*

row	k	val	quality
8	8	2	35
7	7	3	35
...	...	...	...
4	4	7	31
3	3	12	27
2	2	25	2
1	1	48	1

In our approach we separate combination and selection of DPS into two separate steps: The combination step generates N ($100 \leq N \leq 500$) binary DPSs. Although N is very small compared to the number of possible combinations, the selection step always found SR-realizations which were optimal or near to the (theoretical possible) optimum for the investigated machines.

4. **Selection of binary DPSs**

Input data for this step are the binary DPSs created in the previous step. Each basic DPS of the V-Graph has minimal $\|PP(.)\|$ for a certain length. The selection mechanism is best explained by means of a simple example: Table 2 describes a V-Graph of a Machine M_1 with 121 states. Row 2 tells us that in each SR-realization of M_1 $log_2(2)$ binary discrete memory elements are absolutely necessary. Row 4 denotes a binary DPS whose partition blocks need not be combined, but it leaves 115 (out of 121) states within one block. Row 3 suggests a shift register of length 3 but there will always be 7 states within the largest block. This block has to be split up by additional shift registers. As the basis partitions of row 3 contain these 7 states in one common block all partitions derived from it contain this block, too.

Hence this block has to be split up by DPSs of length 2 or length 1. A 3-2-1-1 realization (3 bits are necessary to encode 7 states and one single memory element is absolutely necessary, as DM_2 equals 1) needs less memory elements than a 3-2-2-1 realization. So an optimal SR-realization has been theoretically identified and is used to evaluate the quality of obtained SR-realizations.

The algorithm repeatedly selects binary DPS into an initially empty set D. The selection is performed using a greedy strategy, as always that DPS of a given length is selected and inserted into D which achieves the smallest total partition product $\Theta(D)$. The algorithm stops when $\Theta(D) = \underline{0}$. The selection mechanism may be advised to prefer a small total amount of memory elements or to prefer longer shift registers, even if the total number of memory cells increases. If the quality of the achieved shift register structure is acceptable the algorithm continues with step 5, otherwise it continues with step 3 using an increased N.

Table 2. Description of typical DPSs

row	k	val	quality
4	4	2	115
3	3	17	7
2	2	20	2
1	1	121	1

5. **State-assignment**
   The selected set of DPSs fixes an unique state coding and saves efforts for state assignment.
6. **Logic optimization**
   This step is performed using ESPRESSO [3] followed by MIS [4] for multi-level implementations.

6 Extensions of the Algorithm

For non-(1,3)-complete state transition functions ($nr_\delta = \emptyset$ or $ns_\delta = \emptyset$) the application of m^+ and m^- do not necessarily yield partitions on S, but only sets of disjoint blocks, whose set union may not equal S. For a non-empty set nr_δ (ns_δ), m^+ (m^-) results in a set of blocks, where the elements of nr_δ (ns_δ) are not included. Hence this set of blocks is not suited for state encoding purposes. The straightforward application of m^+ (m^-) for the construction of V-Graphs does not produce basic DPSs any longer.

In [2] it is proposed to achieve (1,3)-completeness by defining additional state transitions for states $s \in nr_\delta \cup ns_\delta$. However, these transitions will restrict possible SR-realizations. Therefore, we suggest modified operations $m^{+\prime}$ and $m^{-\prime}$ which allow the manipulation of non-(1,3)-complete machines without the need of changing δ. In this paper the modified operations are discussed only informally, a thorough definition is given in [18].

The basic idea is that $m^{+\prime}$ ($m^{-\prime}$) must not drop states of nr_δ (ns_δ). Thus $m^{+\prime}$ ($m^{-\prime}$) uses the operation m^+ (m^-) and reinserts dropped states to achieve again partitions on S. The problem that remains is to select a partition block to insert a state $s \in nr_\delta \cup ns_\delta$. We make use of m^+ and m^- being inverse operations on the elements of a DPS: A state $s \in nr_\delta$ ($s \notin nr_\delta \cap ns_\delta$) which is not included in $m^+(\underline{0})$ will not be dropped by $m^-(\underline{0})$. Hence m^- determines where to insert s for $m^{+\prime}$. m^+ is used for the definition of $m^{-\prime}$. A thorough discussion is given in [18].

The application of $m^{+\prime}$ ($m^{-\prime}$) to construct a V-Graph may result in a non symmetric graph (see Fig. 3). However, the DPSs are suited for the combination step and in consequence induce valid state assignments for shift register realizations. In [18] another modification of $m^{+\prime}$ ($m^{-\prime}$) is discussed which preserves symmetric V-Graphs, too.

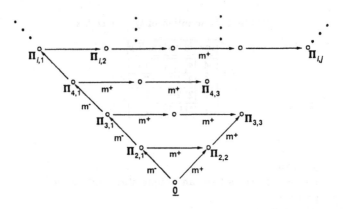

Fig. 3. V-Graph of a non-(1,3)-complete FSM

A further extension of our approach allows partial shift register realizations of machines whose state transition functions are not suited for successful realizations. This extension of the algorithm identifies and extracts state transitions that implement a reset-functionality and state transitions which map states into themselves. The remaining (essential) state-behavior of M is input for the shift register realization algorithm. If a shift register structure is successfully generated, the accomplished state codes are substituted into the original machine description, which now implements the original behavior but the extracted state-transitions. Fig. 4 shows an additional output, which is used to cause a reset of shift registers memory contents for a full implementation of M.

7 Experimental Results

Though the main goal of the proposed state assignment algorithm is the realization of shift register structures the achieved area compares favorably well to area optimizing results.

Table 3 summarizes successful applications of the presented algorithm and its extensions. The minimal number of bits for state encoding are shown in column $nr - of - b$. $SR - structure$ illustrates the achieved shift register structure and $Runtime$ shows the algorithm's runtime on a Sun ELC, running Lucid CommonLisp. The results of two-level implementations are given in the last two columns. $Jedi - Esp$ shows PLA area after state assignment using JEDI, followed by ESPRESSO 2.3 and $SR - Esp$ describes area after shift register realization and logic minimization.

As stated in Section 5 we illustrate that a reset-functionality and state transitions which map states into themselves prevent good shift register realizations. The machines scf^*, $kirkman^*$, $s1488^*$, and $s1494^*$ are marked because their reset-functionality is separated and not considered here. For $s208$ and $s420$ a

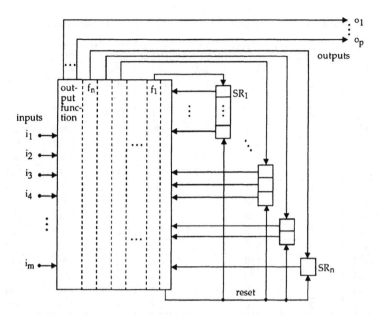

Fig. 4. Shift Register Structure with separated reset-circuitry

reset state r is defined, but there exists another state r', which also shows reset-functionality. Hence, in $s208^{**}$ and $s420^{**}$ not the transitions into the specified reset state r but the state transitions into r' have been excluded. $train11+$ is generated from $train11$ by extracting state transitions mapping a state s into itself. The original machines had no shift register structure at all, the results for the modified machines are shown in Table 3. The proposed realization algorithm has been applied to the modified machines and all area results refer to the essential state transition functions.

Six machines of Table 3 have non-(1,3)-complete state transition functions, but could be manipulated using the $m^{+'}$ ($m^{-'}$) operators presented above.

In Table 4 some results of multi-level implementations (MisII 2.2) are shown. The achieved shift register structures are compared to a state assignment calculated with JEDI. *Lits* are given in factored form prior and after library mapping using *mcnc.genlib* and *script.mcnc*.

The results can be summarized as follows:

- About 25-30 % of the benchmark machines allow a shift register realization. The quality of the achievable shift register structure is determined at a very early stage of the algorithm.
- If a given FSM permits a shift register realization the proposed state assignment procedure compares very well with standard state assignment tools. For 10 of 17 (12 of 17) machines our procedure achieved better or equal results for two-level-implementations (multi-level-implementations).

Table 3. Experimental Results for Two-level Implementations

FSM	$nr - of - b$	LL_6/DM_2	SR-structure	Runtime (sec.)	Jedi-Esp (Area)	SR-Esp (Area)	AreaChange (%)
ex4	4	5/1	3-1	9.0	561	660	118
planet	6	8/1	2-2-1-1	83.2	4947	4845	98
shiftreg	3	3/0	3	1.1	72	48	67
tav	2	-/0	2	1.0	198	180	91
scf*	7	6/1	3-3-1-1	146.8	20567	18850	92
kirkman*	4	15/0	4	90.3	3906	4320	110
train+	4	-/0	3-2	5.8	153	140	92
s1488*	6	8/1	2-2-1-1	26.7	4823	5141	107
s1494*	6	8/1	2-2-1-1	39.3	5088	5300	104
s208**	5	-/0	5	39.8	819	819	100
s420**	5	-/0	5	18.2	1155	1155	100
tma	5	3/2	2-1-1-1	9.3	1155	1190	103
fsm01	5	8/1	2-2-1	16.8	1176	1176	100
fsm02	2	3/0	3	0.4	147	168	114
fsm09	2	2/0	2	0.4	64	64	100
fsm13	4	2/2	2-1-1	3.5	810	810	100
fsm15	2	2/0	2	0.6	612	561	92

Table 4. Experimental Results for Multi-level Implementations

FSM	Jedi − MisII		SR − MisII		Lits(fac)Change %/%
	Lits(fac)	Area	Lits(fac)	Area	
ex4	66/95	95	69/105	106	105/109
planet	564/886	892	555/842	853	98/95
shiftreg	9/11	11	2/2	2	22/18
tav	26/35	35	22/31	31	85/89
scf*	829/1316	1348	887/1205	1189	107/92
kirkman*	154/249	250	170/261	262	110/105
train+	21/30	30	7/9	9	33/30
s1488*	543/820	834	523/807	815	96/98
s208**	49/73	74	42/65	65	86/89
s420**	50/72	72	48/72	73	96/100
fsm01	155/248	251	155/223	225	100/90
fsm02	28/36	37	18/25	25	64/69
fsm09	12/17	17	9/11	11	75/65
fsm13	89/127	128	84/118	118	94/93
fsm15	34/48	50	33/45	46	97/94

- The shift-functionality of memory cells increases final FSM implementation area. But the comparison of memory area has to take into account the area which is necessary for a conventional scan path. Because memory cells with scan path functionality are not included in the benchmark library no concrete numbers of implementation area are available.
- The results do not consider the reduced number of connections between combinational logic and memory which will cause further improvements of final implementation area.

8 Conclusion

Our results confirm the value of theory founded estimations for shift register realizations of finite state machines. An algorithm has been presented to obtain the optimal partial shift register structure for arbitrary FSMs. Although the main goal of the proposed state assignment strategy is supporting Design For Testability, two-level and multi-level implementations of shift register realizations compare well with results from conventional area optimizing state assignment tools.

References

1. P. Ashar, S. Devadas, and A.R. Newton. Optimum and heuristic algorithms for an approach to finite state machine decomposition. *IEEE Trans. on CAD*, 10(3):1062–1081, 1991.
2. K. Boehling. Zur Theorie der Schieberegister-Realisierung von Schaltwerken. Technical Report 1, Gesellschaft fuer Mathematik und Datenverarbeitung, Bonn, 1968.
3. R. Brayton, G. Hachtel, C. McMullen, and A. Sangiovanni-Vincentelli. *Logic Minimization Algorithm for VLSI Synthesis*. Kluwer Academic Publishers, Dordrecht, 1984.
4. R. Brayton, R. Rudell, and A. Sangiovanni-Vincentelli A. Wang. Mis: A multilevel logic optimization system. *IEEE Trans. on CAD*, CAD-6(6):1062–1081, 1987.
5. S. Devadas and A.R. Newton. Exact algorithms for output encoding, state assignment, and four-level boolean optimization. *Trans. on CAD*, 10:13–27, 1991.
6. P. Duzy, H. Kraemer, M. Pilsl, W. Rosenstiel, and T. Wecker. Callas - conversion of algorithms to library adaptable structures. In *Proc. of the VLSI 89 Conference*, Munich, 1989.
7. A. Friedman. Feedback in synchronous sequential switching circuits. *Trans. Electron. Comput.*, EC-15:354–367, 1966.
8. M. Geiger and T. Mueller-Wipperfuerth. FSM decomposition revisited: Algebraic structure theory applied to MCNC benchmark FSMs. In *Proc. 28th Design Automation Conference*, pages 182–185, San Francisco, 1991.
9. R. Stearns J. Hartmanis. *Algebraic Structure Theory of Sequential Machines*. Prentice Hall, 1966.
10. R. Martin. Studies in feedback-shift-register synthesis of sequential macines. Technical Report 50, MIT, Cambridge, Massachusetts, 1969.

11. T. Mueller-Wipperfuerth and M. Geiger. Algebraic Decomposition of MCNC Benchmark FSMs for Logic Synthesis. In *EURO ASIC '91*, pages 146–151, Paris, 1991.

12. A.J. Nichols. Minimal shift register realizations of sequential machines. *Trans. Electron. Comput.*, EC-14:688–700, 1965.

13. F. Pichler. *Mathematische Systemtheorie*. DeGruyter, Berlin, 1975.

14. F. Pichler. Schieberegister-Realisierung von Endlichen Automaten. Technical Report FSM-DFT PI 86-2, University of Linz, Systems Science, Linz, 1986.

15. F. Pichler and H. Schwaertzel. *CAST - Computerunterstützte Systemtheorie*. Springer, Berlin, 1990.

16. W.D. Roome. Algorithms for multiple shift register realizations of sequential machines. *Trans. on. Computers*, CC-22(10):933–943, 1973.

17. G. Saucier, C. Duff, and F. Poirot. State assignment using a new embedding method based on an intersecting cube theory. In *Proc. of the 26th Design Automation Conference*, pages 321–326, 1989.

18. J. Scharinger. Implementierung der Schieberegister-Realisierung fuer CAST-Systeme. Master's thesis, University of Linz, Austria, 1991.

19. S. Yang. *Logic Synthesis and Oprimization Benchmarks User Guide*. MCNC, 1991. Version 3.0.

Cluster-Based Modelling of Processes with Unknown Qualitative Variables

Raimo Ylinen

Helsinki University of Technology, Control Engineering Laboratory
SF-02150 Espoo, Finland

Abstract

A new method for modelling of complex processes containing unknown qualitative variables is presented. The method is based on clustering the input-output-pairs of the model into a finite number of clusters each representing a different value of unknown qualitative variables. The clustering algorithm is derived from the assumption that each cluster consists of realisations of a normally distributed random vector. The method has been tested by simulated process model and real data from mineral processes.

1 Introduction

The conventional methods for automatic process control are based on process model and sufficient number of measurements. In practice, however, there are often variables which cannot be measured nor even be presented quantitatively. The relationships between the variables can also be such that they cannot be deduced from known laws of nature.

There exist three types of variables: 1) quantitative variables, the values of which can be presented using some measure, 2) qualitative variables, the values of which can be ordered in some natural way, 3) qualitative variables, the values of which cannot even be ordered.

In the model suitable for control design the variables have to be divided into input- and output-variables, the relationships of which are cause-effect-relationships between the input- variables and the output-variables or possibly other relationships between the output-variables.

If the model contains only variables of types 1 or 2, it is at least in principle possible to present the relationships using such general rules or conditions, which are satisfied by more than single values of variables. But in case of variables of type 3 almost the only possibility is to give a list of all values satisfying the relationship.

It is obvious that the models consisting only of lists of input- output-pairs cannot be used e.g. for construction of explicit feedback control loops. Instead they are suitable for design of expert systems, in particular if the number of pairs can be reduced. In this paper a method for this purpose will be presented.

The method is based on clustering the measured input-output pairs into a relatively small number of clusters, each of which corresponds to a different value of the set of qualitative variables. The developed clustering algorithm is tested by simulation and using real data from mineral industry, where ore quality is very important, often unmeasurable qualitative variable.

2 Cluster Based Modelling

A very basic system model is a relation between all possible values of input-variables $u = (u_1, u_2, \ldots, u_r)$ and of output-variables $y = (y_1, y_2, \ldots, y_s)$ represented by a set of pairs (u, y). If the model is dynamic, the values of variables $u_i, i = 1, \ldots, r$ and $y_i, i = 1, \ldots, s$ are signals i.e. functions of time defined on a suitable time interval.

This kind of model could, in principle, be constructed using measured process variables as such. The influence of disturbances is then included in output-signals. However, this is usually not possible in practice, because of its large information content and, therefore the large data processing capacity needed for storing and manipulating the model. In the case of qualitative variables it is, however, often the only possibility.

The traditional way for reducing the information content of the model description, for instance for control design, is based on construction of some rule, usually a mathematical equation, the solutions of which define the set of input-output-pairs. In practice the model equation is constructed choosing first its structure and parametrization, and fitting then the parameters to process measurements.

The most parameter fitting methods are based on models which are linear with respect to parameters, in particular in case of large-scale process models. The influence of disturbances will be taken into account only in some average way. This means that for modelling of real complex processes with qualitative variables different models for different circumstances (values of qualitative variables, operating point etc.) are needed. The problem is, how the circumstances have to be classified and how such a change in circumstances can be detected that a new model is necessary.

An alternative, natural way for reducing the information need is to decrease the number of input-output-pairs belonging to the model grouping them to so-called clusters. This means that the pairs corresponding to the same circumstances are classified to the same cluster. The clustering is depicted by the following simple example.

Suppose that the model represents a process with one input u and one output y, and is of the form

$$S = \{(u_1, y_1), (u_2, y_2), (u_3, y_3), (u_1, y_1'), (u_2, y_2'), (u_3, y_3')\} \tag{1}$$

where signals u_i, y_i, y_i' are shown in Fig. 2 a). If the pairs are drawn in (u, y) coordinates of Fig. 2 b) it is seen that they can be grouped e.g. into three clusters

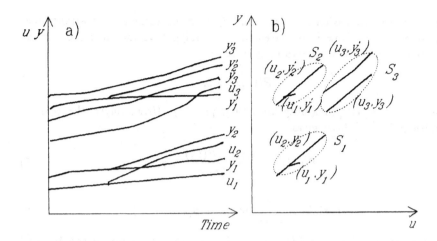

Figure 1. Input- and output-signals of the model a) as functions of time and b) as clusterized

$$S_1 = \{(u_1, y_1), (u_2, y_2)\}, \quad S_2 = \{(u_1, y_1'), (u_2, y_2')\}, \quad S_3 = \{(u_3, y_3'), (u_3, y_3')\} \tag{2}$$

In order to reduce the information need it is not sensible nor even possible to give lists of the measured points belonging to the clusters, but they have to be modelled by the use of some suitably parametrized descriptions.

Very useful is the assumption that the measured data consist of realizations of different random vectors, the differences caused by the influences of qualitative variables or disturbances. The problem is to construct these vectors, i.e. their probability density functions. This kind of clustering can be accomplished applying the methods of Cluster Analysis developed in Statistics.

3   Clustering Algorithm

From process control point of view it is important that the clustering methods are automatic and do not require any experienced statistician as a user. On the other hand, the results have not to be best possible, because the circumstances change all the time and there always are new measurements available. Instead it is important that the clustering and classification are stable so that the errors do not increase in time.

A method which requires a choice of only few parameters is presented in what follows. The basic assumption is that each cluster C_i represents a normally distributed random vector $\bar{x}_i$, i.e. its probability distribution is ([1],[2])

$$p(x|C_i) = \frac{1}{(2\pi)^{n/2}(\det R_i)^{1/2}} e^{-\frac{1}{2}(x-m_i)^T R_i^{-1}(x-m_i)} \tag{3}$$

where n is the dimension of $\bar{z}_i$ and m_i, R_i its mean value and covariance matrix, respectively. Now a measurement x is assigned to the cluster C_i, if the conditional probability satisfies (Bayes' rule)

$$P(C_i|x) > p_i \tag{4}$$

where p_i is a given probability limit $(0 < p_i < 1)$. Using Bayes' theorem for conditional probabilities gives

$$P(C_i|x) = \frac{p(x|C_i)P(C_i)}{p(x)} > p_i \tag{5}$$

or

$$p(x|C_i) > \frac{p_i}{P(C_i)}p(x) = q_ip(x) = p'_i(x) \tag{6}$$

If the a priori probability $P(C_i)$ is not known, the parameter q_i can be used instead of the parameter p_i. If $p(x)$ is also unknown, a constant parameter p'_i (likelihood) can be used.

Taking the natural logarithm of both sides of the inequality and arranging the terms gives

$$(x - m_i)^T R_i^{-1}(x - m_i) < -\ln(\det R_i) - \ln((2\pi)^n) - 2\ln p'_i(x) \tag{7}$$

This can be used as a condition for assigning x to the cluster C_i as such, or as simplified so that the right side is taken as a constant parameter d^2 independent on the cluster and x [1]. The left side of the condition is the squared, generalized Euclidean distance weighted by the matrix R_i^{-1} of the point x from the point m_i.

If the point x could be assigned to more than one clusters, the cluster can be chosen so that

$$(x - m_i)^T R_i^{-1}(x - m_i) + \ln(\det R_i) \tag{8}$$

or simply

$$(x - m_i)^T R_i^{-1}(x - m_i) \tag{9}$$

is a minimum.

In practice the means m_i and covariance matrices R_i are not usually known, so that they have to be replaced by their estimates calculated from the measurements

$$m_i(k_i) = \sum_{j=1}^{k_i} x_i(j)/k_i \tag{10}$$

$$R_i(k_i) = \sum_{j=1}^{k_i} (x_i(j) - m_i(k))(x_i(j) - m_i(k))^T/k_i \tag{11}$$

which are updated when a new point is assigned to a cluster.

The previous equations can be connected to each other. Define the matrix

$$
\begin{bmatrix}
1 & x_i(1)^T \\
1 & x_i(2)^T \\
\cdots & \cdots \\
1 & x_i(k_i)^T
\end{bmatrix}
= [1 \ X_i(k_i)^T] = k_i^{1/2} \tilde{X}_i(k_i)^T \tag{12}
$$

Then the extended covariance matrix

$$
\tilde{R}_i(k_i) = \tilde{X}(k_i)\tilde{X}_i(k_i)^T =
\begin{bmatrix}
1 & m_i(k_i)^T \\
m_i(k_i) & R_i(k_i) + m_i(k_i)m_i(k_i)^T
\end{bmatrix} \tag{13}
$$

is obtained. The matrix $\tilde{X}_i(k_i)$ is updated as follows

$$
\tilde{X}_i(k_i+1)^T =
\begin{bmatrix}
(\frac{k_i}{k_i+1})^{1/2}\tilde{X}_i(k_i)^T \\
\cdots \quad \cdots \\
(\frac{1}{k_i+1})^{1/2} \ (\frac{1}{k_i+1})^{1/2}x_i(k_i+1)^T
\end{bmatrix} \tag{14}
$$

In practice it is useful to weight the latest measurements more than the old ones, for instance like

$$
\tilde{X}_i(k_i+1)^T =
\begin{bmatrix}
\lambda^{1/2}\tilde{X}_i(k_i)^T \\
\cdots \quad \cdots \\
(1-\lambda)^{1/2} \ (1-\lambda)^{1/2}x_i(k+1)^T
\end{bmatrix} \tag{15}
$$

where λ, $0 < \lambda < 1$ is a parameter to be chosen. A similar weighting is obtained by defining an upper limit k_{max} to k_i, i.e. when $k_i > k_{max}$, then k_i is set to k_{max}.

Because the size of X_i increases when a new measurement is obtained, it cannot be used in continuous process modelling. This problem can be avoided if the covariance matrix $R_i(k_i)$ is factored to so-called Cholesky-factors, i.e. it is written to the form

$$
R_i(k_i) = U_i(k_i)^T U_i(k_i) \tag{16}
$$

where $U_i(k_i)$ is of upper triangular form, with non-negative diagonal entries. The use of Cholesky-factors makes also easier the calculation of the inverse of $R_i(k_i)$ in condition (7).

Cholesky-factorization can be applied also to $\tilde{R}_i(k_i)$ giving

$$
\tilde{R}_i(k_i) =
\begin{bmatrix}
1 & 0 \\
m_i(k_i) & U_i(k_i)^T
\end{bmatrix}
\begin{bmatrix}
1 & m_i(k_i)^T \\
0 & U_i(k_i)^T
\end{bmatrix}
= \tilde{U}_i(k_i)^T \tilde{U}_i(k_i) \tag{17}
$$

Cholesky-factorization is accomplished by using so-called orthogonal Householder-transformations to $\tilde{X}_i(k_i)$.

When a new point is assigned to a cluster, the Cholesky-factors can be updated directly applying the Householder-transformations to the matrix

$$
\tilde{V}_i(k_i+1)^T =
\begin{bmatrix}
\lambda^{1/2}\tilde{U}_i(k_i)^T \\
\cdots \cdots \\
(1-\lambda)^{1/2} \ (1-\lambda)^{1/2}x_i(k_i+1)^T
\end{bmatrix} \tag{18}
$$

giving

$$\begin{bmatrix} 1 & m_i(k_i+1)^T \\ 0 & U_i(k_i+1) \\ 0 & 0 \end{bmatrix} = \begin{bmatrix} \tilde{U}_i(k_i+1) \\ 0 \end{bmatrix} \qquad (19)$$

This procedure is applicable even to an empty cluster by the choice

$$\tilde{U}_i(0) = \begin{bmatrix} 0 & 0^T \\ 0 & \sigma_i I \end{bmatrix} = \begin{bmatrix} 0 & m_i(0)^T \\ 0 & U_i(0) \end{bmatrix} \qquad (20)$$

where σ_i is an assumed standard deviation in cluster C_i. If an estimate of the mean is available, it can be used instead of 0.

Now a measurement x is assigned to the cluster C_i if (c.f. (7))

$$(x - m_i(k_i))^T U_i(k_i)^{-1} (U_i(k_i)^T)^{-1} (x - m_i(k_i)) < d_i^2 \qquad (21)$$

For numerical accuracy it is useful that the values of the measured variables are of same size class. This can be realized for instance by calculating the mean $m(k)$ and the covariance matrix $R(k)$ or the Cholesky-factor $U(k)$ of all measurements, and by normalizing the measured vectors to

$$z = (U(k)^T)^{-1}(x - m(k)) \qquad (22)$$

There can be variables which are not as important as the others from the clustering point of view. This can be taken into account by defining a diagonal weighting matrix $W = \text{diag}[w_1, w_2, \ldots, w_n]$. The normalization changes the weighting matrix to the form

$$W_n(k) = (U(k)^T)^{-1} W U(k)^T \qquad (23)$$

The use of the weighting matrix changes the equation (21) to the form

$$(x - m_i(k_i))^T W U_i(k_i)^{-1} (U_i(k_i)^T)^{-1} W (x - m_i(k_i)) < d_i^2 \qquad (24)$$

On the other hand, it is often interesting to know which variables are most significant from the clustering point of view i.e. indicate most strongly the existence of different clusters. This can be analyzed using so-called Canonical Analysis [2]. This leads to the solution of the generalized eigenvalue problem

$$R_b e = \lambda R_w e \qquad (25)$$

where R_b is the between clusters covariance matrix and R_w the within clusters covariance matrix. The significance of the variables can be read from the corresponding components of the eigenvectors associated to the largest eigenvalues. The components have to be presented in the same scale which can be done by multiplying them by the standard deviations of the corresponding variables.

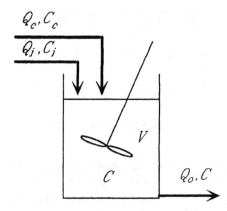

Figure 2. Perfect mixer

4 Simulation Example

Consider the perfect mixer depicted by Fig.2 with volume V, two input flows Q_c, Q_i and one output flow Q_o. The corresponding concentrations are C_c, C_i and C. Suppose that the output concentration C is controlled by the flow Q_c and the problem is to construct a model for their relationship.

The system is governed by the equations

$$V\frac{dC}{dt} = C_i Q_i + C_c Q_c - C Q_o \tag{26}$$

$$Q_i + Q_c = Q_o \tag{27}$$

(V is constant) In discrete time

$$C(k+1) = a(k)C(k) + b(k)Q_c(k) + c(k) \tag{28}$$

where

$$
\begin{aligned}
a(k) &\triangleq \left(1 - \Delta t \frac{Q_i(k) + Q_c(k)}{V}\right) \\
b(k) &\triangleq \Delta t \frac{C_c(k)}{V} \\
c(k) &\triangleq \Delta t \frac{Q_i(k)C_i(k)}{V}
\end{aligned}
\tag{29}
$$

The flow Q_i and the concentrations C_i and C_c are varying in an unknown way. Q_c, c and the measurement of C are disturbed by white noises distributed uniformly in intervals $[-\sigma_i/2, \sigma_i/2]$, $i = 1, 2, 3$.

The system was simulated using the sample time $\Delta t = 1$, the control flow

$$Q_c(k) = \bar{Q}_c(k) + 0.05\sin(0.2k) \tag{30}$$

	$1 \leq k \leq 100$	$101 \leq k \leq 200$	$201 \leq k \leq 300$
$\bar{Q}_c(k)$	0.2	0.1	0.1
$a(k)$	0.9	0.9	0.9
$b(k)$	1.0	1.0	1.0
$c(k)$	0.0	0.1	0.2

Table 1. Parameters in simulation

	cluster 1	cluster 2	cluster 3	total
$\hat{a}$	0.9000	0.9000	0.9000	1.0117
$\hat{b}$	1.0000	1.0000	1.0000	0.2759
$\hat{c}$	0.0000	0.1000	0.2000	-0.0605

Table 2. Parameter estimates in Run 1

and other parameters presented in Table 1. In simulation run 1 the noise parameters were $\sigma_1 = \sigma_2 = \sigma_3 = 0$. Using the variables C, ΔC and Q_c and the parameters $d = 3$, $\sigma_0 = 1$, $W = \text{diag}[1,1,1]$, $k_{max} = 100$ gave 3 clusters. The results of simulation and clustering are presented in Figs.3 and 4. The parameters a,b and c for each cluster and for the total data were estimated using least squares estimation and the estimates are in Table 2. It is seen that the clustering is completely based on the values of the parameter c and the results are completely correct.

In Run 2 the noise parameters were $\sigma_1 = \sigma_2 = \sigma_3 = 0.01$. The clustering parameters $d = 3.75$, $\sigma_0 = 2.5$, $W = \text{diag}[1,1,1]$, $k_{max} = 100$ gave again 3 clusters which are presented in Figs.5 and 6. The corresponding parameter estimates are in Table 3. The clustering algorithm found also in this case the changes in parameter c almost correctly and therefore the parameter estimates are relatively good. Finally in Run 3 the noise parameters were $\sigma_1 = \sigma_2 = \sigma_3 = 0.05$. The clustering parameters $d = 3.75$, $\sigma_0 = 2.5$, $W = \text{diag}[1,1,1]$, $k_{max} = 100$ gave still 3 clusters which are presented in Figs.7 and 8. In this case the disturbances are too high in order to obtain a correct clustering. Therefore the corresponding parameter estimates presented in Table 4 are also erroreous.

	cluster 1	cluster 2	cluster 3	total
$\hat{a}$	0.8750	0.9528	0.8613	1.0053
$\hat{b}$	1.0836	1.0452	1.0845	0.2880
$\hat{c}$	0.0368	-0.0026	0.3050	-0.0476

Table 3. Parameter estimates in Run 2

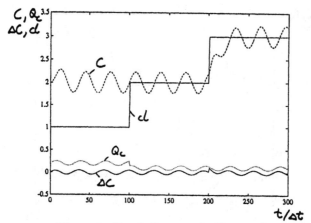

Figure 3. Simulation run 1. Time series

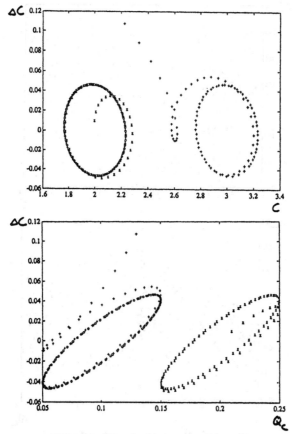

Figure 4. Simulation run 1. Clusters

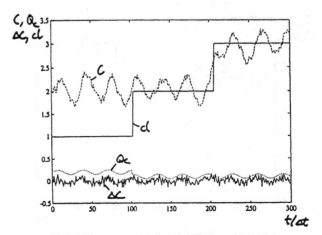

Figure 5. Simulation run 2. Time series

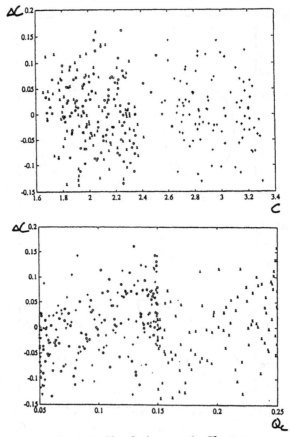

Figure 6. Simulation run 2. Clusters

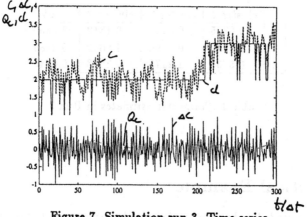

Figure 7. Simulation run 3. Time series

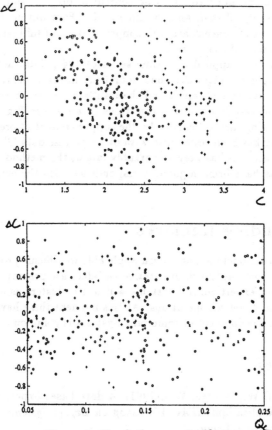

Figure 8. Simulation run 3. Clusters

	cluster 1	cluster 2	cluster 3	total
$\hat{a}$	0.8705	0.3427	0.1812	0.7306
$\hat{b}$	-2.2382	0.8894	1.5002	-0.2983
$\hat{c}$	0.5632	1.2278	2.3078	0.6649

Table 4. Parameter estimates in Run 3

5 Practical Examples

Cluster-based modelling has been tested also by using the data from real mineral processes. Both a grinding process and a flotation process has been modelled.

A copper flotation plant is presented here as an example. The model consists of eight input-variables: the copper content of ore, the xanthate feed, the zinc sulphate feed, the cyanid feed, the zinc content of ore, the sulphur content of ore, the conductivity of feed, and the lime feed . Four output-variables were: the copper content of concentrate, the copper content of tailings, the recovery, and the zinc content of concentrate.

The clustering was applied to a time series of 104 points each again representing averages of variables over one hour time interval. Five clusters with 1,45,30,14,14 points were obtained and the results have been presented by time series (Fig.9) and two dimensional projections (examples in Fig.10). The most significant clustering factor seems to be the copper content of ore.

The clustering has been tested also with other process data [3], [4]. Although the number of variables has been remarkably larger, the method has performed well. The results have been in good correspondence with the empirical process knowledge.

6 Concluding Remarks

The cluster-based modelling has proven applicable to difficult process modelling problems where some unknown qualitative variables are present. It is also possible to construct expert control systems which use clusters as a basic process model for design of control operations. At this moment these have not yet been realized but the development of some pilot systems is going on.

References

[1] Ginsberg, D.W., Whiten, W.J.(1991). A data based expert system for engineering applications. *IFAC Workshop on Expert Systems in Mining and Metal Processing*, Espoo.

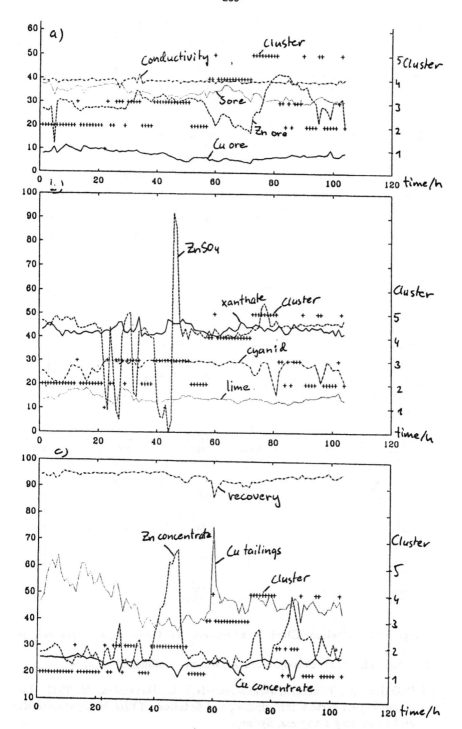

Figure 9. Flotation. Time series: a)non-manipulated inputs, b)manipulated inputs, c)outputs

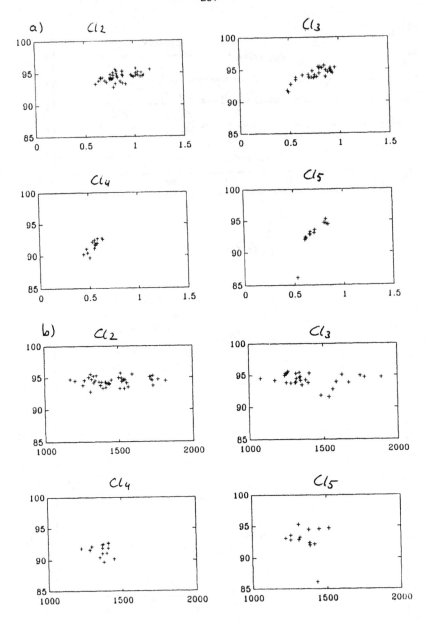

Figure 10. Flotation. Clusters: a) recovery vs. Cu ore b) recovery vs lime

[2] James, M.(1985). *Classification Algorithms*. Collins, London.

[3] Pulkkinen,K., Ylinen,R., Jämsä-Jounela,S.-L., Järvensivu, M.(1993). Integrated expert system for grinding and flotation. *XVIII International Mineral Processing Congress*, Sydney.

[4] Ylinen,R., Jämsä-Jounela,S-L.,Miettunen,J.(1993). Use of cluster analysis in process control. *XII World Congress of IFAC*, Sydney.

The Role of Partitions and Functionals in Descriptor Computation for Data Receptive Fields

J.A. Muñoz-Blanco, J.C. Quevedo-Losada and O. Bolivar-Toledo

Departamento de Informática y Sistemas
Universidad de Las Palmas de Gran Canaria
Las Palmas de Gran Canaria
Spain

Abstract. A fundamental concept in Artifitial Vision and Image Processing is that of Complete Description of a data field. From the analitycal point of view completness requires the conservation in the number of degrees of freedom of the visual environment. This constancy has allowed us to establish a special conservation principe, which is determinated by the receptive fields and the functionals performed on them.

The purpose of this work is the use of an alternative representation domain, based on the above concepts, and to convergein a spetial non orthonormal transform which combine classic functionlas and data field partitions in order to generate a complete description that, as usual, can be truncated for a particular visual task like a clasification problem.

We have design a clasification system, which is based on a reduced inference system, and we have study the influence of different parameters characteristics of these algebraic analitycal transforms, on the goodness of the clasification system.

1. Introduction

A fundamental concept in Artificial Vision and Image Processing is that of Complete Description [1]. A desciption is admitted to be complete, for a visual environment, if it contains all the data and their properties needed to cover the given objetives. For that, in the general problems of Artificial Vision is not posible, a priori, to establish the necessary and suficient conditions to decide whether a representation is complete or not, based only on the visual data an the nature of the representation nature. In other words, it is necessary to add the representation purpose, and even in this case is imposible to decide, in general, about the complete character of the representation, mainly because the objetives and/or the purposes are espressed in a fuzzy way and even the degree of the fuzzyness is changing.

We have developped different alternatives for data representation spaces based on completness requeriments, based mostly on orthonormal transforms and emphasizing the functional characteristics of each transformation, which allow us for implementing visual systems of recognition to solve some concrete practical situations.

This fact is based on the conjeture according to which given a set of images (or Data Fields) to be clasified, in a probabilistic or fuzzy way, the classification accuracy, in the sense of minimum error probability, or fuzzyness, in the asignation to the corresponding clases, can be increased when the original data are represented in an alternative space.

In this contex the Data Field concept appears as the consequent generalization of the image representation in one, two or more dimensions. An unidimensional data field, with length N and resolution R, is a ordered set of N places, i, such thata number I_i can be assigned to each place with resolution R.

In artificial vision, the system is embodied into a computer where typically the resolution is a constant. Thus, we may restrict the problem to that of transforms of "places" of constant resolution. According to this completness of a transform may be looked at as a problem of reversible transformation of places and stated as follows:

From a data field D, addresable by index i, a complete transformation is a rule to construct a second field D´, adressable by index j, from which field D can be recovered. The associated series of problems are related to the necessary and sufficient completness conditions for families of transforms.

Normally, artificial vision adds very strong restrictions tending to the reduction of the original dimension of the field, that take advantage of additional knowledge of the family of visual problems under attention. Thus, an again typical problem can be stated as follows:

From a data field D of dimension N, is it possible to build a new field D´ of M dimension, such that M < < N and such that a visual task can be performed by queries only to D´?. A possible solution consists in generating a D´of the same dimension that D by a complete transform, and then truncate it.

Here the question is faced by the stating of a theorem [2] on complete descriptions that may be truncated for visual tasks. The theorem shows that for a family of algebraic partitions of the data field, the computation of certain number of analytical descriptors in each partition provides for a complete description that can be truncated. As usual the validity of the truncation rest in heuristics proofs.

A general characteristic of this algebraic analytical transformation, named **Functional Receptive Field Transformation** is the essentialy parallel character of the computational structure that performs them. This fact has allowed the development of fast algorithms to implement it.

2. Application to Visual Recognition

An objective of this work is the generation of an alternative representation domain based on the above non-orthonormal transforms which combine not only classic functionals but data field partitions, for a concrete practical situation .

In this way, we have designed a classification system based on a reduced inference system [3] shown in the following figure.

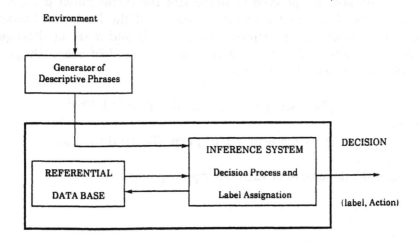

Following this scheme, we have generated the descriptor phrases from the above transformation and developped diferent strategies in order to compare the efficiency of classifier system as a function of the different parameters involved in the transformation defined previously. The parameters considered has been: the dimension of the data field, the number of partitions and their extension, the shifting interpartition and the different type of functionals chosen.

We have adquired a set of 90 images. These consist of 10 different acquisitions from each of the 9 different shapes. We apply a segmentation process to these images by making a space of monodimentional measure where there are only two cluster. Finally, we achieved a process to normalize and make them invariant against traslations, rotations and homotecias.

For the generation of the descriptive phrases that must label each class containing the forms, both patterns and unknown, for posterior recognition, we started from value N (freedom degrees or resolution). To obtain the values of L (partitions number) and d (freedom degrees of the receptive fields), the unique restriction must be impossed by the following expression:

$$L = [(N - d) / \delta] + 1$$

being δ the interpartitions shift.

This means that if we fix freedom degrees and the shift, we obtain the necessary number of partitions.

At last, we proceed to implement the classification process. To achieve this have used a modified version of the Euclidean Distance. Given an unknown descriptive phrase, $D'(i)$ and a set of descriptive phrases corresponding to each referential data field class, $D(i,j)$, the Euclidean Distance is defined as:

$$[\text{Distance } (j)]^2 = \Sigma [(D'(i) - D(i,j) / 10^{E(j)}]^2$$

where:

$$E(j) = \min (\exp. (D'(i)), \exp. (D(i,j)) \quad (i=1,...,n)$$

being n the dimension of each descriptive phrase.

Once computed every distance between unknown descriptive phrase and each one of the descriptive phrases which conform the referential data base, the inference system associates the unknown pattern to that whose distance between them be lowest.

3 Experimental Results

In our referential data base we consider 63 images as pattern and 27 images as unknown.

According to Alt [4], the best results for purposes of classification are given by low order moments. For this reason, the moments used as functionals in every cases are $M_{0,3}$, $M_{0,4}$, $M_{1,2}$, $M_{1,3}$, in addition to the moments $M_{0,6}$, $M_{0,8}$, $M_{0,10}$, $M_{1,4}$, $M_{1,6}$, $M_{1,8}$, $M_{1,10}$ arbitrarily chosen.

In addition, to test the influence of the functional we try with Fourier Descriptor, from which we define 3 different functionals:

1) Energy Total

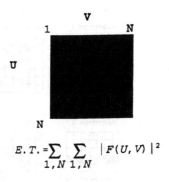

$$E.T. = \sum_{1,N} \sum_{1,N} |F(U,V)|^2$$

2) Energy of the center

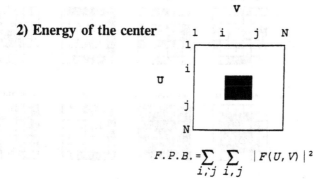

$$F.P.B. = \sum_{i,j} \sum_{i,j} |F(U,V)|^2$$

3) Energy of the periphery.

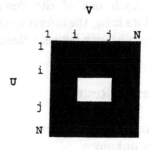

$$F.P.A. = \sum_{1,i} \sum_{1,N} |F(U, V)|^2 + \sum_{j,N} \sum_{1,N} |F(U, V)|^2 +$$
$$\sum_{1,N} \sum_{1,i} |F(U, V)|^2 + \sum_{1,N} \sum_{j,N} |F(U, V)|^2$$

CASE 1: In this case, we increase the Resolution of the Data Field.

Functionals: Moments

Partitions:

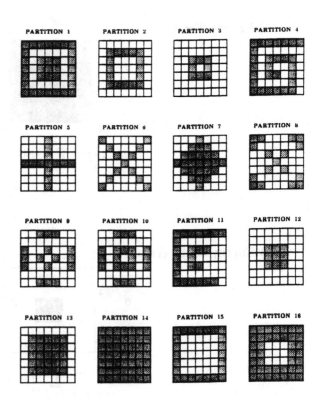

Results with Resolution 128 X 128

MOMENTS

		$M_{0,3}$	$M_{0,4}$	$M_{0,6}$	$M_{0,8}$	$M_{0,10}$	$M_{1,2}$	$M_{1,3}$	$M_{1,4}$	$M_{1,6}$	$M_{1,8}$	$M_{1,10}$
P	1	1	1	1	1	1	1	1	1	1	1	1
	2	3	3	3	3	2	4	4	3	3	3	3
A	3	2	0	0	0	0	2	1	0	0	0	0
R	4	3	2	3	3	3	2	2	2	2	3	3
	5	1	1	1	1	1	2	1	1	1	1	1
T	6	4	4	4	4	3	3	0	0	1	1	2
	7	2	2	2	2	2	2	2	2	2	2	2
I	8	1	1	3	3	2	0	0	0	3	2	2
	9	1	1	2	4	2	1	2	1	1	2	1
T	10	1	1	1	3	2	1	1	1	1	2	1
I	11	2	1	1	1	3	1	2	1	1	3	3
	12	2	0	0	0	0	2	0	0	0	0	0
O	13	3	3	2	3	3	4	4	3	3	3	3
N	14	2	2	2	2	2	2	2	2	2	2	2
	15	1	1	1	1	1	1	1	1	1	1	1
S	16	2	2	2	2	2	2	2	2	2	2	2

Results with Resolution 256 X 256

MOMENTS

		$M_{0,3}$	$M_{0,4}$	$M_{0,6}$	$M_{0,8}$	$M_{0,10}$	$M_{1,2}$	$M_{1,3}$	$M_{1,4}$	$M_{1,6}$	$M_{1,8}$	$M_{1,10}$
P	1	1	3	1	2	5	0	1	2	1	3	1
	2	0	2	1	1	1	1	1	1	1	1	1
A	3	1	0	0	0	0	0	1	0	0	0	0
R	4	2	1	2	4	2	0	3	1	3	2	3
	5	0	0	0	0	0	0	0	0	0	0	0
T	6	2	2	2	1	1	2	2	2	2	1	1
	7	1	1	1	2	2	1	1	1	1	2	2
I	8	1	1	0	3	0	0	0	0	0	0	0
	9	1	4	1	2	3	1	2	2	1	3	1
T	10	1	4	1	2	3	0	2	2	1	3	1
I	11	0	0	1	1	0	0	0	0	0	0	0
	12	0	0	0	0	0	0	1	0	0	0	0
O	13	0	2	1	1	1	1	1	1	1	1	1
N	14	0	1	0	0	0	1	0	0	0	0	0
	15	1	3	2	4	5	0	3	6	1	5	1
S	16	0	1	0	0	0	1	0	0	0	0	0

CASE 2: We study the dependences with respect a different functionals.

Functionals: Moments and Fourier

Partitions:

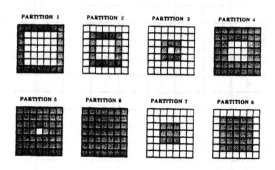

Results:

		$M_{0,3}$	$M_{0,4}$	$M_{0,6}$	$M_{0,8}$	$M_{0,10}$	$M_{1,2}$	$M_{1,3}$	$M_{1,4}$	$M_{1,6}$	$M_{1,5}$	$M_{1,10}$	E.T.	F.P.B.	F.P.A.
P	1	1	0	4	0	10	1	0	3	1	3	0	7	9	6
A	2	0	0	0	0	0	0	0	0	0	0	0	0	1	1
R	3	1	1	1	3	3	0	0	0	1	0	1	0	1	0
T	4	0	0	0	0	0	0	0	0	0	0	0	0	1	1
I	5	0	0	0	1	1	0	0	0	0	0	1	0	1	0
T	6	0	0	1	3	1	0	0	0	1	0	1	0	0	0
I	7	0	0	1	3	2	0	0	0	1	0	1	0	0	0
O N S	8	0	0	1	3	1	0	0	0	1	0	1	0	0	0

MOMENTS   FOURIER

CASE 3: In this experiment, we reduce the dimension of the Receptive Field, so we have to generate more of them, for the completness

Functionals: Moments and Fourier

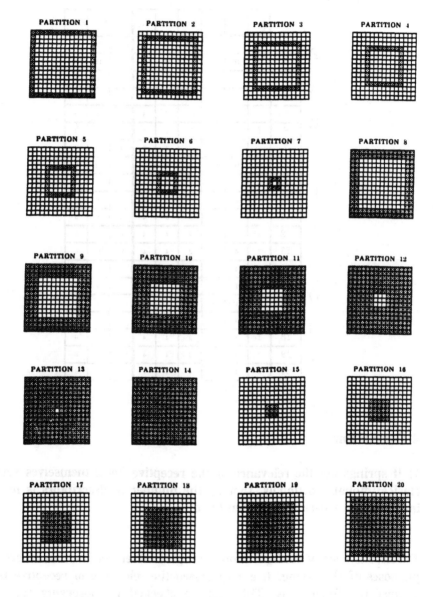

Partitions:

Results:

		MOMENTS						FOURIER		
		$M_{0,3}$	$M_{0,4}$	$M_{0,5}$	$M_{1,2}$	$M_{1,3}$	$M_{1,6}$	E.T.	F.P.B.	F.P.A.
P	1	24	24	24	24	24	24	20	20	20
	2	17	16	16	16	16	16	11	18	13
A	3	1	0	3	2	0	0	1	8	1
	4	0	0	0	1	1	0	1	3	1
R	5	3	3	5	3	3	4	6	6	6
	6	1	1	1	1	1	1	3	3	3
T	7	2	2	2	1	2	2	4	7	4
	8	17	16	16	16	16	16	9	13	9
I	9	2	0	3	2	0	0	1	9	1
	10	0	0	0	1	0	0	0	3	0
T	11	3	3	3	3	3	4	5	5	5
	12	3	3	3	4	4	4	6	6	5
I	13	3	4	4	3	4	4	6	6	6
	14	3	4	4	3	4	4	6	6	6
O	15	2	2	2	1	1	2	4	5	4
	16	1	1	1	2	1	1	4	3	4
N	17	3	3	4	3	4	4	6	6	6
	18	3	4	4	3	4	4	6	6	6
S	19	3	4	4	3	4	4	6	6	6
	20	3	4	4	3	4	4	6	6	6

4 Conclusions

1) It springs out the relevance of the receptive fields themselves versus clasical global functionals, that is, the number of descriptors is not so important as where are choosen from.

2) There is a set of configurations that give better results than others for purposes of classifying. It is enphasized the relevance of receptive field against the functionals. This can be checked by observing that the variability of partitions versus the functionals is smaller than the functionals versus the partitions.

3) In general, an increase of resolution causes an increase in the efficiency of the classification system.

4) We think that in the future, the research must be focused on that optimal configuration, analysing the dependence of the system with respect the different parameters involving in the description: the dimension of de receptive field, the complexity of the data field and the functional selected.

REFERENCES

[1]   Candela Solá, S.: "Transformaciones de Campo Receptivo Variable en Proceso de Imágenes y Visión Artificial". Tesis Doctoral. Universidad de Las Palmas de Gran Canaria. 1.987

[2]   Bolivar Toledo, O.: "Hacia una Teoría de las Transformaciones en Campos de Datos Receptivos y Campos de Datos. Implicaciones en Teoría y Proceso de Imágenes". Tesis Doctoral. Universidad de las Palmas de Gran Canaria. 1.989.

[3]   Muñoz Blanco, J.A.:"Jerarquización de Estructuras de nivel bajo y medio para Reconocimiento Visual. Aplicaciones a texturas de Formas". Tesis Doctoral. Universidad de Las Palmas de Gran Canaria. 1.987

[4]   Alt, F.L.: Digital pattern recognition by moments. J. - ACM9, 240-258 (1.962).

On Some Algorithmic and Computational Problems for Neuronal Diffusion Models

Virginia Giorno[1], Amelia G. Nobile[2] and Luigi M. Ricciardi[3]

[1] Dipartimento di Informatica e Applicazioni dell'Università, Salerno, Italy

[2] Dipartimento di Matematica e Informatica dell'Università, Udine, Italy

[3] Dipartimento di Matematica e Applicazioni dell'Università, Napoli, Italy

Abstract. In this work we consider some one–dimensional diffusion processes arising in single neurons' activity modelling and discuss some of the related theoretical and computational first passage time problems. With reference to the Wiener and the Ornstein–Uhlenbeck processes, we outline some theoretical methods and algorithmic procedures. In particular, the relevance of the computational methods to infer about asymptotic trends of the firing pdf is pointed out.

1 Introduction

The purpose of this paper is to point out certain theoretical and computational results that we have obtained in order to provide a quantitative description of the input–output behaviour of single neurons subject to a diffusion–like dynamics. The literature on neuronal modelling is too wide and well known to be reviewed here. We limit ourselves to mentioning that if a neuron is viewed as a black box characterized by an input and by an output the following two distinct problems can be posed: i) to determine the output for a given input or ii) to analyze the output in order to gather information about the input. In both cases the class of input functions must be preliminarily specified. Hereafter, we shall assume that the neuron's membrane potential is modelled by a one–dimensional diffusion process. Namely, we shall assume (however, this can be proved in some cases, as shown for instance in [23] and [11]) that the neuron's membrane potential can be described by a scalar diffusion process $\{X(t), \ t \geq t_0; \ t_0 \in \Re\}$ generated by a stochastic differential equation

$$dX(t) = \mu\left[X(t), t\right] dt + \sigma\left[X(t), t\right] dW(t), \tag{1}$$

where $W = \{W(t); \ t \geq t_0\}$ is a standard Wiener process and μ and σ are real–valued functions of their arguments satisfying certain regularity conditions (cf., for instance [15]). More precisely, the process $X(t)$ represents changes in the membrane potential between two consecutive neuronal firings. The reference level for the membrane potential is taken equal to the resting potential, viz. $X(t_0) = x_0$. The threshold potential, denoted by $S = S(t)$ $[S(t_0) > x_0]$, will be assumed to be a deterministic function of time. In the sequel we shall focus our attention on the properties of the first passage time (FPT) random variable T, defined as

$$T = \inf_{t \geq t_0} \{t : X(t) > S(t) | X(t_0) = x_0\}. \tag{2}$$

Hence, T is the theoretical counterpart of the interspike interval. The importance of interspike intervals is due to the generally accepted hypothesis that the information transferred within the nervous system is usually encoded by the timing of occurrence of neuronal spikes. Therefore, the reciprocal relationship between the firing frequency and the interspike interval naturally leads to the problem of determining the probability density function (pdf) of T, namely the function

$$g[S(t), t | x_0, t_0] = \frac{\partial}{\partial t} P\{T \leq t\}. \tag{3}$$

When this density cannot be obtained analytically (which is practically the rule) or when it is too difficult to give sufficiently precise estimations of it, the analysis is restricted to its moments, primarily mean and variance. In some cases it can finally be useful to work out asymptotic estimates of the FPT pdf for large times or for large thresholds, as it will be indicated in the sequel.

An alternative description of the process $X(t)$ is sometimes obtained via the so called "diffusion equations approach". First of all one defines the pdf of $X(t)$ conditional on $X(t_0) = x_0$:

$$f(x, t | x_0, t_0) = \frac{\partial}{\partial x} P\{X(t) \leq x | X(t_0) = x_0\}. \tag{4}$$

It can then be seen that f satisfies the Fokker–Planck equation

$$\frac{\partial f}{\partial t} = -\frac{\partial}{\partial x} [A_1(x, t) f] + \frac{1}{2} \frac{\partial^2}{\partial x^2} [A_2(x, t) f] \tag{5}$$

and the Kolmogorov equation

$$\frac{\partial f}{\partial t_0} + A_1(x_0, t_0) \frac{\partial f}{\partial x_0} + \frac{1}{2} A_2(x_0, t_0) \frac{\partial^2 f}{\partial x_0^2} = 0, \tag{6}$$

where the coefficients $A_1(x, t)$ and $A_2(x, t)$ are the "drift" and "infinitesimal variance" of the process defined as

$$A_i(x, t) = \lim_{\Delta t \downarrow 0} \frac{1}{\Delta t} \int (y - x)^i \, f(y, t + \Delta t | x, t) \, dy$$

$$\equiv \lim_{\Delta t \downarrow 0} \frac{1}{\Delta t} E\{[X(t + \Delta t) - X(t)]^i | X(t) = x\} \qquad (i = 1, 2). \tag{7}$$

It is essential to mention that the quantities $A_1(x, t)$ and $A_2(x, t)$ just defined are related in the following way to the functions μ and σ earlier considered when talking about the differential equations approach (however, see [23]):

$$\begin{cases} A_1(x, t) = \mu(x, t) \\ A_2(x, t) = \sigma^2(x, t). \end{cases} \tag{8}$$

To determine f via the diffusion equations the following initial conditions must be considered:

$$\lim_{t \downarrow t_0} f(x, t | x_0, t_0) = \delta(x - x_0) \tag{9a}$$

$$\lim_{t_0 \uparrow t} f(x, t | x_0, t_0) = \delta(x_0 - x). \tag{9b}$$

They express the circumstance that initially the whole probability mass is concentrated at the initial value x_0. However, these conditions are not always sufficient to determine uniquely the transition pdf, so that suitable boundary conditions may have to be imposed ([7], [8], [15]).

The neuronal models based on the applications of diffusion processes are predominantly time homogeneous and thus the functions μ and σ appearing in (1), (5) and (6) do not depend explicitly on t, i.e., $\mu(x, t) = \mu(x)$, $\sigma(x, t) = \sigma(x)$. Consequently, one has $f(x, t | x_0, t_0) = f(x, t - t_0 | x_0, 0) = f(x, t - t_0 | x_0)$. Time non–homogeneous diffusion models have been considered by Matsuyama et al. [18], Matsuyama [17], Ricciardi [24], Lánský [16]. However, since very few analytical results are available, simulation methods are required.

2   The Firing pdf

The determination of the firing pdf for a neuron modelled by a diffusion process is an FPT problem in which the unknown is the FPT pdf through a preassigned boundary $S(t)$. Let us now define the transition pdf, $\alpha(x, t | x_0, t_0)$, of $X(t)$ in the presence of an absorbing boundary $S(t)$. We set

$$\alpha(x, t | x_0, t_0) = \frac{\partial}{\partial x} P\{X(t) \leq x; \; X(\vartheta) < S(\vartheta) \; \forall \vartheta \in [t_0, t] | X(t_0) = x_0\}. \tag{10}$$

It can be shown that such a function may be obtained by solving Eqs. (5) and (6) subject not only to initial conditions (9) but also to the boundary condition $\alpha[S(t), t | x_0, t_0] = 0$. A second boundary condition, expressing the behaviour of $X(t)$ at the other hand of the interval where diffusion takes place, must be added. This may be a regularity condition, a reflection condition or other kinds of conditions depending upon the model one is dealing with.

The functions (3) and (10) are intimately related. Indeed, denoting by ν the lower bound of the state space ($-\infty$ if we assume that the membrane potential can be hyperpolarized without bounds), the following identity clearly holds:

$$P\{\nu < X(t) < S(t); \; X(\vartheta) < S(\vartheta) \; \forall \vartheta \in [t_0, t] | X(t_0) = x_0\} = P\{T > t\}. \tag{11}$$

In terms of definitions (3) and (10), identity (11) can be re–written as

$$\int_\nu^{S(t)} \alpha(x, t | x_0, t_0) \, dx = \int_t^\infty g[S(\tau), \tau | x_0, t_0] \, d\tau. \tag{12}$$

Hence,

$$g[S(t), t | x_0, t_0] = - \frac{\partial}{\partial t} \int_\nu^{S(t)} \alpha(x, t | x_0, t_0) \, dx. \tag{13}$$

The left–hand–side is the firing pdf. Its determination thus requires the determination of the transition pdf of $X(t)$ in the presence of an absorbing boundary at $S(t)$.

Let us now make the assumption (to be removed later) that the neuronal threshold is a constant, i.e. $S(t) = S$, with $S > x_0$ a constant. Also, let us assume that $X(t)$ is time–homogeneous. From (13) we then obtain:

$$g(S, t | x_0) = - \int_\nu^S \frac{\partial}{\partial t} \alpha(x, t | x_0) \, dx = \left\{ A_1(x) \, \alpha - \frac{1}{2} \frac{\partial}{\partial x} \left[A_2(x) \, \alpha \right] \right\} \Big|_\nu^S \tag{14}$$

where we have used the fact that α satisfies Eq. (5) with A_1 and A_2 independent of t.

An alternative approach is the following. Since the sample paths of $X(t)$ are continuous functions, any sample path that reaches a state $x > S$ at time zero must necessarily cross S for the first time at some intermediate time τ $(0 < \tau < t)$. We can then write

$$f(x, t | x_0) = \int_0^t f(x, t - \tau | S) \, g(S, \tau | x_0) \, d\tau \qquad (x > S). \tag{15}$$

This is a first–kind Volterra integral equation in the unknown firing pdf $g(S, t | x_0)$. Its solution is made complicated by the circumstance that the kernel $f(x, t - \tau | S)$ exhibits a singularity of the type $1/\sqrt{t - \tau}$ as $\tau \uparrow t$. Hence, the problem of determining $g(S, t | x_0)$ from Eq. (15) via numerical methods is by no means trivial. However, we shall see that an alternative integral equation with continuous kernel can be obtained from (15) and profitably used to investigate the behaviour of the firing pdf (cf. [2]). An analytic approach to the solution of Eq. (15) is based on the Laplace transform (LT). Set

$$\begin{cases} g_\lambda(S | x_0) = \mathcal{L}_t \{ g(S, t | x_0) \} \\[2mm] f_\lambda(x | y) = \mathcal{L}_t \{ f(x, t | y) \} \\[2mm] f_\lambda(x | x_0) = \mathcal{L}_t \{ f(x, t | x_0) \}, \end{cases} \tag{16}$$

where $\mathcal{L}_t$ denotes the LT with respect to t of the function in curly brackets. Henceforth, we shall safely assume $\lambda > 0$. Since a convolution integral appears on the right–hand–side of (15), passing to the LT we obtain:

$$g_\lambda(S | x_0) = \frac{f_\lambda(x | x_0)}{f_\lambda(x | S)} \qquad \forall x > S. \tag{17}$$

Note that the right–hand–side of (17) is actually independent of the arbitrarily chosen state x. Hence, if the transition pdf of $X(t)$ is known and if its LT can be calculated, the right–hand–side of (17) can be written down. The function $g(S, t | x_0)$ can then be obtained as an inverse LT. Alternatively, we can obtain $g_\lambda(S|x_0)$ as a solution of an ordinary differential equation. Indeed, the Kolmogorov equation (6) for a time–homogeneous process reads

$$\frac{\partial f}{\partial t} = A_1(x_0)\,\frac{\partial f}{\partial x_0} + \frac{1}{2}\,A_2(x_0)\,\frac{\partial^2 f}{\partial x_0^2}.$$ (18)

Taking the LT of both sides we obtain:

$$\mathcal{L}_t\left\{\frac{\partial f(x,t|x_0)}{\partial t}\right\} = A_1(x_0)\,\frac{\partial f_\lambda(x|x_0)}{\partial x_0} + \frac{1}{2}\,A_2(x_0)\,\frac{\partial^2 f_\lambda(x|x_0)}{\partial x_0^2}.$$ (19)

We now have:

$$\mathcal{L}_t\left\{\frac{\partial f(x,t|x_0)}{\partial t}\right\} = \int_0^\infty e^{-\lambda t}\,\frac{\partial f(x,t|x_0)}{\partial t}\,dt$$

$$= \lim_{t\downarrow 0}\left[e^{-\lambda t}\,f(x,t|x_0)\right] + \lambda\int_0^\infty e^{-\lambda t}\,f(x,t|x_0)\,dt.$$ (20)

Since the limit in (20) is zero by virtue of the initial condition (9a) and of the assumption $x > S > x_0$, from (17), (19) and (20) we finally obtain:

$$\frac{A_2(x_0)}{2}\,\frac{d^2 g_\lambda(S|x_0)}{dx_0^2} + A_1(x_0)\,\frac{dg_\lambda(S|x_0)}{dx_0} - \lambda\,g_\lambda(S|x_0) = 0.$$ (21)

This is a second order ordinary differential equation that must be solved with the conditions

$$g_\lambda(S|S) = 1$$ (22a)

$$g_\lambda(S|x_0) < \infty, \qquad \forall x_0 < S.$$ (22b)

Condition (22a) expresses the circumstance that the FPT through S starting from S itself is a degenerate random variable that takes the value 0 with probability 1 (instantaneous first–passage). Hence, $g(S,t|S) = \delta(t)$ and hence its LT is unit. Condition (22b) is instead a regularity condition whose role is to discard solutions of (21) that blow up and thus cannot be pdf's.

It is worth to point out that even though the inverse LT of the function $g_\lambda(S|x_0)$, obtained either by means of (17) or by solving Eq. (21) with condition (22), cannot be calculated it can nevertheless provide useful information on the FPT. Indeed, $g(S,t|x_0)$ is the moment generating function of T, so that

$$t_n(S|x_0) \equiv E(T^n) = \left.\frac{d^n g_\lambda(S|x_0)}{d\lambda^n}\right|_{\lambda=0} \qquad (n = 1,2,\ldots).$$ (23)

As we shall see, the knowledge of the moments can play an essential role in suggesting some general features of the firing pdf.

3   Neuronal Models

Gerstein and Mandelbrot [10] have been the first authors to postulate that for certain neurons subject to spontaneous activity the firing pdf could be modelled by the FPT pdf of a Wiener process characterized by the constant infinitesimal moments

$$A_1(x) = \mu, \qquad \mu \in \Re \tag{24a}$$

$$A_2(x) = \sigma^2, \qquad \sigma \in \Re^+. \tag{24b}$$

As is well known, in this case the transition pdf (4) is normal, with mean $\mu\,(t - t_0) + x_0$ and variance $\sigma^2\,(t - t_0)$. Since this is a temporally homogeneous diffusion process, we can take $t_0 = 0$. By means of the methods outlined in Section 2, one can then prove that the FPT pdf of such a process is given by

$$g(S, t | x_0) = \frac{S - x_0}{\sigma\sqrt{2\pi t^3}}\,\exp\left[-\frac{(S - x_0 - \mu t)^2}{2\sigma^2 t}\right], \qquad x_0 < S. \tag{25}$$

For $\mu \geq 0$, firing is a sure event as (25) is normalized to unity. If one takes $\mu < 0$, the FPT pdf (25) can be interpreted as the firing pdf conditional upon the event "firing occurs". The case $\mu = 0$ is also of interest since (25) is a stable law ([9]). This case provides an interpretation of numerous experimental results indicating that the shapes of histograms are sometimes preserved when the adjacent interspike intervals are summed (see, for instance, [25], [10] and [14]). Although the assumptions underlying Gerstein–Mandelbrot's model are undoubtedly oversimplified as some well-known electrophysiological properties of neuronal membrane are not taken into account, it must be stressed that the fitting of some experimental data by the FPT pdf (25) is truly remarkable. This is probably the reason why some authors have attempted to generalise this model in various respects.

Note that in (25) the firing threshold is assumed to be a constant. If, more realistically, a time–varying threshold is introduced to account for relative refractoriness – which is relevant in high firing rate conditions – the determination of the firing pdf for the Wiener neuronal model cannot be accomplished analytically unless the firing threshold is unrealistically assumed to be linearly time–dependent. For other types of threshold functions, ad hoc numerical methods had to be devised (cf., for instance, [5] ,[1], [21] and [22], [2] and references therein).

In conclusion, despite the excellent fitting of some data, Gerstein and Mandelbrot's neuronal model based on the Wiener process has been the object of various criticisms. We limit ourselves to pointing out that this model does not include the well-known spontaneous exponential decay of the neuron's membrane potential that occurs between successive PSP's. Hereafter we shall thus discuss another diffusion model for neuronal activity that includes this specific feature. This is customarily denoted as the Ornstein–Uhlenbeck (OU) neuronal model because of its analogy with the model used by these authors to describe

Brownian motion. The OU neuronal model is the diffusion process characterized by the following drift and infinitesimal variance:

$$A_1(x) = -\frac{x}{\vartheta} + \mu \tag{26a}$$

$$A_2(x) = \sigma^2, \tag{26b}$$

where $\mu \in \Re$ while ϑ and σ are positive constants. Comparing (26) with (24) we see that now the drift is state–dependent. Recalling (1) and (8) we can interpret the OU model as generated by the following SDE:

$$dX(t) = \left(-\frac{X}{\vartheta} + \mu\right) dt + \sigma\, dW(t), \tag{27}$$

where $W(t)$ is a standard Wiener process. Eq. (27) is taken as representative of the time course of the membrane potential. In the absence of randomness ($\sigma = 0$), Eq. (27) yields the spontaneous exponential decay of the membrane potential towards the resting potential $\rho = \mu\vartheta$. Hence, ϑ can be viewed as the time–constant of the neuron's membrane (approximately 5 msec).

Let us now return to the OU model with infinitesimal moments (26). Its transition pdf is obtained by solving either the Fokker–Planck equation (5) or the Kolmogorov equation (6) with the initial conditions (9). One thus finds (cf., for instance, [23]):

$$f(x, t|x_0) = \left[2\pi V(t)\right]^{-1/2} \exp\left\{-\frac{\left[x - M(t|x_0)\right]^2}{2V(t)}\right\}, \tag{28}$$

where

$$M(t|x_0) = \mu\vartheta\left(1 - e^{-t/\vartheta}\right) + x_0\, e^{-t/\vartheta} \tag{29a}$$

$$V(t) = \frac{\sigma^2\vartheta}{2}\left(1 - e^{-2t/\vartheta}\right). \tag{29b}$$

Hence, at each time t the transition pdf is normal with mean $M(t|x_0)$ and variance $V(t)$. It must be pointed out that the OU model differs from the Wiener model in some relevant respects. For instance, it possesses an equilibrium regime in that in the limit as $t \to \infty$ the pdf (28) becomes normal with mean $\mu\vartheta$ and variance $\sigma^2\vartheta/2$. Furthermore, the crossing of the firing threshold is a sure event. However, differently from the Wiener model, the FPT problem is in general unsolvable, even in the case of constant thresholds and analytical solutions can only be determined in very particular cases, of no interest in the neuronal modelling context.

To give an idea of the difficulties one meets when trying to solve the FPT problem for the OU neuronal model for arbitrary thresholds, let us make use of the approach outlined in Section 2 for the case of a constant threshold S. Eq. (21) then reads:

$$\frac{\sigma^2}{2}\frac{d^2 g_\lambda}{dx_0^2} + \left(\mu - \frac{x_0}{\vartheta}\right)\frac{dg_\lambda}{dx_0} - \lambda g_\lambda = 0. \tag{30}$$

Eq. (30) can be solved with conditions (22) to yield:

$$g_\lambda(S|x_0) = \exp\left[\frac{(x_0 - \mu\,\vartheta)^2 - (S - \mu\,\vartheta)^2}{2\,\sigma^2\,\vartheta}\right] \frac{D_{-\lambda\,\vartheta}\left[\sqrt{\frac{2}{\sigma^2\,\vartheta}}\,(\mu\,\vartheta - x_0)\right]}{D_{-\lambda\,\vartheta}\left[\sqrt{\frac{2}{\sigma^2\,\vartheta}}\,(\mu\,\vartheta - S)\right]}, \qquad (31)$$

where $D_\nu(z)$ is the Parabolic Cylinder Function, or Weber function:

$$D_\nu(x) = \sqrt{\pi}\,2^{\nu/2}\,e^{-x^2/4}\left[\frac{1}{\Gamma[(1-\nu)/2]}\,\Phi\left(-\frac{\nu}{2}, \frac{1}{2}; \frac{x^2}{2}\right)\right.$$
$$\left. - \frac{\sqrt{2}}{\Gamma(-\nu/2)}\,x\,\Phi\left(\frac{1-\nu}{2}, \frac{3}{2}; \frac{x^2}{2}\right)\right],$$

where $\Phi(a, c; x)$ denotes the Kummer function

$$\Phi(a, c; x) = \sum_{n=1}^{\infty} \frac{a\,(a+1)\,\ldots\,(a+n-1)}{c\,(c+1)\,\ldots\,(c+n-1)}\,\frac{x^n}{n!}.$$

An alternative expression for $g_\lambda(S|x_0)$ can be found in[19]. In any case, the function g_λ is too complicated to lead to any useful approximation for its inverse LT (which is not known in closed form). Hence, the firing time problem for the OU neuronal model must be investigated by different means.

We conclude this Section by stressing the fundamental differences existing between the Wiener and the OU neuronal models. The Wiener model is unrealistically lacking the presence of the finite time–constant of the neuronal membrane, but it is susceptible of a closed form solution. Indeed, the firing pdf has the simple expression (25) (from which one can also calculate mean, variance, skewness, etc. of the firing time in the presence of a constant threshold). The OU model, although more realistically depicting the neurophysiological reality, does not allow us to obtain any closed form expression for the firing pdf. However, the moments of the firing time for the case of a constant threshold can be obtained analytically. Eventhough their expressions are too complicated to allow for any intuitive understanding in terms of the values assigned to the involved parameters, they can be evaluated by standard computational methods.

4 Numerical Evaluation of the Firing pdf

In Section 3 we have stressed the lack of a closed form solution to the firing pdf for the OU neuronal model. We shall now briefly indicate an efficient procedure to obtain accurate numerical evaluations for the general case of time–varying thresholds and for arbitrary one–dimensional diffusion models (viz. not necessarily of the OU type).

As mentioned in the foregoing, the calculation of the FPT pdf is essentially equivalent to the determination of a transition pdf in the presence of an absorbing boundary. This, in turn, is a very complicate task. Hence, efficient numerical algorithms are desirable, especially if one wishes to deal with the case of a time–varying neuronal threshold. It is fair to say that the first fundamental contribution to the solution of this problem is due to Durbin [5] for the case of a standard Wiener process and a time–dependent boundary. A purely numerical method was successively developed by Anderssen et al. [1]. The idea is based on the remark that Eq. (15) also holds in the limit as $x \to S(t)$, i.e. when the constant boundary S is changed into a continuous function $S(t)$ and the arbitrary state x at time t indefinitely approaches the value that the threshold takes at that time. Hence, Eq. (3.6) becomes:

$$f[S(t),t|x_0,t_0] = \int_0^t f[S(t),t|S(\tau),\tau] \, g[S(\tau),\tau|x_0,t_0] \, d\tau, \qquad (32)$$

with $x_0 < S(t_0)$. The determination of $g[S(t),t|x_0,t_0]$ for a standard Wiener process was accomplished by Anderssen et al. [1] by a method based on the expansion of g in terms of Legendre polynomials and on the solution of the resulting system of linear equations. Such a method was then modified by Favella et al. [6] and applied to the numerical solution of the FPT problem for the OU process. However, all these algorithms necessitate of the use of large computation facilities and of sophisticated library programs. As a consequence, they are expensive to run and not suitable to suggest the modeler in real time how to pick the various parameters to fit the recorded data or to decide whether to reject the postulated model. Some progress along this line was made by adopting an entirely different approach ([2]). The guiding idea was to remove the singularity of the kernel of Eq. (32) (cf. also Section 2) by trying to modify the equation itself. While referring to the quoted paper for a complete treatment, here we limit ourselves to mentioning the essential results. Let

$$F(x,t|y,\tau) = P\{X(t) \le x|X(\tau) = y\} \equiv \int_{-\infty}^x f(z,t|y,\tau) \, dz \qquad (33)$$

be the transition distribution function of the membrane potential $X(t)$ and let $S(t)$ be a sufficiently smooth threshold function. Then, for $x_0 < S(t_0)$ the FPT pdf $g[S(t),t|x_0,t_0]$ of $X(t)$ through $S(t)$ satisfies the following second–kind Volterra integral equation:

$$g[S(t),t|x_0,t_0] = -2\,\psi[S(t),t|x_0,t_0] + 2 \int_{t_0}^t g[S(\tau),\tau|x_0,t_0] \, \psi[S(t),t|S(\tau),\tau] \, d\tau, \qquad (34)$$

where we have set

$$\begin{cases} \psi[S(t),t|y,\tau] = \varphi[S(t),t|y,\tau] + k(t)\,f[S(t),t|y,\tau] \\[2mm] \varphi(x,t|y,\tau) = \dfrac{d}{dt}\,F(x,t|y,\tau), \end{cases} \qquad (35)$$

with $k(t)$ an arbitrary continuous function. Apparently, Eq. (34) is more complicate than Eq. (32). However, it possesses an extra degree of freedom, viz.

the arbitrary function $k(t)$. Its specification can be made in a way to remove the singularity of the kernel or, when possible, to make the integral on the right–hand–side of (34) vanish, thus obtaining the closed form solution $g[S(t), t|x_0, t_0] = -2\ \psi[S(t), t|x_0, t_0]$. To illustrate this method, let us separately consider the Wiener and the OU neuronal models.

We start with the Wiener model. One has (cf. Section 3):

$$f(x, t|y, \tau) = \frac{1}{\sqrt{2\pi\sigma^2 (t - \tau)}}\ \exp\left\{-\frac{[x - y - \mu (t - \tau)]^2}{2\sigma^2 (t - \tau)}\right\}. \tag{36}$$

Hence,

$$F(x, t|y, \tau) = \frac{1}{2}\left[1 + \text{Erf}\left\{\frac{[x - y - \mu (t - \tau)]}{\sqrt{2\sigma^2 (t - \tau)}}\right\}\right]. \tag{37}$$

From (35) and (37) we then have:

$$\psi[S(t), t|y, \tau] = f[S(t), t|y, \tau]\ h(t, \tau, y), \tag{38}$$

where the function $h(t, \tau, y)$ is defined as follows:

$$h(t, \tau, y) = S'(t) - \frac{\mu}{2} - \frac{S(t) - y}{2(t - \tau)} + k(t), \tag{39}$$

with $S'(t) = dS(t)/dt$. The following result holds. Let $S(t)$ be twice differentiable with continuous derivatives. Then,

$$\lim_{\tau\uparrow t} \psi[S(t), t|S(\tau), \tau] = 0 \tag{40}$$

if and only if

$$k(t) = \frac{1}{2}\ [\mu - S'(t)]. \tag{41}$$

Furthermore, one has:

$$\psi[S(t), t|S(\tau), \tau] = 0 \quad \forall t, \tau\ :\ t_0 \le \tau < t \quad and \quad \lim_{\tau\uparrow t} \psi[S(t), t|S(\tau), \tau] = 0 \tag{42}$$

if and only if

$$\begin{cases} S(t) = a\,t + b \\[2mm] k(t) = \dfrac{\mu - a}{2} \end{cases} \tag{43}$$

with a and b arbitrary real numbers. In other words, if a smooth threshold $S(t)$ is assigned, the firing pdf for the Wiener neuronal model satisfies Eq. (34) in which the kernel $\psi[S(t), t|S(\tau)]$ is made continuous (actually vanishing) at $\tau = t$ with the choice (41) of $k(t)$ in (38). Hence, a straightforward numerical solution of (34) is possible without need to resort to main frame computers or program libraries (cf. [2]).

Similar considerations can be made for the case of a diffusion process with linear drift and constant infinitesimal variance:

$$\begin{cases} A_1(x) = \alpha\, x + \beta, \\ A_2(x) = \sigma^2, \end{cases} \tag{44}$$

α, β and $\sigma \neq 0$ being arbitrary real constants. (Note that for $\alpha = 1/\vartheta$ and $\beta = \mu$ this process identifies with the OU neuronal model of Section 3). Then, under the initial condition $P\{X(t_0) = x_0\} = 1$, for all $y \in \Re$ and $\tau < t$ one has:

$$f(x,t|y,\tau) = \left\{ \frac{\exp\left[-2\,\alpha\,(t-\tau)\right]\alpha}{\pi\,\sigma^2\,\left[1 - \exp\left(-2\,\alpha\,(t-\tau)\right)\right]} \right\}^{1/2}$$

$$\times \exp\left\{ \frac{-\alpha\,\left[(x + \beta/\alpha)\,\exp\left[-\alpha\,(t-\tau)\right] - (y + \beta/\alpha)\right]^2}{\sigma^2\,\left[1 - \exp\left(-2\,\alpha\,(t-\tau)\right)\right]} \right\} \tag{45}$$

and

$$F(x,t|y,\tau) = \frac{1}{2}\left\{ 1 + \mathrm{Erf}\left[\left\{ \frac{\alpha}{\sigma^2\,\left[1 - \exp\left(-2\,\alpha\,(t-\tau)\right)\right]} \right\}^{1/2} \right. \right.$$

$$\left. \left. \times \left[(x + \beta/\alpha)\,\exp\left[-\alpha\,(t-\tau)\right] - (y + \beta/\alpha)\right] \right] \right\}. \tag{46}$$

From (35) and (46) we thus have:

$$\psi[S(t),t|y,\tau] = f[S(t),t|y,\tau]\,H(t,\tau,y)\,, \tag{47}$$

with

$$H(t,\tau,y) = S'(t) - \alpha\left[S(t) + \frac{\beta}{\alpha}\right] - \frac{\alpha\,\exp\left[-\alpha\,(t-\tau)\right]}{1 - \exp\left[-2\,\alpha\,(t-\tau)\right]}$$

$$= \left\{ \left[S(t) + \frac{\beta}{\alpha}\right]\exp\left[-\alpha\,(t-\tau)\right] - \left(y + \frac{\beta}{\alpha}\right) \right\} + k(t). \tag{48}$$

Let now again $S(t)$ be a twice differentiable function with continuous derivatives. Then [2],

$$\lim_{\tau \uparrow t} \psi[S(t),t|S(\tau),\tau] = 0 \tag{49}$$

if and only if

$$k(t) = \frac{1}{2}\left[\alpha\,S(t) + \beta - S'(t)\right]. \tag{50}$$

Furthermore:

$$\psi[S(t),t|S(\tau),\tau] = 0 \quad \forall t,\tau\,:\,t_0 \leq \tau < t \quad and \quad \lim_{\tau \uparrow t} \psi[S(t),t|S(\tau),\tau] = 0 \tag{51}$$

if and only if

$$\begin{cases} S(t) = -\dfrac{\beta}{\alpha} + A\,e^{\alpha\,t} + B\,e^{-\alpha\,t} \\[2mm] k(t) = B\,\alpha\,e^{-\alpha\,t}, \end{cases} \tag{52}$$

with A and B arbitrary constants. Therefore, also for the OU neuronal model the firing pdf is a solution of Eq. (34) in which the kernel is continuous at

$\tau = t$ if the choice (50) of $k(t)$ is made. The numerical solution of Eq. (32) under the continuity conditions (40) and (49), and hence with the choices (41) and (50), respectively, of the originally arbitrary function $k(t)$, becomes a rather straightforward and simple task [2]. Therefore, by using a personal computer and without any need of library programs, the Wiener and the OU neuronal models can be investigated for arbitrary time–varying thresholds and hence parameters identification problems can be effectively approached.

One might wander whether the regularization procedure of Eq. (32) is limited to the above two neuronal models. Actually, a more general result can be proved. Indeed, under mild assumptions one can prove that for any time–homogeneous diffusion processes characterized by drift $A_1(x)$ and infinitesimal variance $A_2(x)$, the kernel of Eq. (32) can be made vanishing at $\tau = t$ by selecting [12]

$$k(t) = \frac{1}{2}\left\{ A_1[S(t)] - S'(t) - \frac{A_2'[S(t)]}{4} \right\}, \tag{53}$$

a condition that includes (41) and (50) as particular cases.

We conclude this Section by providing an example in which eq. (34) is implemented to evaluate the FPT pdf for the standard Wiener process originating at $x_0 = 0$ at time $t_0 = 0$ for a Daniel's boundary [4]

$$S(t) = \frac{H}{2} + \mu t - \frac{\sigma^2 t}{H} \ln \frac{[c_1 + \sqrt{\Delta(t)}]}{2}, \tag{54}$$

with $\Delta(t) = c_1^2 + 4c_2 \exp(-H^2/\sigma^2 t)$, $H \neq 0$, $c_1 > 0$ and c_2 real numbers such that $c_1^2 + 4c_2 > 0$. Indeed, for such a boundary a closed form solution to the FPT problem is known:

$$g[S(t), t|0, 0] = |H|[2\pi\sigma^2 t^3]^{-1/2}$$
$$\times \left\{ \exp\left[-\frac{[S(t) - \mu t]^2}{2\sigma^2 t} \right] - \frac{c_1}{2} \exp\left[-\frac{[S(t) - H - \mu t]^2}{2\sigma^2 t} \right] \right\}. \tag{55}$$

On the other hand, from Eq. (34) we have:

$$g[S(t_0 + kh), t_0 + kh|x_0, t_0] = -2\psi[S(t_0 + kh), t_0 + kh|x_0, t_0]$$
$$+ 2\int_{t_0}^{t_0+kh} d\tau\, g[S(\tau), \tau|x_0, t_0]\, \psi[S(t_0 + kh), t_0 + kh|S(\tau), \tau], \tag{56}$$
$$(k = 1, 2, \ldots)$$

where $h > 0$ is the integration step. Note that by taking $k(t)$ as in (41), condition (40) is satisfied. Hence, from (56) we obtain

$$g_1[S(t_0 + h), t_0 + h|x_0, t_0] = -2\,\psi[S(t_0 + h), t_0 + h|x_0, t_0],$$
$$g_1[S(t_0 + kh), t_0 + kh|x_0, t_0] = -2\,\psi[S(t_0 + kh), t_0 + kh|x_0, t_0]$$
$$+ 2h \sum_{j=1}^{k-1} g_1[S(t_0 + jh), t_0 + jh|x_0, t_0]$$
$$\times \psi[S(t_0 + kh), t_0 + kh|S(t_0 + jh), t_0 + jh], \quad (k = 2, 3, \ldots) \tag{57}$$

where use of a composite trapezium rule has been made.

Table 1. Standard Wiener process and Daniels boundary with $\mu = 0$, $\sigma^2 = H = 2$, $c_1 = 3$ and $c_2 = 1$. For the integration steps $h = 0.01$ and $h = 0.001$ the evaluated first-passage-time p.d.f. g_1, absolute error $\Delta = g - g_1$, relative error $\rho = g^{-1}(g - g_1)$ and the computed probability mass m are listed for $t_j = 0.1 + 0.2i$ ($i = 0, 1, \ldots, 14$).

	t	g_1	Δ	ρ	m
$h = 0.01$	0.1	$0.1230595E+01$	$-0.6587342E-07$	$-0.5352972E-07$	$0.4298450E-01$
	0.3	$0.1180677E+01$	$-0.7456284E-06$	$-0.6315265E-06$	$0.3224750E+00$
	0.5	$0.7226209E+00$	$-0.1413565E-05$	$-0.1956168E-05$	$0.5082466E+00$
	0.7	$0.4759008E+00$	$-0.1014729E-05$	$-0.2132232E-05$	$0.6256584E+00$
	0.9	$0.3342791E+00$	$-0.6164188E-06$	$-0.1844028E-05$	$0.7054863E+00$
	1.1	$0.2458567E+00$	$-0.3506283E-06$	$-0.1426151E-05$	$0.7628643E+00$
	1.3	$0.1869531E+00$	$-0.2070343E-06$	$-0.1107414E-05$	$0.8057754E+00$
	1.5	$0.1457657E+00$	$-0.1340091E-06$	$-0.9193472E-06$	$0.8388170E+00$
	1.7	$0.1158833E+00$	$-0.7940904E-07$	$-0.6852505E-06$	$0.8648307E+00$
	1.9	$0.9356882E-01$	$-0.5101326E-07$	$-0.5451954E-06$	$0.8856724E+00$
	2.1	$0.7651656E-01$	$-0.3506552E-07$	$-0.4582738E-06$	$0.9026077E+00$
	2.3	$0.6323746E-01$	$-0.2302033E-07$	$-0.3640299E-06$	$0.9165299E+00$
	2.5	$0.5273317E-01$	$-0.1197061E-07$	$-0.2270035E-06$	$0.9280873E+00$
	2.7	$0.4431297E-01$	$-0.1273337E-07$	$-0.2873510E-06$	$0.9377618E+00$
	2.9	$0.3748645E-01$	$-0.7072146E-08$	$-0.1886588E-06$	$0.9459185E+00$
$h = 0.001$	0.1	$0.1230595E+01$	$-0.6584439E-07$	$-0.5350614E-07$	$0.4288606E-01$
	0.3	$0.1180676E+01$	$-0.4559044E-07$	$-0.3861383E-07$	$0.3224995E+00$
	0.5	$0.7226196E+00$	$-0.9183196E-07$	$-0.1270820E-07$	$0.5082599E+00$
	0.7	$0.4758998E+00$	$-0.4728551E-07$	$-0.9936024E-07$	$0.6256654E+00$
	0.9	$0.3342786E+00$	$-0.3431984E-07$	$-0.3431984E-07$	$0.7054902E+00$
	1.1	$0.2458564E+00$	$-0.1386206E-07$	$-0.5638277E-07$	$0.7628664E+00$
	1.3	$0.1869529E+00$	$-0.8431609E-07$	$-0.4510018E-07$	$0.8057766E+00$
	1.5	$0.1457655E+00$	$-0.1903176E-07$	$-0.1305642E-07$	$0.8388175E+00$
	1.7	$0.1158832E+00$	$-0.9313979E-08$	$-0.8037383E-07$	$0.8648308E+00$
	1.9	$0.9356878E-01$	$-0.7203846E-08$	$-0.7698985E-07$	$0.8856723E+00$
	2.1	$0.7651653E-01$	$-0.2879011E-08$	$-0.3762600E-07$	$0.9026074E+00$
	2.3	$0.6323746E-01$	$-0.5258141E-08$	$-0.8314916E-07$	$0.9165294E+00$
	2.5	$0.5273316E-01$	$-0.2016633E-08$	$-0.3824223E-07$	$0.9280868E+00$
	2.7	$0.4431296E-01$	$-0.6087457E-08$	$-0.1373742E-06$	$0.9377612E+00$
	2.9	$0.3748644E-01$	$-0.1826937E-08$	$-0.4873594E-07$	$0.9459179E+00$

In Table 1 approximation g_1 to g, errors Δ and relative errors ρ are listed for the case of the standard Wiener process originating in the origin at time 0 for the boundary (54) with $\mu = 0$, $\sigma^2 + H + 2$, $c_1 = 3$ and $c_2 = 1$. The last column shows the integrated probability mass m up to the considered time.

5 The Moments of the Firing Time

The computational method outline in Section 4 to evaluate the firing pdf via the numerical solution of the integral equation is useful as it can be implemented also for time time–dependent thresholds. As we shall soon see, it has also proved to be precious to discover certain types of features exhibited by the firing pdf that would have otherways passed unnoticed. However, we shall now indicate an interesting computational result with reference to the OU neuronal model with a constant threshold S. For the sake of simplicity, let us consider the OU process characterized by infinitesimal moments (26) in which we set $\mu = 0$. Then, making use of (23) and (31) – or, alternatively, by means of a method due to Nobile et al. [19] – the following results can be proved:

$$t_1(S|x_0) = \vartheta \left\{ \sqrt{\pi} \left[\varphi_1 \left(\frac{S}{\sigma\sqrt{\vartheta}} \right) - \varphi_1 \left(\frac{x_0}{\sigma\sqrt{\vartheta}} \right) \right] + \psi_1 \left(\frac{S}{\sigma\sqrt{\vartheta}} \right) - \psi_1 \left(\frac{x_0}{\sigma\sqrt{\vartheta}} \right) \right\}$$

$$(58a)$$

$$t_2(S|x_0) = 2\,\vartheta\,t_1(S|x_0) \left[\sqrt{\pi}\,\varphi_1 \left(\frac{S}{\sigma\sqrt{\vartheta}} \right) + \psi_1 \left(\frac{S}{\sigma\sqrt{\vartheta}} \right) \right]$$
$$+ 2\,\vartheta^2 \left\{ \sqrt{\pi}\,\ln 2 \left[\varphi_1 \left(\frac{S}{\sigma\sqrt{\vartheta}} \right) - \varphi_1 \left(\frac{x_0}{\sigma\sqrt{\vartheta}} \right) \right] \right.$$
$$\left. - \sqrt{\pi} \left[\varphi_2 \left(\frac{S}{\sigma\sqrt{\vartheta}} \right) - \varphi_2 \left(\frac{x_0}{\sigma\sqrt{\vartheta}} \right) \right] - \psi_2 \left(\frac{S}{\sigma\sqrt{\vartheta}} \right) + \psi_2 \left(\frac{x_0}{\sigma\sqrt{\vartheta}} \right) \right\}$$
$$(58b)$$

$$t_3(S|x_0) = 3\,\vartheta\,t_2(S|x_0) \left[\sqrt{\pi}\,\varphi_1 \left(\frac{S}{\sigma\sqrt{\vartheta}} \right) + \psi_1 \left(\frac{S}{\sigma\sqrt{\vartheta}} \right) \right]$$
$$+ 6\,\vartheta^2\,t_1(S|x_0) \left[\sqrt{\pi}\,\ln 2\,\varphi_1 \left(\frac{S}{\sigma\sqrt{\vartheta}} \right) - \sqrt{\pi}\,\varphi_2 \left(\frac{S}{\sigma\sqrt{\vartheta}} \right) - \psi_2 \left(\frac{S}{\sigma\sqrt{\vartheta}} \right) \right]$$
$$+ 3\,\vartheta^3\,\sqrt{\pi} \left(\ln^2 2 + \frac{\pi^2}{12} \right) \left[\varphi_1 \left(\frac{S}{\sigma\sqrt{\vartheta}} \right) - \varphi_1 \left(\frac{x_0}{\sigma\sqrt{\vartheta}} \right) \right]$$
$$- 6\,\vartheta^3\,\sqrt{\pi}\,\ln 2 \left[\varphi_2 \left(\frac{S}{\sigma\sqrt{\vartheta}} \right) - \varphi_2 \left(\frac{x_0}{\sigma\sqrt{\vartheta}} \right) \right]$$
$$+ 6\,\vartheta^3\,\sqrt{\pi} \left[\varphi_3 \left(\frac{S}{\sigma\sqrt{\vartheta}} \right) - \varphi_3 \left(\frac{x_0}{\sigma\sqrt{\vartheta}} \right) \right]$$
$$+ 6\,\vartheta^3 \left[\psi_3 \left(\frac{S}{\sigma\sqrt{\vartheta}} \right) - \psi_3 \left(\frac{x_0}{\sigma\sqrt{\vartheta}} \right) \right],$$
$$(58c)$$

where we have set:

$$\varphi_1(z) = \int_0^z e^{t^2}\,dt \qquad (59a)$$

$$\varphi_2(z) = \sum_{n=0}^{\infty} \frac{z^{2n+3}}{(n+1)!\,(2n+3)} \sum_{k=0}^{n} \frac{1}{2k+1} \qquad (59b)$$

$$\varphi_3(z) = \sum_{n=0}^{\infty} \frac{z^{2n+5}}{(n+2)!\,(2n+5)} \sum_{k=0}^{n} \frac{1}{2k+3} \sum_{j=0}^{k} \frac{1}{2j+1} \qquad (59c)$$

$$\psi_1(z) = \sum_{n=0}^{\infty} \frac{2^n}{(n+1)\,(2n+1)!!}\, z^{2n+2} \qquad (59d)$$

$$\psi_2(z) = \sum_{n=0}^{\infty} \frac{2^n\, z^{2n+4}}{(2n+3)!!\,(n+2)} \sum_{k=0}^{n} \frac{1}{k+1} \qquad (59e)$$

$$\psi_3(z) = \sum_{n=0}^{\infty} \frac{2^n\, z^{2n+6}}{(2n+5)!!\,(n+3)} \sum_{k=0}^{n} \frac{1}{k+2} \sum_{j=0}^{k} \frac{1}{j+1}. \qquad (59f)$$

The above expressions are much too complicated to shade any light on the quantitative behaviour of the moments as a function of the involved parameters and of the neuronal threshold. Nevertheless, some unexpected features have been discovered as a result of systematic computations in which the mean t_1 and the variance $V = t_2 - t_1^2$ of the firing time have been obtained. As shown in Nobile et al. [19], for positive boundaries of the order of a couple of units or more, whatever the initial state the variance of the firing time equals the square of its mean value to an excellent degree of approximation. This is, for instance, shown by Table 2 in which t_1 and V are listed in the first two columns for various thresholds and initial states. In the third column the values of t_1^2 are listed. Note that the goodness of the agreement between t_1^2 and V increases as the threshold S moves away from the starting point x_0. All this is clearly suggestive of an exponential approximation to the firing pdf for large thresholds. However, if such an approximation holds, for large thresholds the skewness of the firing time, viz. $t_1^3/(t_2 - t_1^2)^{3/2}$, should approach the value 2. As the fourth column of Table 2 clearly shows, this is indeed the case, to a very high degree of precision. Putting all this information together, the conjecture emerges that the firing pdf is susceptible of an excellent exponential approximation for a wide range of thresholds and of initial states. As a matter of fact, these "experimental" results have led us to formulate theorems on the asymptotic exponential behaviour of the FPT pdf not only for the OU process but also for a wider class of diffusion processes, both for constant and for time–varying boundaries (Nobile et al. [20], Giorno et al. [13]).

As is to be expected, the convergence to the exponential distribution for increasing thresholds is accompanied by a large increase of the mean firing time. Indeed, as the threshold moves away from the reset voltage, spikes become more and more rear and hence the mean interval between successive neuronal firings grows larger and larger. This kind of behaviour is in agreement with the finding that for some neurons the histograms of the recorded interpike intervals become increasingly better fitted by exponential functions as the firing rates decrease (see, for instance, Škvařil et al. [26]).

Table 2. Mean t_1, variance V and skewness Σ of the first–passage time through the boundary S for an Ornstein–Uhlenbeck process originating at $x_0 = 3$, 0 and -2, respectively (unpublished data. Note that some of the present data are slightly different from those of [3], due to an improvement of the computation procedure). The values of t_1^2 have been also reported to show that they equal those of V to a degree of approximation that improves as the boundary increases. Note that as the boundary increases the values of t_1, V and Σ become insensitive to the starting point of the process.

	S	t_1	V	t_1^2	Σ
$x_0 = 3$	4	$0.193146E+04$	$0.406163E+07$	$0.373054E+07$	2.00533
	5	$0.140654E+06$	$0.198076E+11$	$0.197836E+11$	2.00000
	6	$0.282674E+08$	$0.799052E+15$	$0.799047E+15$	2.00000
	7	$0.159800E+11$	$0.255359E+21$	$0.255359E+21$	2.00000
$x_0 = 0$	1	$0.209341E+01$	$0.584203E+01$	$0.438235E+01$	2.29375
	2	$0.104284E+02$	$0.105275E+03$	$0.108752E+03$	2.01050
	3	$0.869316E+02$	$0.742438E+04$	$0.755711E+04$	1.99998
	4	$0.201839E+04$	$0.406906E+07$	$0.407391E+07$	2.00000
	5	$0.140741E+06$	$0.198076E+11$	$0.198080E+11$	2.00000
	6	$0.282675E+08$	$0.799052E+15$	$0.799052E+15$	2.00000
	7	$0.159800E+11$	$0.255359E+21$	$0.255359E+21$	2.00000
$x_0 = -2$	-1	$0.523297E+00$	$0.215861E+00$	$0.273839E+00$	2.21670
	0	$0.142520E+01$	$0.106694E+01$	$0.203121E+01$	1.84970
	1	$0.351861E+01$	$0.690897E+01$	$0.123806E+02$	1.89575
	2	$0.118536E+02$	$0.106342E+03$	$0.140508E+03$	1.98218
	3	$0.883568E+02$	$0.742545E+04$	$0.780693E+04$	1.99955
	4	$0.201982E+04$	$0.406906E+07$	$0.407966E+07$	2.00000
	5	$0.140742E+06$	$0.198076E+11$	$0.198084E+11$	2.00000
	6	$0.282675E+08$	$0.799052E+15$	$0.799052E+15$	2.00000
	7	$0.159800E+11$	$0.255359E+21$	$0.255359E+21$	2.00000

The OU neuronal models considered in the foregoing can be modified in a way to include a time–dependent drift. For instance, it is conceivable that the model (26) be changed into

$$
\begin{cases}
A_1(x) = -\dfrac{x}{\vartheta} + \mu + P(t) \\[2mm]
A_2(x) = \sigma^2
\end{cases}
\tag{60}
$$

if a time–dependent extra–effect is induced by some kind of DC external stimulation acting on the neuron. It is, in particular, interesting to investigate the behaviour of the model (60) when $P(t)$ is a periodic function of period T that reflects some oscillatory action of the environment. This situation quite naturally leads us to considering an FPT problem for the OU process (26) and a periodic boundary. To give a hint of why this is so, let us refer to the OU neuronal model with infinitesimal moments (60), where $P(t)$ is periodic with period T. It is then intuitive that by a suitable transformation an FPT problem through a constant threshold S_0 can be changed into an FPT problem for a time–independent OU model through a periodic boundary. Putting it the other way around, an FPT problem through a periodic boundary $S(t)$ for the OU process defined via

$$
\begin{cases}
dX(t) = -\dfrac{1}{\vartheta} X(t)\, dt + \sigma\, dW(t) \\[2mm]
P\{X(0) = x_0\} = 1,
\end{cases}
\tag{61}
$$

is equivalent to the FPT problem through the constant boundary $S(0) = S_0$ for the process

$$
Y(t) = X(t) - S(t) + S_0
\tag{62}
$$

that satisfies

$$
\begin{cases}
dY(t) = \left[-\dfrac{1}{\vartheta} Y(t)\, dt - \dfrac{S(t)}{\vartheta} + \dfrac{S_0}{\vartheta} - S'(t) \right] dt + \sigma\, dW(t) \\[2mm]
P\{Y(0) = x_0\} = 1.
\end{cases}
\tag{63}
$$

In other words, both $X(t)$ and $Y(T)$ start at x_0 at time 0. However, $Y(t)$ is an OU process containing a periodic drift and the threshold is constant while $X(t)$ is a time–independent linear–drift process but the corresponding threshold is a periodic function. To approach the problem of determining the firing pdf, use of the method outlined in Section 4 has been made. We have thus been able to prove that the periodicity of the threshold induces an oscillatory behaviour of the firing pdf which, in turn, approaches a time non–homogeneous exponential limit as the threshold moves farther and farther away from the reset value of the process. For instance, for the normalized OU process, viz. for $A_1 = -x$ and $A_2 = 2$, as the periodic threshold $S(t)$ of period T moves away from the equilibrium point (the origin, in this case) in the positive direction, the following asymptotic relation holds for the FPT pdf $g(t)$:

$$
\tag{64}
g(t) \sim Z(t)\, e^{-\lambda t},
$$

where λ is a positive constant and $Z(t)$ is a periodic function of period T. It is quite impressive to see to what degree of accuracy relation (64) emerges as a consequence of the use of the computational method of Section 4 (cf. Giorno et al. [13]). Figures 1–6 show the FPT pdf and corresponding functions $Z(t)$ for the OU process with drift $-x$ and infinitesimal variance 2 through some oscillatory boundaries. Note that (64) is satisfied to a very high degree of accuracy.

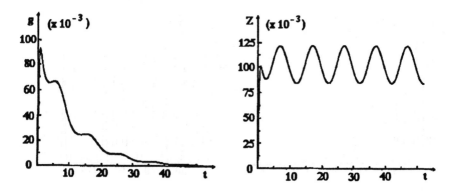

Fig. 1. OU process with $A_1(x) = -x$, $A_2 = 2$, originating at $x_0 = 0$ and boundary $S(t) = A + C \ sin(2\pi t/T)$. FPT density and function $Z(t)$ in (64) with $\lambda = 0.98032053E - 1$ are plotted for $A = 2$, $C = 0.1$, $T = 10$.

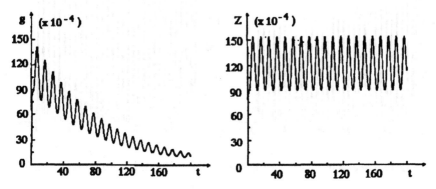

Fig. 2. Same as in Figure 1 with $A = 3$, $C = 0.1$, $T = 10$, $\lambda = 0.11810255E - 1$.

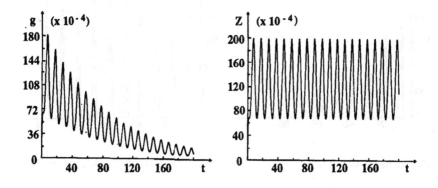

Fig. 3. Same as in Figure 1 with $A = 3$, $C = 0.2$, $T = 10$, $\lambda = 0.12343852E - 1$.

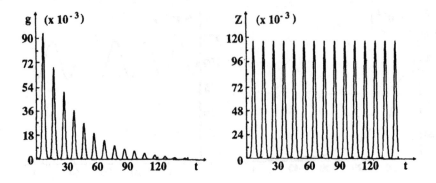

Fig. 4. Same as in Figure 1 with $A = 3$, $C = 1$, $T = 10$, $\lambda = 0.31411586E - 1$.

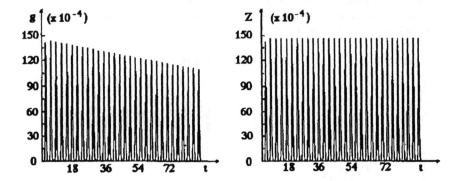

Fig. 5. Same as in Figure 1 with $A = 4$, $C = 1$, $T = 3$, $\lambda = 0.3297975E - 2$.

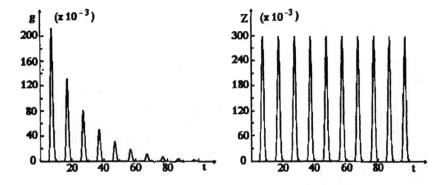

Fig. 6. Same as in Figure 1 with $A = 4$, $C = 2.5$, $T = 10$, $\lambda = 0.48428087E - 1$.

References

1. Anderssen, R.S., DeHoog, F.R. and Weiss, R.: On the numerical solution of Brownian motion processes. J. Appl. Prob. **10** (1973) 409–418

2. Buonocore, A., Nobile, A.G. and Ricciardi, L.M.: A new integral equation for the evaluation of first–passage–time probability densities. Adv. Appl. Prob. **19** (1987) 784–800

3. Cerbone, G., Ricciardi, L.M. and Sacerdote, L.: Mean, variance and skewness of the first passage time for the Ornstein–Uhlenbeck process. Cybern. Syst. **12** (1981) 395–429

4. Daniels, H.E.: The minimum of a stationary Markov process superimposed on a U–shaped trend. J. Appl. Prob. **6** (1969) 399–408

5. Durbin, J.: Boundary–crossing probabilities for the Brownian motion and Poisson processes and techniques for computing the power of the Kolmogorov–Smirnov test. J. Appl. Prob. **8** (1971) 431–453

6. Favella, L., Reineri, M.T., Ricciardi, L.M. and Sacerdote, L.: First passage time problems and some related computational problems. Cybernetics and Systems **13** (1982) 95–128

7. Feller, W.: Parabolic differential equations and semigroup transformations. Ann. Math. **55** (1952) 468–518

8. Feller, W.: Diffusion processes in one dimension. Trans. Amer. Math. Soc. **77** (1954) 1–31

9. Feller, W.: An Introduction to Probability Theory and its Applications vol.2. Wiley, New York, 1966

10. Gerstein, G.L. and Mandelbrot, B.: Random walk models for the spike activity of a single neuron. Biophys. J. **4** (1964) 41–68

11. Giorno, V., Lánský, P., Nobile, A.G. and Ricciardi, L.M.: Diffusion approximation and first–passage–time problem for a model neuron. III. A birth–and–death process approach. Biol. Cybern. **58** (1988) 387–404

12. Giorno, V., Nobile, A.G., Ricciardi, L.M. and Sato, S.: On the evaluation of the first–passage–time densities via non–singular integral equations. Adv. Appl. Prob. **21** (1989) 20–36

13. Giorno, V., Nobile, A.G. and Ricciardi, L.M.: On the asymptotic behaviour of first–passage–time densities for one–dimensional diffusion processes and varying boundaries. Adv. Appl. Prob. **22** (1990) 883–914

14. Holden, A.V.: A note on convolution and stable distributions in the nervous system. Biol. Cybern. **20** (1975) 171–173

15. Karlin, S. and Taylor, H.M.: A Second Course in Stochastic Processes. Academic Press, New York, 1981

16. Lánský, P.: On approximations of Stein's neuronal model. J. Theor. Biol. **107** (1984) 631–647

17. Matsuyama, Y.: A note on stochastic modeling of shunting inhibition. Biol. Cybern. **24** (1976) 139–145

18. Matsuyama, Y., Shirai, K. and Akizuki, K.: On some properties of stochastic information processes in neurons and neuron populations. Kybernetik **15** (1974) 127–145

19. Nobile, A.G., Ricciardi, L.M. and Sacerdote, L.: Exponential trends of Ornstein–Uhlenbeck first–passage–time densities. J. Appl. Prob. **22** (1985) 360–369

20. Nobile, A.G., Ricciardi, L.M. and Sacerdote, L.: Exponential trends of first–passage–time densities for a class of diffusion processes with steady–state distribution. J. Appl. Prob. **22** (1985) 611–618

21. Park, C. and Schuurmann, F.J.: Evaluations of barrier–crossing probabilities of Wiener paths. J. Appl. Prob. **13** (1976) 267–275

22. Park, C. and Schuurmann, F.J.: Evaluations of absorption probabilities for the Wiener process on large intervals. J. Appl. Prob. **17** (1980) 363–372

23. Ricciardi, L.M.: Diffusion processes and related topics in Biology.(Lecture Notes in Biomathematics, vol. 14.), Springer, Berlin Heidelberg New York, 1977

24. Ricciardi, L.M.: Diffusion approximation and computational problems for single neurons activity. In: Amari S. and Arbib M.A. (eds.) Competition and Cooperation in Neural Nets. (Lecture Notes in Biomathematics, vol. 45.) Springer, New York. 1982, 143–154

25. Rodieck, R.W., Kiang, N.Y.–S. and Gerstein, G.L.: Some quantitative methods for the study of spontaneous activity of single neurons. Biophys. J. **2** (1962) 351–368

26. Škvařil, J., Radil–Weiss, T., Bohdanecký, Z. and Syka, J.: Spontaneous discharge patterns of mesencephalic neurons, interval histogram and mean interval relationship. Kybernetik **9** (1971) 11–15

Hierarchic Representation for Spatial Knowledge

Dorota H. Kieronska and Svetha Venkatesh

Department of Computer Science
Curtin University of Technology
Perth, PO Box U1987
Australia

1 Introduction

One of the fundamental issues in building autonomous agents is to be able to sense, represent and react to the world. Some of the earlier work [Mor83, Elf90, AyF89] has aimed towards a reconstructionist approach, where a number of sensors are used to obtain input that is used to construct a model of the world that mirrors the real world. Sensing and sensor fusion was thus an important aspect of such work. Such approaches have had limited success, and some of the main problems were the issues of uncertainty arising from sensor error and errors that accumulated in metric, quantitative models. Recent research has therefore looked at different ways of examining the problems. Instead of attempting to get the most accurate and correct model of the world, these approaches look at *qualitative* models to represent the world, which maintain *relative* and *significant* aspects of the environment rather than *all* aspects of the world. The relevant aspects of the world that are retained are determined by the *task* at hand which in turn determines how to sense. That is, task directed or purposive sensing is used to build a qualitative model of the world, which though inaccurate and incomplete is sufficient to solve the problem at hand.

This paper examines the issues of building up a hierarchical knowledge representation of the environment with limited sensor input that can be actively acquired by an agent capable of interacting with the environment. Different tasks require different aspects of the environment to be abstracted out. For example, low level tasks such as navigation require aspects of the environment that are related to layout and obstacle placement. For the agent to be able to reposition itself in an environment, significant features of spatial situations and their relative placement need to be kept. For the agent to reason about objects in space, for example to determine the position of one object relative to another, the representation needs to retain information on relative locations of start and finish of the objects, that is endpoints of objects on a grid. For the agent to be able to do high level planning, the agent may need only the relative position of the starting point and destination, and not the low level details of endpoints, visual clues and so on. This indicates that a hierarchical approach would be suitable, such that each level in the hierarchy is at a different level of abstraction, and thus suitable for a different task. At the lowest level, the representation contains low level details of agent's motion and visual clues to allow the agent to navigate and reposition itself. At the next level of abstraction the aspects of the representation allow the agent to perform spatial reasoning, and finally the highest level of abstraction in the representation can be used by the agent for high level planning.

2 Related Work

In work that aims at building 3D descriptions of environments, Ayache and Faugeras AyF89] use passive vision sensors, stereo. Geometric relations in the environment are computed by recursive prediction and verification algorithms that use the Extended Kalman Filter. In other work, Moravec [Mor83] used different types of stereo vision to model and find paths in the environment. Elfes [Elf90] proposed the use of a tessellated representation of spatial information called "Occupancy grid". The occupancy grid is updated using multi-view, multisensor input.

In parallel to the research done in 3D vision, other researchers have used a more qualitative approach to the representation of spatial environments. Kuipers et. al [KuB90] has proposed a spatial representation for indoor environments based on locally distinctive features. An agent explores the environment creating a map that describes locally distinctive places, and the path taken by the agent in traversing from one distinctive point to another. The agent uses a sonar sensor in the simulation and a local hill climbing algorithm is used to determine locally distinctive places. These features are not necessarily globally distinctive, and a rehearsal procedure is used to distinguish places that appear to have similar locally distinctive features.

An approach to the representation of large-scale, cross-country environments in terms of distinct landmarks can be found in Levitt [Lev87]. The main aim of this approach is navigation in an outdoor environment. The agent establishes a position in the environment in terms of visual landmarks that can be seen, and the relative position of these landmarks in terms of the agent's position.

Yeap and Handley present the notion of a Raw Cognitive Map which is made up of a network of Absolute Space Representations (ASR). An ASR is a representation of an environment that has been visited by the agent. The boundaries in an ASR are surfaces that are perceived as obstacles to the viewer's movement and the gaps between them are exits from an ASR. A hierarchy is then formed by grouping one or more ASR's as a *place*, and then groups of these as a higher level place and so on. The raw cognitive map is then organised as a hierarchy of place representations to form the full cognitive map.

In his work on representing space Hernandez [Her91] has used a topological/orientation pair to describe the position of an object. Topological relations are described in terms of Egenhofer and Franzosa's [EgF91] work on describing topological relations. They described these relations in terms of the four intersections of the boundaries and the interiors of two sets. Orientation is described in terms of qualitative directions: back, right-back, right, right-front, front, left-front, left, left-back. The orientations are defined in terms of a reference frame which can be either intrinsic, extrinsic or deictic.

One common theme to most of these formalisms is that any topographical map must be expressed in terms of some form of observable and distinguishable features. Such features fall into two groups: *situations* - that can be sensed directly in terms of visual or other sensory perception, and *objects* - abstract, high-level structures, for example buildings, doors, trees etc., which involve conceptual understanding and comprehension. The high-level features are easily recognized and dealt with by humans, however this is not so for machine perception.

This paper examines the construction of a hierarchic knowledge representation for spatial environments based on simple primitives that can be sensed. The levels of this hierarchy are useful at different levels of planning. The more abstract levels of the hierarchy can be used for high level planning, and the lower levels of the representation are linked directly to sensory percepts. The layout of this paper is as follows; Section 3 contains the spatial representation, and an example of the construction of this representation follows in section 4. Possible applications and usage of the hierarchic representation are discussed in Section 5, and our conclusions follow in Section 6.

3 The Spatial Representation

The knowledge representation we propose, called *active map*, is produced by an active, sighted agent in a large-scale environment. The agent is active if it is capable of autonomous motion and possibly other physical interaction with its environment.

3.1 Assumptions

We choose the following sensory inputs:

(i) *Proximity*: To be able to follow an upright surface (object) maintaining equal distance from it.

(ii) *Surface discrimination*: The agent has a visual capability limited to the recognition of selected visual clues. At this stage, upright bounding regions of objects, namely *surfaces*, are considered visual clues of importance. It is assumed that any two surfaces can touch but cannot intersect. Thus, in this paper the term "surface" refers to upright surfaces, and not to the terrain the agent is following.

(iii) *Absolute direction*: To determine the direction of an observation and determine the direction of self-motion. The absolute direction is expressed in terms of readings from a compass, as shown in Figure 1.

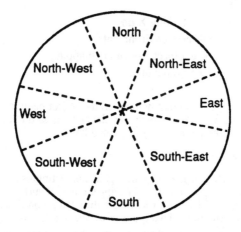

Fig. 1. Panoramic view of an agent in terms of fuzzy directions.

(iv) *Tracking zones* To reason with end-points of boundaries, we have to determine where the boundaries of surfaces start and finish. We propose the following definition for the start and finish of surfaces. When a new surface appears, the agent keeps a look out in four narrow areas aligned with the directional axes: N-S and E-W. These regions, shown in Fig. 2 are termed *tracking zones*.

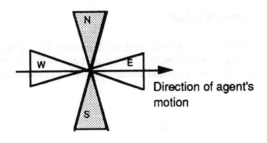

Direction of agent's motion

Fig. 2. Tracking Zones

3.2 Overview

As the agent traverses around an object, it observes the scene. The map is built incrementally as the agent exhaustively traverses the environment according to a simplified rule:

> Choose an unmapped object X
>
> Reach X
>
> **while** circumnavigating X **do**
>
> > Observe environment
> >
> > Note objects for future mapping
> >
> > Expand the map
>
> end

The surface closest to the agent that belongs to the object being circumnavigated, say surface *s*, is termed the followed surface, and the agent is said to be *following(s)*. The five levels of the active map are briefly described below:

Level 1: Sensory level

As the agent circumnavigates an object, the agent can distinguish simple spatio-temporal events relating to the appearance and disappearance of surfaces, and the change in direction of self-motion. The lowest level is a directed graph representing both topographic and visual information. Two types of events are defined:

(a)   *Virtual events*, associated either with a change in the agent's direction of motion, or with junctions marking the connectivity of two surfaces.

(b)   *Visual events*, associated with visual events that are observed by the agent that are caused either by an appearance or a disappearance of a surface.

Each event is represented as a node and the edges are labelled by the direction of self-motion of the agent, and the surface the agent is following.

Level 2: Spatial Event level

To construct spatial relations, the start- and end-points of surfaces must be located. The end-points of intervals form the second level of the knowledge representation. The intervals are established on the basis of observed end-points of respective surfaces, and the surfaces can then be interpreted as intervals. To record the relevant events, the agent responds to the end-points of surfaces appearing in the four tracking zones.

When an end-point is noted, it becomes an event $B[x]_{direction}$ or $E[x]_{direction}$ depending on events observed immediately prior to the current one (B is equivalent to Begin, and E is equivalent to End).

Level 3: Abstract Surface Level

The second level contains multiple observations of the same physical points because the end-points of surfaces can be observed from more than one location. In the third level of hierarchy, multiple observations of the same physical point are unified. This level is once again a directed graph: the nodes represent the physical start- and end-points of surfaces and the arcs are labelled by the direction of the interval between the end points.

Level 3 is equivalent to levels 1 & 2. However it should be noted, that high-level task planning and reasoning about the environment needs to operate on abstract objects and not directly on sensory inputs. Therefore, it is important to include this intermediate level of abstraction before proceeding to more complex representation.

Level 4: Objects as Sequences of surfaces

From the previous layer, it is possible to extract sets of connected surfaces. Typically a set of connected surfaces would correspond to a single object. Thus, although the agent does not explicitly do high level object recognition, abstracting sets of connected surfaces is equivalent to extracting objects.

Level 5: Properties of objects

The association of properties with objects forms the next level of knowledge representation. For example, the property of *mail* can be associated with the post office building. A high level, task planning system would require an appropriate object representation.

3.3 Hierarchic Representation

The active map is a set of relations, which can be represented as a graph. Level 1 serves as the foundation for the active map and can be defined as a directed graph. Each subsequent level is built upon the lower one(s).

3.3.1 The Representation

Level 1 of the active map is represented as a directed graph with the following definition:

Definition 1: *Level 1* of the active map is defined to be a directed graph $G_1 = <S, N_v, N_{nv}, E, \phi_1>$, where

S is a set of surfaces, and each surface is represented as a triple $<s, va, m>$, where s is a unique surface label, *va* is a visual attribute (such as required for surface recognition by the agent's sensory system) and *m* is a marker. Any surface that has been followed by the agent is marked.

N_v is a set of node labels at visual nodes, where each label consists of a set of ordered pairs (s_i, dir), where s_i is a surface label and *dir* is a direction in which this surface can be observed. The ordering is imposed by the sequence in which the agent scans the surroundings.

N_{nv} is a set of node labels at virtual nodes, where each label is either blank or consists of a pair (s_i, s_j) which implies the connectivity of surface s_i with surface s_j.

E is a set of edge labels and each label is a pair *(following(s), dir)*. The surface the agent is following is s and *dir* represents the direction of the agent's motion.

ϕ_1 is a set of triples $<e_i, n_i, n_j>$ where $e_i \in E$, $n_i, n_j \in (N_V \cup N_{nv})$. Each triple defines an edge labelled e_i, which leads from n_i to n_j.

Definition 2: An edge $<(following(s_1), dir_1), n_1, n_2> \in \phi_1$ is *equivalent* to another edge $<(following(s_2), dir_2), n_3, n_4> \in \phi_1$ if:

$s_1 = s_2$ and $dir_1 = dir_2$ and $n_1 = n_3$ and $n_2 = n_4$

or $s_1 = s_2$ and $dir_1 = -dir_2$ and $n_1 = n_4$ and $n_2 = n_3$

(direction as defined in definition 1)

Level 2 builds on level 1, and in fact becomes an augmented directed graph. The initial graph (representing level 1) contains all the directional information, and the only visual information contributed by level 2 are the observations of end-points of surfaces. Level 2 is a relation of end-points of surfaces and the direction in which they can be observed together with the link to level 1, to indicate where, with relation to the rest of the active map, the observation took place.

Definition 3: *Level 2* of the active map (which subsumes level 1) is a directed graph $G_2 = <S, N_v, N_{nv}, E, EP, \phi_2>$, where S, N_v, N_{nv}, and E are defined as before and EP is a set of sequences of observations of end-points, where each sequence of observations is an ordered sequence of pairs of the form $(ep_i(s_i), z)$ where

$ep_i(s_i), j \in \{1, 2\}$ is a function that given a surface s_i and a value j returns the jth end-point of s_i,

z is a tracking zone in which $ep_i(s_i)$ is observed, $z \in \{S, N, W, E\}$.

ϕ_2 is a set of quadruples $<(e_i, ep), n_i, n_j>$ where $e_i \in E$, $ep \in EP$, $n_i, n_j \in (N_v \cup N_{nv})$ and each quadruple represents an edge labelled by (e_i, ep), connecting nodes n_i and n_j.

Once all the information has been gathered, a third level is constructed from the levels 1 and 2. Here, the observations of endpoints are used to abstract out the actual physical points (being observed), and the connectivity of these endpoints, i.e. the surfaces.

Definition 4 *Level 3* is a directed graph $G_3 = <S, N_v, N_{nv}, E, EP, \phi_2, PP, L, \phi_3>$, where $S, N_v, N_{nv}, E, EP, \phi_2$ are defined as before and

PP is a set of physical points of the form: $<ep_j(s_a), O>, j \in \{1, 2\}$ and $O \in \phi_2$.

$\forall (o \in O)$, where $o = <(e_i, ep), n_i, n_j>$

$\exists (ep_i(s_b), z) \in EP$ such that $i = j$ and $a = b$.

L is a set of edges of the form (s_i, dir) where s_i is a surface id if the edge represents surface, or $s_i = \emptyset$ if it is a relation between two end-points of different surfaces, *dir* is the directional relationship,

ϕ_3 is a set of triples $<l, ep_i, ep_j>$ where $l \in L$ and $ep_i, ep_j \in PP$. Each triple defines an edge labelled by l connecting ep_i and ep_j.

Now that the surfaces are defined, we can define cyclic groups of surfaces that represent objects, or on a more general sense we introduce a concept of structure, which may or may not map onto an actual (physical) object, but which represents an entity that can be circumnavigated.

Definition 6: A *structure* is defined to be a cyclic sequence of surfaces, which are represented at level 2 as

$$<<(\textit{following}(s_1), \textit{dir}_1), \textit{seq}_1, n_1, n_2>, <(\textit{following}(s_2), \textit{dir}_2), \textit{seq}_2, n_2, n_3> \dots <(\textit{fol-}$$

$$\textit{lowing}(s_m), \textit{dir}_m, \textit{seq}_m, n_m, n_1>>,$$

where $<(\textit{following}(s_i), \textit{dir}_i), \textit{seq}_i, n_i, n_{i+1}> \in \phi_2$, such that

(1) $s_i = s_{i+1}$ or $\left(s_i \neq s_{i+1} \text{ and } !\exists s_j, i < j \leq m \text{ such that } s_j = s_i\right)$

and

(2) $s_i \neq s_{i+1} \Rightarrow n_i \in N_{nv}$ and $n_i = (s_i, s_{i+1})$

4 Construction of the Active Map

The map construction may begin at an arbitrary point. The agent notices all surfaces visible from the starting location and this observed scene is used to form the first node in the map. Subsequently, the agent approaches the closest structure (any of the visible surfaces). Whenever an event occurs the agent notes the appropriate changes and then continues. Details of the algorithms to build the active map are in [KiV93].

This following example (Fig. 3) demonstrates the construction of the active map while following object A. The agent starts by following surface **a** starting at position P. The agent observes **a** in a SE direction, and **d** in a NE direction. This information is recorded at the visual node 1. The agent maintains a SE direction while following **a**, and this is recorded on the arc 1-2 in the graph. Around position Q, the agent stops seeing surface **d** and thus a new visual node 2 is recorded. Further down the path at R, the agent sees new surfaces **b, h, e** and this is recorded as the new visual node 3. At S, the agent changes direction from a SE to a NE direction and this is recorded as a direction node I. As the agent continues to circumnavigate, the agent continues to construct this directed graph. This directed graph forms Level 1, or the sensory level in the active map.

The endpoints of objects are perceived as the agent circumnavigates the object. For example, at 1, the agent observes the Begin of **a** in a N direction. At position 2 the agent observes the endpoint of **a** in a E direction, and then the end of **a** in a N direction (at 3), and then finally the end of **a** in a W direction (at 4). Thus as the agent moves around the object the agent observes the same physical point in more than one direction. Also, the agent makes observations of all endpoints, that is not only the endpoints of object being followed, but also the endpoints of other visible objects. Thus for example, at position 6, the agent observes the start of surface **h** in a E direction. Recording these endpoints forms the second level of this active map.

At the third level, multiple observations of the same physical point are merged. Thus observations made at positions 2,3 and 4 are related to node X which is the begin of surface **b** and the observations 6, 7 and 8 are related to the point and the observations 6,7 and 8 are related to the node Y which represents the end of surface **b**. The arc between the nodes X and Y is labelled by the direction the agent was travelling in when following surface **b** and thus is labelled (N,E). This level also keeps the information about connectivity of surfaces; for example it maintains that **a** is connected to surface **b** and so on.

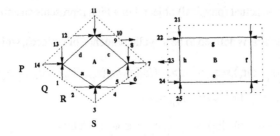

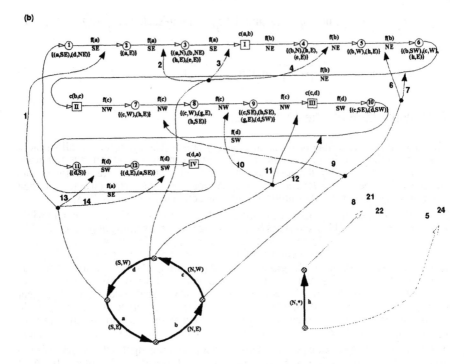

Fig 3. An example of a hierarchic map

In the next level of the active map surfaces **a**, **b**, **c** and **d** are collected together as a single object, Object **a**, and the relationship(s) between **a** and other objects are maintained. At the final level of the hierarchy, properties are associated with individual objects.

5 Using the active map

In the previous sections we have defined an active map and the way that it models some of the qualitative aspects of physical objects located in space. Such a map would be required for many applications involving object manipulation, navigation or spatial understanding (in a form of an intelligent, passive observer).

5.1 Navigation-based applications

The first application of this type of representation is to enable an autonomous agent to reason about points in space. At level 2, the active map contains details of relative observations of endpoints. In other work [VeK92] we have shown how relative observations and agent's motion can be used to build up information about the relative positions of endpoints and intervals. This type of spatial reasoning is suitable not only for understanding the topography of the environment, but also useful for answering spatial queries about objects in space, such as, what is the closest object, to the given object, in a NW direction.

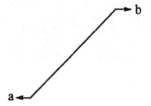

Fig. 4. The agent can relate observations of endpoints

Given that an agent makes the observation **a** in a W direction followed by an observation **b** in the E direction when following a NW path (Fig. 4), two deductions can be made:

1. The physical point causing observation **a** is south of the physical point causing observation **b**.

2. The physical point causing observation **a** is east of the physical point causing observation **b**.

The second application of this representation is in high and low level planning. Levels 3, 4 and 5 allow for the model to be used for high level planning. For example, given a high level tasks such as "Collect mail", Level 5 will indicate which spatial object has the property "mail" associated with it. At a lower level of abstraction, at Level 4 this would lead to an object, defined a set of connected surfaces. Given a starting point and the destination identified as the object connected with 'mail', alternate navigation paths can be planned. To implement a suitable navigation plan the agent would have to use the details at Levels 1-3, to predict what can be seen and what directions need to be followed during the traversal of the path. It can be seen that the low level details of obstacle avoidance and repositioning need not form part of the high level navigation plan constructed from Level 4, since all relevant information is already stored in lower levels.

To enable the agent to do low level planning such as navigation, Level 1 can be used. The graph as constructed in Layer 1, represents all possible navigation plans for an agent. In fact, navigation is a specialized planning problem. To further the analogy, we observe that the nodes in this graph contain knowledge about distinct observable states of the world and edges represent actions leading to changes in those states.

5.2 Manipulation strategy as a planning problem

Another application area in which such a representation can be used is in manipulating and arranging objects in 2D and 3D. To do so, the relationship between the endpoints of spatial intervals represented in Levels 2 and 3 need to be reasoned with. As a basis for this reasoning, we use concepts and methods from planning. The remaining part of this paper introduces the notion of expressing constraints using events and states, and then demonstrates how those constraints can be used to derive a plan for spatial manipulation.

In [KVT91] we have described an event-based planning system based on temporal logic. This planner can be used to generate sequences of actions altering spatial layout of objects. First, we will describe how to express spatial relations in a form suitable for our planner.

5.2.1 Expressing constraints.

Planning is associated with autonomous agents. An intelligent agent is required to achieve some goal, which in turn is achieved via a sequence of actions that the agent performs. An agent is any device capable of performing at least one action which results in some change to the world. This change may be internal or external to the agent. Figure 5 shows a pictorial representation of the relationship between *states* and *events*.

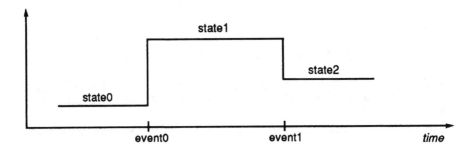

Fig. 5. Relationship between events and states

To model planning, the relevant domain is described in terms of *events* and *states*. Events represent atomic actions that cause the state of the world to change. States represent the intervals of time during which the relevant (i.e. explicitly modelled) aspect of the world remains constant. A collection of all the states is a description of the system at any point in time. Every state change is associated with some event causing this change. And similarly, every event represents some state change (possibly more than one). This implicit mapping between states and events allows the user to freely mix state and event constraints within the specification, as they can all be easily translated into the event-based form required by our plan generator.

5.2.2 The primitive operators.

We define the operators B and E, to derive the starting and terminating events associated with a state as follows:

B[state] - returns the event that starts a state.

E[state] - returns the event that terminates a state.

To represent ordering on events, we define two primitive operations, **coincides_with** and **starts_before**, as follows:

B[state1] *coincides_with* B[state2]

B[state1] and B[state2] represent the simultaneous occurrence of two events. This is mapped onto a single event that causes changes in two states in the following manner. A new event B[state1,state2] is created and each occurrence of B[state1] and B[state2] is replaced with B[state1, state2]. This new event must satisfy the constraints imposed on B[state1] and B[state2].

B[state1] *starts_before* B[state2]

This means that the event B[state1] occurs before B[state2]. It should be noted that an explicit ordering of the terminating events is redundant since those events can be interpreted as beginning events of the following states.

In Table I, we demonstrate how these simple primitives are sufficient to describe all the temporal relations on intervals as described by Allen [All83] (X and Y are intervals). Implicitly, the following condition holds for all intervals:

$B_{[X]}$ *starts_before* $E_{[X]}$

Table 1: The mapping of temporal relations on intervals as defined by Allen into our system primitives.

X before Y	$E_{[X]}$ *starts_before* $B_{[Y]}$
X equal Y	$(B_{[X]}$ *coincides_with* $B_{[Y]})$ and $(E_{[X]}$ *coincides_with* $E_{[Y]})$
X meets Y	$E_{[X]}$ *coincides_with* $B_{[Y]}$
X overlaps Y	$(B_{[X]}$ *starts_before* $B_{[Y]})$ and $(B_{[Y]}$ *starts_before* $E_{[X]})$ and $(E_{[X]}$ *starts_before* $E_{[Y]})$
X during Y	$(B_{[Y]}$ *starts_before* $B_{[X]})$ and $(E_{[X]}$ *starts_before* $E_{[Y]})$
X starts Y	$(B_{[X]}$ *coincides_with* $B_{[Y]})$ and $(E_{[X]}$ *starts_before* $E_{[Y]})$
X finishes Y	$(E_{[X]}$ *coincides_with* $E_{[Y]})$ and $(B_{[Y]}$ *starts_before* $B_{[X]})$

5.2.3. Mapping spatial constraints

In the above discussion, the constraints expressed were temporal, and thus intrinsically one dimensional in nature. Spatial information can be analysed and reasoned with by expressing constraints along orthogonal axes. Thus spatial intervals along a given direction can be considered to be analogous to temporal intervals, and constraints between spatial intervals can be expressed in a method similar to that for describing temporal constraints.

Consider a given initial configuration, as shown in Fig. 6.

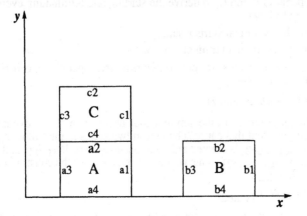

Fig. 6. Three rectangles in 2D.

Some of the constraints along the x and y axes can be expressed as follows:

Along x axis: a_2 before b_4 (1)

                       a_2 equal c_4 (2)

Along y axis: a_1 met-by c_1 (3)

                       a_1 equal b_1 (4)

Let us consider the problem of finding a suitable set of manipulation routines to change the configuration in Fig. 6 to that shown in Fig. 7.

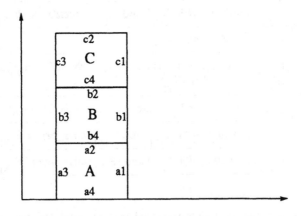

Fig. 7. Final arrangement of three rectangles in 2D.

Several researchers [Fre92, MuJ83] have observed that a continuum exists between two intervals as one interval moves across another. [MuJ83] expresses this continuum in terms of a graph of a progression of relations.

In figure 8, an interval c moves from a state of being *after* B to *met-by* and *overlap* before it becomes *equal* to B. This progression graph can be used to extract the spatial movements that are required in going from an initial relationship between two intervals, to a final relation between the two intervals. For example, if C is *after* B in the initial configuration and C *equals* B in the final configuration, then the interval C has to be moved such that it is *met-by*, *overlap-by* and then *equal* C.

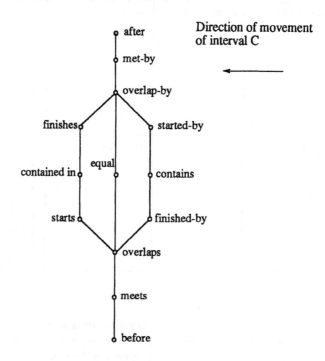

Fig. 8. Transitive relations in 1D, as interval C slides along interval B

Consider the problem of obtaining the configuration in Fig. 7 from the starting configuration in Fig.6. Then the initial conditions are given by (1) - (4) and the final conditions are:

Along x axis:	a_2 equal b_4	(5)
	b_2 equal c_4	(6)
Along y axis:	a_1 met-by b_1	(7)
	b_1 met-by c_1	(8)

To derive the final state, pairs of constraints in the initial and final state are used to compute the spatial movement required to achieve the spatial change. After each movement is computed, any spatial relations that need to be updated or added are processed. The process of propagating each pair of constraints along this direction is sufficient to produce a sequence of movements that will achieve the final spatial configuration.

Table 2: Movements along y-axis

Initial Condition	Final Condition	Movement derived from Fig. 8	
a_1 equal b_1	a_1 met-by b_1	+ve y-axis	y, c1, b1, a1, b1, x
c_1 equal b_1	b_1 met-by c_1	+v1 y-axis	y, c1, c1, b1, a1, x

Table 3: Movement along x-axis

Initial Condition	Final Condition	Movement derived from Fig. 8	
a_2 before b_4	a_2 equal b_4	-ve x-axis	y, c1, b1, b4, a1, x

6 Conclusion

We have presented a method of building a qualitative hierarchical model of space. The levels of abstraction in this model are suitable for performing various tasks. At the lowest level, the representation retains aspects of the environment suitable for navigation and repositioning. At the next level, information about the endpoints of surfaces is stored to enable spatial reasoning to be performed. Information about endpoints is effectively stored at two levels of the hierarchy: at the second level every observation of an endpoint is stored, and the next level multiple observations of the same endpoints are unified as one. Further up in the hierarchy, connected sets of surfaces are identified as structures, with which properties of form or function can be associated. The interesting aspect of the representation is that it can be constructed by a mobile agent in any domain by systematically circumnavigating all the objects. It makes minimal assumptions about the sensors that are required, and the robustness of the representation relies on the fact that in exploring a terrain, one physical point will be visible from more than one position. The unification of these multiple observations forms the foundation for relating points in space.

We have also shown various applications for this type of representation of space. To start with, navigation is an instance of the classical planning problem. Via associating higher level properties with structures represented in level 5, we can also use our model for task planning. Finally, we have shown how the model of the spatial relations can be used for direct extraction of movements required to achieve a goal arrangement of objects, given some initial configuration. The movements are based on the notion of continuity of the spatial relations between intervals.

7 References

[All83]   Allen, J.F.,1983, *Maintaining knowledge about temporal intervals*, Communicatoins of the ACM, vol. 26, no. 11, pp. 832-843.

[AyF89]   Ayache, N. and Faugeras, O. D., 1989, *Maintaining Representations of the Environment of a Mobile Robot*, IEEE Trans. on Robotics and Automation, vol. 5, No. 6, pp. 804-819).

[EgF91]   Egenhofer, M. and Franzosa, R., 1991,*Point-set topological spatial relations*, International Journal of Geographical Information Systems, vol. 5, no. 2, pp. 161-174.

[Elf89]   Elfes, A., 1989, *Occupancy Grids: A Stochastic Spatial Representation for Active Robot Perception*, Proc. of the Sixth Conference on Uncertainty in AI, July 1990.

[Fre92]   Freksa, C., 1992, *Temporal Reasoning Based on Semi-intervals*, Artificial Intelligence, vol. 54, pp. 199-227.

[Her92]   Hernandez, D., 1992, *Qualitative Reperesentation of Spatial Knowledge*, Ph.D. thesis, Faculty of Informatics, Technical University of Munich.

[KVT91]   Kieronska, D.H., Venkatesh S. and C.P. Tsang, 1991, *Planning with Events and States*, 2nd Annual Conference on AI simulation and Planning in High Autonomy systems, Florida.

[KiV92a]   Kieronska, D.H and Venkatesh S., 1992, Deduction of high level spatial relations *IEEE International workshop on Emerging Technologies and Factory Automation* Melbourne, 11-14 August, pp. 684-687

[KiV92b]   Kieronska, D.H and Venkatesh S(1992),Representation of large scale spatial environments, *11th European Meeting on Cybernetics and Systems research*, Vienna, 21-24 April 1992, pp. 1439-1116.

[KuB90]   Kuipers, B. and Byun. Y-T., 1990, *A Robot Exploration and Mapping Strategy Based on a Semantic Hierarchy of Spatial Representation*, AAAI'90 Workshop on Qualitative Vision, pp. 1 - 28.

[LeL90]   Levitt, T.S. and Lawton D.T., 1990, *Qualitative Navigation for Mobile Robots*, Artificial Intelligence, vol. 44, pp. 305 - 360.

[Lev87]   Levitt, T.S., 1990, *Qualitative Navigation*, Proc. DARPA Image Understanding Workshop, Los Angeles, Morgan Kaufmann.

[Mor83]   Moravec, H. P.,1983, *The Stanford Cart and the CMU Rover*, Proceedings of the IEEE, Vol. 71, No. 7, pp. 872-874.

[MuJ90]   Mukerjee, A. and Joe, G., 1990, *A Qualitative Model for Space*, Proc. AAAI-90, pp. 721 - 727.

[VeK92]   Venkatesh S. and Kieronska, D.H.(1992), Spatial Reasoning For Intelligent Navigation, *International Conference on Automation, Robotics and Computer Vision*, pp. CV16.1.1 - CV16.1.5 1992.

[Zad81]   Zadeh,L.A., 1981, PRUF- a meaning representation language for natural languages, in Fuzzy Reasoning and its applications, Eds. Mamdani, E.H. and Gaines, B.R., pp. 1-66.

3 APPLICATIONS

Probabilistic Models in Qualitative Simulation

Wilfried Grossmann, Hannes Werthner

Institut für Statistik, OR und Computerverfahren, Universität Wien,
Austria

Abstract. Qualitative simulation, the main reasoning technique used in qualitative reasoning, necessarily generates spurious solutions [9, 14]. This fact can be viewed as the central problem of an approach which integrates both the fields of AI and systems theory. Works, which exploit concepts of the latter area such as state space diagrams or qualitative cells, are able to classify the entire space of solutions in advance as well [1, 7, 11]. They also show the correspondences between methods developed in the area of system theory and qualitative reasoning. Such similarities will be discussed in a first part of the paper. Since qualitative simulation generates an entire set of solutions, it might be very useful to indicate priorities among them as well. This will be presented in a second part where we apply a probabilistic concept, i.e. markov processes.

1   Introduction

Qualitative reasoning incorporates two central tasks: the modeling task and the derivation of dynamic behavior, providing a causal explanation as well. As a simple example to demonstrate qualitative reasoning features we use a container filled with some liquid, e.g. water, above a heat source. We want to know what in principle can happen in such a situation: the liquid may be transformed into gas, it may disappear or the relationship between the amount of gas and that of liquid may reach an equilibrium. This illustrates following aspects of human reasoning [10]:
- Humans understand basic concepts which can be applied to different situations. The concept of heat transfer, for example, explains how both the container and the water become hotter.
- Humans establish a cause-effect chain of changes. One can increase, for example, the water temperature by increasing the temperature of the heat source.
- Humans realize that interesting things happen around limit points. If the flame is high enough, the water will begin to boil.
- Humans are able to reason about changes over time. We can order events on the time axis, for example, water begins to boil before the container may begin to melt.
- Humans can reason from particular points of views and at different levels of detail. We can identify what is important, e.g. in our example no reference is made to the shape of the container.

We can identify two motivations for the qualitative reasoning research:

- Modeling real world situations in order to support the reasoning activities of persons such as engineers. The scope of the process is a model of the real world and references to numerical models are assumed to exist. However, the necessary information is either not available or these models are too complex or too slow to provide the desired answer. Moreover, there may not even be the need for an exact quantitative solution.
- Imitating human, thinking mainly of engineers. Also in this case a model of reality exists but it is just an instrument to study the reasoning process itself.

Both levels together provide qualitative reasoning with capacities for modeling which are most probably superior to those of other approaches. As a basic characterization we can state that qualitative reasoning deals with dynamic systems; it is based on very coarse description with little information and simple structure; its concepts semantically correspond to methods in system theory; and it relies on techniques used in AI. Moreover, the following basic assumptions can be identified:

Compositionality: The structural description of a device or also a situation is an assembling of single basic descriptional entities - either physical or abstract objects - which can be connected by well defined interconnections. The respective assemblies may be further composed in a hierarchic manner. The basic entities are described in the form of behavioral primitives. The behavior of the overall device or situation follows cogently from the respective individual behaviors.

Locality: Effects are local, the only way an effect can be propagated is via its interconnection, either physical or abstract ones, to other parts. Moreover, such effects can be localized.

Function: Function has to be separated from behavior. Human artefacts have specific functions which can also be described as the mapping from the behavior to the goal (purpose) of a device.

Class wide assumptions: Qualitative reasoning assumes, similar to the object-oriented programming paradigm, that distinct objects can be grouped into a unique class, just the common features are represented. Hence, each qualitative description denotes an entire class.

No-function-in-structure: A model should be described by its structure and behavioral primitives, the overall behavior follows from this description. The description of a device should not anticipate its function. Function is interpreted as a consequence of behavior.

Reduced quantity space: Values of variables are mapped to qualitative values, represented by reduced and ordered - at least partially- sets of values. These sets may vary from small ones with just three members $(+, -, 0)$ - denoting the positive, the negative range and zero - to bigger ones with more members. These qualitative values represent intervals - crisp or overlapping - on the real number line. But the quantity space may also consist of a variable number of members, ordered with respect to quantities defined beforehand, so called landmark values. On these qualitative value sets special algebras have to be defined, resulting into ambiguous situations.

Reduced relationships: Between entities and their variables just a reduced set of relationships is available. However, not only simple algebraic operations are defined, also more complex functional dependencies may be used. Additionally, qualitative derivative relationships are defined as the main mechanism responsible for the changes of states.

Reduced representation of time: Since qualitative reasoning deals with generic descriptions of situations and their time dependent evolution, it just produces a time ordered sequence of well distinguishable events. It is described what may happen before or after something, but no exact information about time is available.

In the following chapter we shortly present different approaches of qualitative reasoning, chapter 3 discusses correspondences between methods of system theory and qualitative reasoning. In the last chapter we present the mapping of one of the discussed approaches onto a representation based on markov chains.

2 Different Approaches in Qualitative Reasoning

Device-centered Ontology[1]

A device is described as a set of individual components and conduits [2]. The latter denote connections between components. Components may only be connected by the means of so called terminals or nodes. Interconnected components constitute a larger component, which for its part may be connected to other components. Material such as liquid or electricity may flow in these devices. Both modeling primitives, components and conduits, have associated sets of confluences which are equations constraining their elementary behavior. Behavior is described in terms of the attributes of the processed material.

The entire set of confluences provides a description of both the static and dynamic features of a device and are based on physical laws. It depicts the single components, their interconnections and the processing of material. The latter one represent the qualitative counterpart of Kirchhoff´s law for current and voltage. The so called no-function-in-structure principle is realized as confluences describe only the behavior of the respective part, i.e. component or conduit. The scope of this approach is - given such structural description of a device and the behavior of its components - to produce a total envisionment which contains the set of all possible future states of the system and the transitions between these states. This result of the reasoning process corresponds to a causal explanation of the overall behavior.

A variable is described by its value, taken from the reduced quantity space {-, 0, +}, and its derivative, also described by its sign. Qualitative values as well as derivatives may be related by the means of confluences, available connectives are just simple algebraic operators such as +, - and *. The reasoning process proceeds as follows: after an enumeration of all states based on the limited values of the state variables, the latter ones serve as starting points for a constraint satisfaction process inside each state. However, based on this value set and the available algebra, ambiguous results may be obtained, i.e. different value assignments satisfy the set of constraints. Afterwards the values of the derivatives determine possible transitions between neighbouring states. This latter procedure is based on rules obtained from mathematical analysis such as the qualitative mean value theorem. The final result, the envisionment graph, corresponds to a state space description with all possible state transitions.

---

[1] Ontology is defined by [13] as the conceptual definition and cataloguing of entities and relationships composing the mind representation of the real world.

Process-centered Ontology

Whereas in the device-centered approach no explicit notion of directed influence or causality exists, the process-centered ontology possesses such a primitive with its concept of a process [4]. Thus, causality is already compiled into the model. This ontology also starts from individual objects and variables, which describe their attributes. However, this approach includes supplementary concepts such as views, processes and histories. A view describes a specific situation in which some objects exist, it indicates the relationship between individuals and their qualitative variables. A process may change these relationships, it represents the only primitive to change the state of a device or a situation. A history on its side describes the time trajectories of the single individuals and indicates the change of the entire system. These concepts introduce an explicit notion of time. Individuals cooperate when their histories meet, i.e. when processes create relationships between them. In difference to the previous approach no explicit necessity exists that two objects must touch each other, it is sufficient that variables of two objects influence each other by the means of processes.

The description of variables is more complicated as in the previous case. Additionally to the signs of a variable's value and its derivative, the magnitude of both values is taken into consideration. The latter are partially ordered with respect to some known landmark values. The process of reasoning in the process-centered ontology can be defined as follows: First the existing individuals, relationships and quantity conditions are identified to create the set of active views and processes. Afterwards the signs of the derivatives are examined where only variables with derivatives ≠ 0 change. Variables may be influenced either directly or indirectly, the first ones are resolved in a first step, the latter ones afterwards. In a next phase it is checked whether some variables pass a threshold, i.e. change valid predicates and their relationships in the quantity space. In such a way the overall structure may change, new processes and views may be activated. The final result, called limit analysis, is similar to the previous approach, it also describes all possible behaviors.

Constraint-based Approach

This approach differs from the others as it refers directly to well known mathematical descriptions of dynamic systems in the form of differential equations. It starts from these equations which exactly describe a given physical situation [8]. Based on this description a qualitative version is deduced, using basic qualitative primitives such as operators, functional and derivative relationships. It does not provide modeling primitives like basic components or processes as in the other approaches. Since the focus of this approach is behavior generation, this method fits best to the notion of qualitative simulation. The set of qualitative values consists of non overlapping intervals which cover the whole range of interest. The limits of these intervals are constituted by so called landmark values, the critical values of a variable, which are assumed to be known.

Variables are described by functions of time, i.e. f(t): [a,b] → **R**′ with t ∈ [a,b] ⊆ **R**′ and **R**′ = **R** ∪{-∞, +∞}.[2] It is assumed that f is continuous and continuously differentiable and has only a finite number of critical points in any interval [a,b], i.e. the set {t | f′(t) = 0, t ∈ [a,b]} is finite. Furthermore it is assumed that both limits $\lim_{t\downarrow a} f'(t)$ and $\lim_{t\uparrow b} f'(t)$ exist and f′(a) and f′(b) are defined to be equal to these limits. Such functions are also called reasonable functions.

The qualitative state of f at some point t is described by a pair (qval, qdir) where

$$qval = \begin{cases} l_i & \text{if } f(t)=l_i \ (a\,landmark\,value) \\ (l_i, l_{i+1}) & \text{if } f(t)\in(l_i, l_{i+1}) \end{cases}$$

$$qdir = \begin{cases} inc & \text{if } f'(t)\rangle 0 \\ dec & \text{if } f'(t)\langle 0 \\ std & \text{if } f'(t)=0 \end{cases}$$

The value of f is either an entire interval for all values of t in the respective domain or a fixed landmark value. Both types of states strictly interchange. Thus, the qualitative behavior of f on [a,b] can be defined as a sequence $QS(f,t_0)$, $QS(f,t_0,t_1)$, $QS(f,t_1)$, ..., $QS(f,t_{n-1},t_n)$, $QS(f,t_n)$ of states at distinguished time points and states between such time points. A system is now a set $F = (f_1, ..., f_m)$ of functions f_i: [a,b] → **R**′, as previously defined. The set of distinguished time points of F is the union of all such time points of the respective functions.

The algorithm QSIM [8] generates behavior, starting from a description based on qualitative constraints. In the dynamic behavior time points and time intervals have to interchange, i.e. $QS(f,t_i) \Rightarrow QS(f,t_i,t_{i+1})$ and $QS(f,t_i,t_{i+1}) \Rightarrow QS(f,t_{i+1})$. Based on rules similar to the other approaches the algorithm generates all behaviors which are consistent with the constraining equations. This approach generates in a first step all possible successor states and filters afterwards all states which are not consistent with the set of constraints. Moreover, it may also produce new qualitative values, i.e. intervals, during this process. Contrary to the other approaches, it produces - or at least tries to produce - just one trajectory in the state space.

QUALSIM - an Approach Based on Fuzzy Sets and Influence Diagrams

In the previous cases it was assumed that intervals representing qualitative values have a crisp form. However, sometimes even this knowledge may not be available. A value, for example, may fall into some region, but it may also have at the same time some other qualitative value. Therefore, intervals representing qualitative values may also overlap. Such form of knowledge has to do with beliefs that exist relative to assumptions about the modelled reality. It is different to probability which only deals with exclusive cases. In the present case the possibility, for example, of being in two intervals at the same moment may be higher than one. Fuzzy sets are a means to implement and compile such a knowledge. Furthermore, this method also provides methods for inferencing [17]. This approach supports facilities to model available knowledge along the dimension of qualitative abstraction as well as the dimension of uncertainty. In the following we present one

---

[2] -∞ and ∞ are added to use them as explicitly defined landmark values.

typical representative of these approaches, i.e. QUALSIM [6]. It has a fixed qualitative value set with the following members:

+/- EXTRALARGE
+/- LARGE
+/- MEDIUM
+/- SMALL
+/- EXTRASMALL
ZERO

QUALSIM assumes that the entire range of possible values of a variable can be defined in advance. All intervals have the same width. Each variable x is identified by a numeric range R_x defined as a subrange of the real number space. Each value of a variable's magnitude is a fuzzy set defined by a characteristic value v_k, k= 1,...,12, (by default the v_k's are evenly spaced in the positive and negative subintervals of **R**) and by a fuzzy membership function $\mu(x)$ of the type, which is similar to [12]:[3]

$$\mu_A(x)=\begin{cases}0 & x < v_k - a\\ \dfrac{x-v_k+a}{d+a} & x \in [v_k-a, v_k+d]\\ 1 & x \in [v_k+d, v_{k+1}-d]\\ \dfrac{v_{k+1}+b-x}{d+b} & x \in [v_{k+1}-d, v_{k+1}+b]\\ 0 & x > v_{k+1}+b\end{cases}$$

for each k and the corresponding $A \in Q$. In the case of the delimiting points v_1 and v_{12} $\mu_A(x)$ remains at a value of 1. The constant d is defined global and determines, together with the variable dependent values a and b, the fuzziness of the qualitative values.

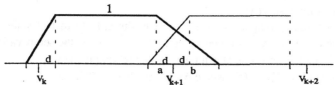

Fig. 1: Value representation based on fuzzy sets

Connections between two variables may possess one of the following labels:

+/- EXTRASTRONG
+/- STRONG
+/- MEDIUM
+/- WEAK
+/- EXTRAWEAK

---

[3] This is different to the original notation used in [6], however, no principle difference exists and the presented form will ease the following discussion.

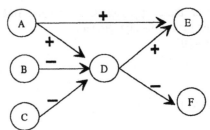

Fig. 2: Influence graph of QUALSIM

In QUALSIM a model is described by a causal influence graph. However, these influences denote the strength of influence and not its probability. The simulation algorithm is fundamentally based on the conversion of qualitative values into quantitative ones and vice versa. The propagation of uncertainty during the simulation is represented by the bifurcation of the simulation tree. The algorithm works as follows, these steps are repeated until some time counter is reached:

- At the beginning the initial qualitative value for state variable x is transformed into its numeric equivalent $x°$ being the characteristic point of the relative fuzzy set, i.e. to the point with the highest membership grade.
- Assuming that $I_i = \{x_j : x_j$ is connected to $x_i, j \in 1.. n\}$ is the set of variables that exert an influence on the variable x_i, the new numeric value is computed as

$$x_i^0(t+1) = x_i^0(t) + \sum_{j \in I_i} \Delta x_j(t) \qquad \text{where } \Delta x_j(t) = c_{ji}^0 \frac{x_j^0(t)}{R_{x_j}} R_{x_i}$$

and $c°_{ij}$ is the numerical mean value of the connection between x_i and x_j and R is the range of the respective variable.
- The calculation of output variables proceeds similarly to that of state variables. Assuming that $K_i = \{x_j : x_j$ is connected to $y_i, j \in 1.. n\}$ is the set of variables that exert an influence on y_i, the new numeric value is computed as

$$y_i^0(t+1) = \sum_{j \in K_i} \Delta x_j(t) \text{ where } \Delta x_j \text{ is calculated as before.}$$

- The numeric value of a variable is transformed again into its qualitative counterpart:
  + If the membership grade of $x°(t+1)$ exceeds a predefined threshold $\mu'(x)$ at least in one of the sets (i.e. if $x°(t+1)$ belongs to the α -cut of at least one of the sets), then the new qualitative value assigned to x_i is the label of the set where $x°(t+1)$ has the highest membership grade.
  + If all the membership grades are lower than α', then an ambiguity occurs and either the user is asked to assign the qualitative value by choosing between the one at time t and the higher (lower) value depending on the sign of $\Sigma_j \Delta x_j$ or a fixed strategy like depth - or breadth - first is applied.
- If the qualitative value of a state variable computed with the above procedure is different from that one at time t, then the corresponding numeric value $x°(t+1)$ is equal to the characteristic value of the relative set. Otherwise, the qualitative derivative (i.e. the sign of the term $\Sigma_j \Delta x_j$) is taken into account and the term

$$\partial x_i = \frac{(1+p)\,v_{k+1} + (1+m)\,v_{k-1}}{p+m+2}$$

is added to $x°_i(t+1)$. p denotes the number of time steps with the variable in the same qualitative state with positive increments, and m is the number of time steps with negative increments. This leads to different results as in the case of order of magnitude methods where a qualitative operation of the form EXTRALARGE - EXTRASMALL will for ever produce the result EXTRALARGE. In QUALSIM a frequently repeated application of this operation will force the variable to change value.

A qualitative model, as already noted, represents an entire set of quantitative models. Thus, its results correspond to the set of solutions of these models. However, due to the limited representation and restricted capacities of the applied reasoning techniques, spurious solutions, which to not correspond to any "real" model, may be produced. Furthermore, no possibility exists to identify these wrong solutions unless external knowledge is applied.

3 Correspondence to Methods in System Theory

In the following we will discuss the correspondence between methods to describe dynamic systems in qualitative reasoning and in system theory, based on figure 3:

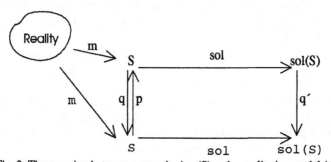

Fig. 3: The mapping between a quantitative (S) and a qualitative model (s)

S denotes a dynamic system based on continuous time and real valued variables which will be defined in the following, s the corresponding qualitative model. s may be otained directly by the mapping m or by q(S). sol and sol are the applied behavior generation methods, we will show that similar concepts are used. A qualitative model can be build either by applying m or m(q(S)). It has been shown by [14] and [15] that q´(sol(S)) $\cap$ sol(s) $\neq$ 0, but not q´(sol(S)) $\subseteq$ sol(s). Moreover, due to the limited capabilities of s p(q(S)) $\neq$ S. A dynamic system S is described by the following tuple:

S = (U, X, Y, δ, γ, P)

where U denotes the set of input variables (or input mapping function), X the set of the state and Y the set of the output variables. P is the parameter set, δ and γ are the state transition and the output function. In this description the values of the members of the respective sets U, X, Y and P are elements of **R**. This is a fundamental difference to qualitative reasoning with its very limited set Q of possible values for qualitative quantities. An exception is [8] and all further interval based approaches which allow for an iterative refinement. Given the definition of a dynamic system S it can be examined which of its sets and functions can be mapped onto a qualitative counterpart. Different levels can be distinguished, where in the general case of qualitative reasoning all sets and functions are mapped to their qualitative counterparts. In this case the set P disappears completely and the other elements of S express slightly different meanings. The qualitative counterpart of S is

$$S = (U, X, Y, \delta_Q, \gamma_Q).$$

In this qualitative model the distinction into the sets U, X and Y is not straightforward, on a descriptional level it does not exist. Qualitative reasoning describes a system being initially at equilibrium and the reasoning process starts by disturbing this situation. Such "input", however, may be applied either by some input variable or by some initial state. The distinction between U and X is thus not a static one, it depends on the specific situation. Normally, the set U has just one member. But if this variable is further used in the reasoning process and changes its value depending on this process rather than on some external condition, it represents a member of X, with a known initial value X_0. On the other side the distinction between the sets X and Y depends on whether the variables are directly influenced. All not directly influenced variables are members of the output set Y, but normally these variables participate again in the calculation of the members of X, i.e. they are input to the qualitative functions δ_Q and γ_Q as well. Moreover, in the device-centered approach a distinction into X, Y and U is not possible at all. The two functions can be stated as follows:

γ_Q represents the intrastate behavior: it is responsible for calculating the values of all variables inside one state, i.e. completing a state, once this state is reached by a state transition.

δ_Q is responsible for the interstate behavior, it forces variables to change from known values to new ones at the next time step and it checks also whether some state variables may pass some threshold value which forces the system to pass to a new state. The two qualitative functions may be written as

$\delta_Q: U \times X \times Y \Rightarrow X$

$\gamma_Q: U \times X \Rightarrow Y$

In the following we will look at the correspondences between the two behavior generation techniques sol(S) and sol(S), as shown in figure 3. Especially, we will examine discrete event simulation, difference equations based simulation and finite state automata which show nice similarities with the main techniques used in qualitative reasoning, at least on a semantic level. However, although we distinguish between model representation and simulation techniques, the latter and the former determine each other. In such a way the mentioned techniques can be interpreted as further specializations of the formalism S to describe dynamic systems.

A discrete event system changes state when a well identified event changes the value of a variable, this variable is part of the vector which describes the state of a system. Therefore, such an event may "touch" only some variables and it may or may not generate a new event. Only events can change things and create or destroy existing objects. Events may occur at not equidistant time points and, secondly, they are not always known in advance. A discrete event system can be described by the tuple (X, U, O, E, ta):

X: the set of state variables (or vector respectively),
U: the set of input variables, i.e. a function $t \to U$, $t \in \mathbf{R}$,
O: the set of output variables,
E: the set of events, and
ta: the time advance function.

An event $e \in E$ changes X, i.e. $e: X_i \to X_{i+1}$ where i denotes the sequence of states with respect to time, i.e. $X_i = X(t_i)$. The time between two consecutive states is not equidistant, i.e. $\exists$ i, that $t_{i+1} - t_i \neq t_i - t_{i-1}$. Each event has an associated time point of its occurrence (e, t_e). Furthermore, an event may either be external, i.e. defined by the environment as a kind of input (e^{ext}) or internal, as a result of some previous event (e^{in}). Two consecutive events are partially ordered:

if (e_i, t_{e_i}) generates (e_j, t_{e_j}) then $i < j$ and $t_{e_i} \leq t_{e_j}$.

A state change can be written as

$(e_i^{in \ or \ ext}, t_{e_i}): (U_i, X_i', t_{e_i}) \to (X'_{i+1}, O'_{i+1}, E')$, where

$E' = E \cup E_1 \setminus E_2 \setminus \{(e_i^{in \ or \ ext}, t_{e_i})\}$ with

$E_1 = \{(e_{j_k}^{in}, t_{e_{j_k}}) \mid k \geq 0, i < j, t_{e_i} \leq t_{e_{j_k}}\}$, i.e. new events may be generated, and

$E_2 = \{(e_{l_m}, t_{e_{l_m}}) \mid m \geq 0, i < l, t_{e_i} \leq t_{e_{l_m}}\}$, i.e. some events may be deleted from the event set. Thus, each event produces a new state with corresponding output values and a probably empty set of new internal events. Furthermore, the activated event and probably some more events are deleted from the set of events. $X' \subseteq X$ and $O' \subseteq O$ which indicate the fact that not all variables are engaged in the actual calculation of new values. The time advance function ta is responsible for shifting the time to the respective time point of the next event in the event list. The ordering relation in this set determines the advancement of time.

In the case of continuous system simulation the system is described by a set of difference equations based on a fixed discretization of the time axis.[4] The set of all state variables is necessary to determine the next state. Furthermore, the distance between any two adjacent time points is equidistant, i.e.

$\delta: (U(t_i), X(t_i)) \to X(t_{i+1})$

$\gamma: X(t_{i+1}) \to O(t_{i+1})$

and $\forall$ i: $t_{i+1} - t_i = t_i - t_{i-1}$.

In this approach a permanent iterative cycle calculates all values at equidistant time points, the calculation of the new values is inherently described by the state transition function δ and no extra devoted primitive such as an event is necessary.

Finally, a finite state automaton (FSA) is described by a tuple (I, X, ϕ, γ), where
I denotes the set of inputs,
X the set of states,
γ a function that maps I to X, and
ϕ a function that maps X to X, i.e. $X_{t+1} = \phi(\gamma(I_t), X_t)$.

---

[4] In this case the formalism S is directly used, in the other two cases either the functions γ and δ are substituted (discrete event simulation) or the set of states is directly represented (finite state automaton).

This is a slightly different point of view since states are directly represented, no direct reference to state variables is made.

We claim that the constraint-based approach resembles the difference equation based methodology whereas the process-centered approach corresponds to the discrete event simulation approach. QUALSIM, based on influence diagrams and fuzzy sets, obviously corresponds to the difference equations based simulation. It follows straightforward from its definition. The device-centered ontology will be considered similar to a FSA.

In contrast to [3] and [16] we classify QSIM and its qualitative differential equation based ontology as a difference equations based approach. The contrary would be only true if the calculation of a specific value can already be classified as an event.[5] The approach of [8] bears a resemblance with the discrete event simulation scheme in so far as time points are not equidistant but nearly arbitrarily spaced on the number line.[6] However, QSIM and his successors and relatives generate new states in a fixed sequence and, even more important, in all these calculations all variables are engaged. The interchanging mode of states with an instantaneous duration on the one hand and states with an interval duration on the other virtually places the time points in an equidistant manner, i.e. { ..., t_i, (t_i, t_{i+1}),t_{i+1},... (t_j, t_{j+1}), t_{j+1}, ...}. Qualitatively, these states just differ in the rules that determine the state transitions, while in all other aspects they can be treated as equidistant. Furthermore, they are always totally ordered which is not the case in discrete event simulation.

However, the most important difference between QSIM and the discrete event approach is that there is no dedicated representational primitive such as an event which is responsible for all state changes. Together with the lack of such an ontological primitive the algorithm works in a completely different way, at least on a semantic or constructive level. Whereas in discrete event simulation a scheduler, which implements the time advance function, always has to check and to look for the event with the smallest occurrence time and to update the event list every time an event has been activated, QSIM permanently "recalculates" the values of all state variables with previously fix defined equations. The generation of new events contains the idea of unforeseen things. Discrete event simulation may introduce new quantities and change the structure of the model during simulation. QSIM does not possess this feature.

The approach represented by the qualitative process theory can be classified as a discrete event simulation approach. States are represented by active processes and views, i.e. specific relationships between variables have to hold. Given such a specific view structure, one or more processes are active. These processes change the values of some variables. These, in turn, may change the values of others. Processes are the only primitives which can change values over time and force the system into a new state. Hence, new views and processes may become active. This representation possesses with the concept of a process a dedicated entity which is responsible for changes. Such processes can be interpreted as an assembly of events. In the process-

---

[5] Both [3]and [16] do not distinguish between [8] and [4], though both differ significantly in the point of view, the representational primitives and in the inferencing algorithms.

[6] Note that this can also be the case when solving differential equations numerically: "intelligent" algorithms, i.e. integration methods, set the time points as they need them, i.e. close to discontinuities the integration step size Δt becomes smaller, in linear ranges larger.

centered view the time distances between two states are not equidistant, neither under a semantic nor under an implementational aspect. Processes touch just some variables and not all of them. Variables are influenced until their values reach some threshold, the remaining ones can be neglected. Another difference to [8] becomes clear: In discrete event simulation events can happen simultaneously. Several processes of the process-centered ontology may also work simultaneously, indeed, in most cases more than one process is active. In the difference equations based approach, on the other hand, the time advancement represents also a sequentialisation of the behavior.

The classification of [2] with respect to the above defined techniques is different. The device-centered ontology corresponds well with the concept of finite state automata. Both sets I and X have limited and beforehand known values. This is similar to the device-centered ontology, where all states are known and can be created as well as enumerated in advance and where the task of the envisionment process is to identify ϕ. The function ϕ represents the causal interpretation that is looked for. On the highest level an envisionment graph looks like a non-deterministic automaton. However, when the values of the respective derivatives in each state are taken into consideration, their different value sets - corresponding to different interpretations - deterministically define the next transition. While the other methods may be classified to some extent as being constructive algorithms which produce new states and values and thus explore the state space, this is not valid in the case of the device-centered ontology. This approach analyses and defines possible state transitions, their preconditions - on the basis of the topical state of the system and some input - and their sequential and causal order.

Based on these considerations we can show the following correspondences:

sol	sol
discrete event simulation	process-centered ontology
difference equations based simulation	constraint-based ontology
difference equations based simulation	influence diagram-based approach
finite state automaton	device-centered ontology

4 An Interpretation based on Markov Chains

The markov property[7] of a system says that the state of a system just depends on its predecessor state. $X_1, .., X_n$ being random variables and S the finite set of states, then a markov chain is a sequence of the random numbers where $P[X_n = s_j \mid X_1,...,X_{n-1}] = P[X_n = s_j \mid X_{n-1}]$. Homogeneous markov chains do not depend on n, i.e. $p_{ij} = P[X_n = s_j \mid X_{n-1} = s_i]$ with $s_i, s_j \in S$ and p_{ij} is independent from time. It denotes the probability that the system changes from state s_i to state s_j. This can be written in the form of a quadratic matrix P indicating the state transition probabilities for all combinations of states. Furthermore, P^n describes the transition probabilities of applying n transitions, i.e. the situation of the modelled system after

---

[7] We use the concept of finite markov chains. Moreover, we do not discuss the case of possibilistic automata, which can also be described by the means of markov processes.

n steps. This transition matrix P has some nice properties, it partitions, for example, the states into equivalence classes. Therefore, a transition matrix not only defines probabilities of specific behaviors, but also indicates the sequence of visited states or the existence of cycles.

In the following we apply this concept onto QUALSIM, which requires the consideration of several problems:
- the given fuzzy number representation of QUALSIM has to be transformed into a probabilistic form to guarantee the discussed markov properties,
- QUALSIM only denotes the strength of an influence between variables, no state description is given in advance,
- the entire set of possible system states is not known.

The solution of the first problem requires a "normalized" representation of qualitative values, i.e. $\Sigma_{A \in Q} \mu_A(x) = 1$ for every $x \in R_x$, i.e. fuzzy numbers are modified in order to represent probabilities.

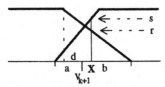

Fig. 4: Relationship between two qualitative fuzzy values

Such a transformation has to maintain the relationship between the two fuzzy numbers, i.e. if x is the current value of the respective state variable the following has to be satisfied:

$r : s = \dfrac{b-x}{d+b} : \dfrac{a+x}{d+a}$ and $r + s = 1$. The important feature is the relationship between

both values and not their magnitudes.

The second and the third of the above mentioned problems can be discussed together: The set S of system states can be constructed in advance based on the limited value set of the qualitative quantity space. If $w = \#(W)$ denotes the number of state variables - W is the set of state variables - and $u = \#(Q)$ the number of qualitative values, then the number of all possible states is u^w. Therefore, the set S can be generated by enumerating all possible value combinations of each element of W. Each element of S represents a vector, the first element of the vector denotes the value of the first variable, the second element denotes the value of the second state variable, and so on. Every value combination represents a state and for every combination of values the respective state can be identified. In this way a finite state space can be constructed in advance.

The function succ(m | m $\in$ Q) denotes the neighbour which follows a member of Q, and pred(m | m $\in$ Q) the neighbour which precedes a member of Q. LARGE is succ(MEDIUM) and SMALL is pred(MEDIUM). succ(EXTRALARGE) is defined to be EXTRALARGE, pred(-EXTRALARGE) is -EXTRALARGE again. The transition probabilities $p_{S_i S_j}$ can now be derived (S_{ik} denotes the kth component of vector S_i - with $S_{ik} \in Q$ - and G denotes the distribution of the numerical correspondents R_{ik} of a qualitative value S_{ik}:[8]

$$p_{S_iS_jk} = \begin{cases} 0 & \text{if } S_{jk} \notin \{ pred(S_{ik}), S_{ik}, succ(S_{ik}) \} \\ \int \mu_{(S_{jk})} (\sum_q R_{iq} \times c_{qk} + R_{ik}) dG & \text{otherwise} \end{cases}$$

The probability to change from S_i to S_j - the system is in S_i at time t and in S_j at time t+1 - is zero when the law of continuity is violated. R_{iq} correspond to those components q that influence the state variable k. The second part represents the original formula to calculate the new value of a state variable where the quantitative correspondent of a qualitative value is weighted by the respective probability distribution. Based on the fact that variables just vary continuously, it follows that P is very sparse. Every state S_i may just have 3^w successors since in each transition just three possibilities of change for each state variable exist. $p_{S_iS_jk}$ represents the kth component of a vector of probabilities and several approaches can be identified to approximate the overall probability value $p_{S_iS_j}$:

a) Since the decisions of choosing a specific range for each component of a state variable are independent, in cases where the state variables are only loosely coupled the multiplication of the single components of the probability vector is a reasonable operation, i.e. $p_{S_iS_j} = \Pi_k\, p_{ijk}$. Multiplication is a reasonable operation since it eliminates those transitions where one or more components of the probability vector are zero, i.e. the values of these components of S_j cannot be reached.

b) Instead of multiplying the components of the probability vector, the minimum of the components could be chosen. This means that the most unlikely case inside each column determines the overall probability of this specific transition or that the inertest variable determines the behavior of the system. In this case a strong dependence between the state variables is assumed. To obtain the markov property every single value has to be weighted by the row sum of all minima, i.e.

$$p_{ij} = \frac{\min_k (p_{ijk} | k \in 1..w)}{\sum_j \min_k (p_{ijk} | k \in 1..w)}.$$

c) As a final approach the single components of each vector could be added. Also in this case we assume a strong dependence between the model's variables. A possible interpretation would be some kind of averaging the transition probabilities which of course makes only sense if no component has a probability of zero. In this approach the value can be calculated according to the following formula $p_{S_iS_j} = 1/F_w * \sum_k p_{ijk}$. The factor F_w represents the sum over all components of $p_{S_iS_j}$ for all successor states S_j of S_i, i.e. $F_w = \sum_k \sum_{j \in succ(S_i)} p_{ijk}$. F_w can be calculated in a recursive way: $F_0 = 0$ and $F_i = 3 * F_{i-1} + 3^{i-1}, \forall i > 0$.

It is demonstrated in [5] that all three approaches satisfy the markov property, i.e. $\sum_j p_{ij} = 1$. However, the method to compute the overall probability of a state transition has to be further explored. It can be assumed that the concept of reachability and velocity of reaching specific states will depend on this choice which, in addition, has

---

[8] G can be obtained by applying the model equations to produce the trajectory of the system. Depending on the c_{ij} this trajectory takes only a finite number of quantitative values inside each qualitative range. Each of these points has thus a well defined probability of occurrence.

to be compared with the simulation results of the original QUALSIM approach. The matrix P represents the state space of the system. The simulation of a system can now be performed by multiplications of matrices and vectors, the latter describing the initial state of the system. These are simple operations. However, in realistic cases the dimension of the matrix - even if it is very sparse - will be enormous. An advantage is that it can be built in advance and that it provides a richer description of the state space, indicating privileged behaviors as well.

5 Conclusions

In this paper we discussed correspondences between concepts of qualitative reasoning and those used in system theory. We think that, based on such similarities, both areas may benefit from each other: AI by introducing techniques to deal with time, simulation methods in system theory by the usage of elaborated knowledge representation methods. In a second part we used the concept of markov processes and discussed their respective application. Such probabilistic approaches may be advantageous in choosing among different solutions generated by qualitative simulation.

References

1.   Aubin J.P. Dynamical Qualitative Simulation. Lecture at Comett Matari Programme, Pavia, Italy, p.41, 1992.

2.   deKleer J., Brown J.S. A Qualitative Physics based on Confluences. Artificial Intelligence 24, 7-83, 1984.

3.   Fishwick P.A., Zeigler B.P. Qualitative Physics: towards the automation of systems problem solving. J. Expt. Theor. Artif. Intell. 3, 219-246, 1991.

4.   Forbus K.D. Qualitative Process Theory. Artificial Intelligence 24, 85-168, 1984.

5.   Grossmann W., Werthner H. A Stochastic Approach to Qualitative Simulation using Markov Processes. To appear in Proc. IJCAI'93, Chambery, France, 1993

6.   Guariso G., Rizzoli A., Werthner H. Identification of Model Structure via Qualitative Simulation. IEEE Trans. of SMC 22 (5), 1075-1086, 1992.

7.   Ishida Y. Using Global Properties for Qualitative Reasoning: A Qualitative System Theory. Proc. of IJCAI-89, Detroit, Mich., 1174-1179, 1989.

8.   Kuipers B.J. Commonsense Reasoning about Causality: Deriving Behavior from Structure. Artificial Intelligence 24, 169-203, 1984.

9.   Kuipers B.J. The limits of Qualitative Simulation. Proc. IJCAI-85, Los Angeles, Cal., 128-136, 1985.

10. Rajagopalan R. Qualitative modeling and simulation: A survey. In: Kerckhoffs E.J.H., Vansteenkiste G.C., Zeigler B.P. (eds.) AI applied to Simulation. Simulation Series 18, No.1, 9-26, 1986.

11. Sacks E. A Dynamic Systems Perspective on Qualitative Reasoning. Artificial Intelligence 42, 349-362, 1990.

12. Shen Q., Leitch R. Integrating Commonsense and qualitative simulation by the use of fuzzy sets. 4th Int. Workshop on Qualitative Physics, Lugano, CH, 221-232, 1990.

13. Strawson F. Individuals: An Essay of Descriptive Methaphysics. Methuen, London, 1959.

14. Struss P. Global Filters for Qualitative Behaviors. Proc. AAAI-88, Saint Paul, Minn., 275-279, 1988.

15. Struss P. Problems of Interval Based Qualitative Reasoning. Technical Report Siemens #INF2ARM-87, 18, 1987.

16. Travé-Massuyès L. Qualitative reasoning over time: history and current prospects. The Knowledge Engineering Review 7 (1), 1-18, 1992.

17. Yager R.R., Ovchinnikov S., Tong R.M., Nguyen H.T. Fuzzy Sets and Applications: Selected Papers by L.A. Zadeh. Wiley, New York, 1987.

STIMS-MEDTOOL: Integration of Expert Systems with Systems Modelling and Simulation

R.P. Otero[1], A. Barreiro[1], H. Praehofer[2], F. Pichler[2] and J. Mira[3]

[1]Dept. Computación, Facultad de Informática, Universidad de A Coruña
E-15071 A Coruña - Spain
[2]Dept. of Systems Theory and Informations Eng.
Institute of Systems Science, Johannes Kepler University
A-4040 Linz - Austria
[3]Dept. Informática y Automática, Facultad de Ciencias, UNED
E-28040 Madrid - Spain

Abstract. A feasibility study concerning the possibility of integrating Medtool (an expert system shell) with STIMS (a object-oriented modelling and simulation environment) is presented. The relationship between the formalisms implemented by the two environments has resulted in the possibility to relate one formalism to each other. The generalized magnitudes formalism on which Medtool is based, appears as a formalism that may be considered for systems modelling and simulation introducing new Artificial Intelligence features into Computer Aided Systems Theory. Finally, having associated the concepts in one formalism to the concepts in the other, the integration of the two shells can be accomplished in an easier fashion and with more meaning.

1 Introduction

In this paper a feasibility study concerning the possibility of integrating Medtool (an expert system shell) with STIMS (a object-oriented modelling and simulation environment) is presented. This way, analytic and logic knowledge is complemented with Artificial Intelligence (AI) representations resulting in a more powerful simulation tool. The aim is also to relate the methods of AI with those of Systems Theory.

The knowledge representation scheme used by Medtool, called *generalized magnitudes* (GM) [1], has been developed for the programming of expert systems - mainly in Medicine [2]- paying special attention to ease of programming and quick operation. Medtool is coded in C under Unix-OpenWindows and under MS-Windows as a complete environment and as a library. From the beginning, the modelling of the domain evolution appeared to us as the hardest part of the expert system to build, therefore particular attention has been paid to time-related knowledge. We have used the modelling of the evolution of physical systems as the field in which we found concepts that helped us in the development of the time-related features of the representation scheme. The concept of physical magnitude has played a central role in the establishment of the GM scheme. As an AI formalism strongly inspired by classical concepts, the GM scheme can be related to the formalisms of Systems Theory.

STIMS is a new object-oriented modelling and simulation environment [3, 4] supporting discrete event, continuous and combined modelling and simulation. It facilitates modular, hierarchical modelling, i.e., complex models can be constructed by coupling together modular building blocks. STIMS is coded in Common Lisp and CLOS and it is intended to serve as a kernel to experiment with several modelling approaches, in particular, to model event-based intelligent control systems [5, 6, 7] and to explore automatic, knowledge-based model construction methods.

Due to the similarities between the DEVS formalism, implemented by the STIMS evironment, and the GM representation used by Medtool, we have begun the study of the integration of the two environments through the relationship between the formalisms. If these formalisms can be related to each other, the integration of the two shells will be accomplished in an easier fashion and with more meaning as the concepts in one scheme are associated to concepts in the other. This has been the approach we have followed, and we will show how the fundamentals of the DEVS formalism can be translated to the GM scheme at the point of bulding DEVS models for systems in GMs.

2 Introduction to the Formalisms

We will assume some familiarity with the DEVS formalism, outlined in figure 1, and we will therefore introduce the GM scheme trying to use the ST terminology.

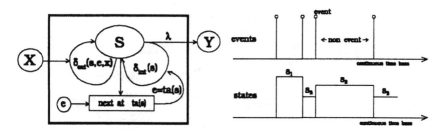

$$DEVS = (X, Y, S, \delta_{ext}, \delta_{int}, \lambda, ta)$$

X, the set of external events; Y, the set of outputs; S the set of sequential states

$\delta_{ext} : Q \times X \to S$ the external transition function

$\delta_{int} : S \to S$ the internal transition function

$\lambda : S \to Y$ the output function

$ta : S \to \mathbf{R}^{+}_0 \cup \infty$ the time advance function

the set of total states: $Q := \{ (s,e) \, / \, s \in S, \ 0 \leq e \leq ta(s) \}$

Fig. 1. Outline of the DEVS formalism

The generalized magnitudes formalism is intended for temporal representation and temporal reasoning, but the aim is to allow reasoning *in* time, making inferences about the evolution of the modelled domain, and not to deal with reasoning about time itself as it appears in other AI paradigms. This is a modular representation in which the basic unit is a GM (figure 2). A GM is defined by giving it a name and a set of possible values it can take. According to the physical concepts of magnitude and state variable, which inspired us, a GM can only take a single value at any particular moment in time. Thus when defining the GMs necessary in order to model a system we must take into account that no more than one value can be simultaneously assigned to a GM. The generalization we consider at this point is that there is no need for cuantification in the values, since symbolic values can be defined for symbolic GMs together with numerical values for numerical GMs.

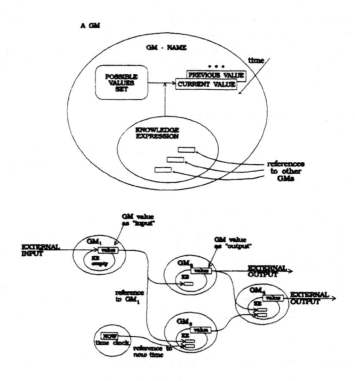

Fig. 2. The models structure in generalized magnitudes

A GM can have a *knowledge expression* (KE) associated with it. This is the way relationships between the GMs can be expressed and the behaviour of the system can be modelled. As a KE is associated with a single GM, it will only be used as an expression that can be evaluated -at any moment- in order to obtain the current value of a single GM. Relational knowledge is strongly associated to structural knowledge thus avoiding many of the problems in applying it, that may arise in other AI schemes. Figure 3 outlines the syntax we use for the KE. They are directly evaluated expressions so there is no need for symbolic processing in the expressions. Several GMs have no KE associated with them and the scheme is unable to obtain a value for these GMs. These GMs are used in order to present to the system the environment facts interacting with it, just like the input variables in the ST formalisms.

Knowledge expression syntax
[*new-value-assigment-expression* [if *condition* ;]*]⁺

Allowed operators
  Math: + - * / mod int abs maximum minimum
        ln exp sin cos tan arctan ...
  Relational: > < = >= <= <>
  Logical: or and not unknown
  Temporal: timeof previous
Temporal constants (pseudoGM):   now begin

Fig. 3. The syntax of the knowledge expressions in the GM formalism

The evaluation of the KEs occurs at discrete time instances in a continuous time base established in the scheme. KE evaluation starts if a non-KE GM has been externally given a new current value or if a special time instant is reached. The mechanism for the application of knowledge (MAK) which evaluates the KE, always tries to apply all the KEs of the model in order to obtain the new value for every GM. Each value a GM takes in each evaluation of its KE has a time point in the continuous time base associated with it. The current time (*now*) is defined dynamically at the moment the KEs are evaluated and this time is associated with the values obtained. In a KE is permitted not only to reference the current value of another GM, but also its previous value and the times (*time-of* operator) those values have associated with them.

With this short introduction to the GM scheme we think the reader can perceive the similarity between the GM and DEVS formalisms, nevertheless we must point out that the GM scheme has other features, not explained here, that are necessary in AI systems, although some of them will appear below in the discussion.

3 DEVS Modelling and Simulation in the GM Scheme

We will now introduce the methodology a designer must follow in order to build a DEVS-equivalent model for a system in the GM formalism. Table 1 shows the equivalence we will establish between the basic concepts in both formalisms. We must point out that the transition functions will need to be modularly expressed for each state variable currently represented as a GM. The application of the knowledge will accomplish the state transitions and will also obtain the output of the system as values for the 'output' GMs.

Table 1. Proposed equivalence between basic concepts in the DEVS and GM formalisms

generalized magnitudes concepts	DEVS concepts
generalized magnitude	*state variable* *input variable* *output variable*
knowledge expression	*state transition functions* *output function*
knowledge expression evaluation *(application of knowledge)*	*state transition* *output*

3.1 The Time Advance Function

The DEVS formalism defines an internal event which is used to fire state transitions without any need for an input to the system. The time advance function *ta(s)* is defined as a number that represents the next activation time for δ_{int}, the internal transition function. A numerical GM we denote as *ta* will represent the time advance function in the GM scheme. We must associate to it a KE whose contents are dependent on the particular system we are modelling, but, in general, could be a conditional expression as follows:

> GM: *ta*
> KE: $ta_{function}(s)$ if $ta_{condition}(s)$;

where *s* stands for the 'state variable' GMs. For example, a typical definition of the time advance function is *ta(s):=sigma*, *sigma* being a state variable. In the GM scheme, this function would be expressed as follows:

> GM: *ta*
> KE: *sigma* ;

where *sigma* is another GM that represents the *sigma* state variable.

Each time a state transition occurs, the time advance function must be applied to the resulting state in order to obtain the time of the next internal event. In the GM scheme, the KE of the *ta* GM will be applied when a new value has been given to any of the s GMs that represent the state variables. During the whole knowledge application process, a particular KE is actually evaluated only when at least one of the referenced GMs in this KE has taken a new value.

3.2 The Elapsed Time

An elapsed time variable e is always defined in DEVS. Its value is a number representing the time elapsed since the last state transition. To represent the e variable in GMs we must define two GMs we call e and *tlast* with associated KEs:

GM: e
KE: timeof(*now*) - previous(*tlast*) ;

GM: *tlast*
KE: timeof(*ta*) ;

Notice that these GMs have a fixed KE, independently of the system to be modelled. The GM *tlast* holds the time (a number) of the state transition, i.e., the time at which the GM *ta* took new value. The GM *e*, representing variable *e*, takes a new value every time knowledge is applied because its KE has a reference to *now* which is an internally defined pseudoGM that is evaluated every time. The value the *e* GM has in any moment of time is the difference between the time of *now* and the time of the previous state transition, i.e., the previous value of *tlast*.

In the GM scheme, the knowledge application will start when a special time is reached. We must indicate which GM will mark these special times. In order to represent the internal state transition we define the GM *tnext* whose value will be the next time for the application of knowledge:

GM: *tnext*
KE: *tlast* + *ta* ;

3.3 The Internal Transition Function

The transition functions must be modularly defined for every state variable of the model. Thus a set of s_i GMs are defined for the representation of the state variables. The internal transition function appears in the KEs associated with these GMs. The contents of these KEs depend on the system to be modelled, but a fixed condition must always be present:

GM: s_i
KE: $s_{i\text{-}function}$ (previous(s_i), s_j) if e=previous(ta) and $s_{i\text{-}condition}$ (previous(s_i),s_j) ;

An internal state transition occurs only when there is an internal event. Thus the 'state variable' GMs s_i only take a new value if the condition e=previous(ta) is true. The reference made to the *ta* GM and the s_i GM through the *previous* operator is necessary in the GM scheme. The name of a GM in a KE will be interpreted as the value the GM will take in the current application of knowledge. Therefore, it would be impossible to evaluate a KE containnig a direct reference to the GM associated with this KE. This behaviour also holds in ST formalisms although it is hidden. For example, *previous(ta)*, that was obtained at the moment of the previous state transition, is the time of the current internal event, but the *ta*, which will be obtained on the current application of knowledge, is the time for the next internal event.

3.4 The Output Function

Similarly to the transition functions we must modularly define the output function for every output variable. A set of y_i GMs are defined for the representation of the output variables:

GM: y_i
KE: $y_{i\text{-}function}$ (previous(s_i)) if e=previous(ta) and $y_{i\text{-}condition}$ (previous(s_i)) ;

The condition e=*previous(ta)* and the reference to the previous state of the system ensures that the output is produced before the internal transition occurs.

3.5 The External Event

The external transition function will produce a state transition when an external event is presented to the system. An external event appears as a new value for some 'input' GMs x_i which have no KE. This is one of the reasons for starting the application of knowledge in the GM scheme, so all the GMs that depend directly or indirectly on the 'input' GMs will also take new value. To deal with all the possible external events from each input variable, we must define a GM *thereisX*:

GM: *thereisX*
KE: x_1 or x_2 or ... x_N ;

where x_i are the 'input' GMs. This GM takes a new value only when any one of the input GMs takes a new value.

3.6 The External Transition Function

The external transition function is applied when there is an external event. The same reasons for a modular definition and reference through previous value we have made for the internal transition function apply here:

GM: s_i
KE: $s_{i\text{-}function}$(previous(s_i),s_j,e,x_k) if *thereisX* and $s_{i\text{-}condition}$(previous(s_i),s_j,e,x_k) ;

The *thereisX* GM ensures that the 'state' GMs take a new value as a result of applying the KEs which represent the external transition function only when there has been an external event.

3.7 DEVS Networks in GM

The GM scheme at the level we have introduced does not consider any structure over the GMs level. Thus when building modular coupled systems in GMs, the designer must constrain himself to not making references involving relationships between the GMs of different subsystems, except for those references representing the connections between component subsystems.

All the connections in a network are made by giving a KE to 'input' GMs. For example, if we want to connect output yA of subsystem A to the input xB of the B component, we must associate a KE to the xB GM as follows:

GM: xB
KE: yA ;

each time the output GM yA takes a new value, this value will be immediately given to the xB GM.

As we now have more than one system modelled we must define a new GM in order to take care of special times for the application of knowledge over all the subsystems. We define a GM *minimumtnext* which will mark the nearest internal event time for the whole network:

GM: *minimumtnext*
KE: minimum (*tnextA*, *tnextB*, ... *tnextZ*) ;

where *tnextA* stands for the *tnext* GM of subsystem A, and so on.

4 Discussion

Two main problems appeared when translating from DEVS to GMs. The first one is due to the existence of two different state transition functions -δ_{int} and δ_{ext}- that deal with the same variable set. As a GM can only have a single KE, this DEVS feature is not directly represented in the basic GM formalism. Nevertheless we can represent it by defining three separate groups of 'state' GMs. For each state variable of the system, we must define three GMs: the actual state GM s_k, the state GM $s_{k,int}$ whose KE contains the knowledge related to the internal transition function and the state GM $s_{k,ext}$ whose KE has the form we have explained for the representation of the external transition function. Each actual state GM s_k must be defined as follows:

GM: s_k
KE: $s_{k,int}$ if e=previous(ta) ;
    else $s_{k,ext}$ if *thereisX* ;

This problem is related to the critical condition problem that appears in DEVS when the internal event and the external event are simultaneous. DEVS define a *select* function to solve the conflict, which may be represented also in GMs, but the solution to this problem in GMs can benefit from particular features of the scheme that allow us, for example, to deal with the problem internally in each subsystem.

The second problem is due to the possible existence in a DEVS system of multiple transitory values for variables in the same moment in time. Although state transitions occur instantaneously, they may involve multiple intermediate values for state variables. Figure 4 shows an example of this situation when $ta(s)=0$.

$$x \rightarrow \delta_{ext} \rightarrow s \rightarrow ta(s)=0 \rightarrow \delta_{int} \rightarrow s' \rightarrow \ldots$$

Fig. 4. Example of two state transitions (s and s') at the same time in a DEVS system

As a GM cannot have more than one value in a time point, this DEVS feature is not representable in the GM scheme. For making this possible we might allow multiple KE evaluations, for each GM, in the same moment in time, giving to all the values obtained the same associated time, but keeping them internally ordered. This proposition, though affordable, would confront basic principles of the GM scheme, mainly the KE firing rule because no new value nor new *now* time exist in the second and following applications of knowledge at the same time.

5 Conclusions

This interdisciplinary study of two different formalisms from two different fields has been very productive. With respect to ST, the GM scheme results in another formalism for systems modelling and simulation that overlaps other existing formalisms, and that introduces in the ST field new features which may be explored. With respect to AI, new features can be added to the representation schemes, inspired in those of the ST formalisms, in order to increase their representational power. This new features may be useful in expert systems for domains far from the typical domains of ST.

Particularly to the GM scheme, the comparison with a formalism that deals with time in a similar way has resulted in new possible features to add to the scheme. For example, we are currently studying the benefit of introducing an *activation expression* associated with every GM, so that KE firing would be driven internally in the GM unit..

We are also currently developing a new version of Medtool intended for systems modelling and simulation which will be more adequate for the integration with the STIMS environment.

Acknowledgements

Ramón P. Otero and Alvaro Barreiro personally want to thank all the members of the Dept. of Systems Theory of the University of Linz-Austria who have made possible their research stay for this work and also acknowledge the financial help of the 1992 AID programme of the Xunta de Galicia-Spain

References

1. R.P. Otero: Medtool, An environment for the development of expert systems. PhD Thesis (in Spanish). Universidad de Santiago de Compostela, Santiago-Spain (1991)
2. J. Mira, R.P. Otero, S. Barro, A. Barreiro, R. Ruíz, R. Marín, A.E. Delgado, A. Amaro, M. Jacket: On knowledge based systems in CCU's: monitoring and patients follow-up. In: J.L. Talmon, J. Fox (eds.): Knowledge based systems in Medicine: methods, applications and evaluation. Lecture Notes in Medical Informatics 47. Berlin: Springer 1991, pp. 174-192
3. H. Praehofer: STIMS - A prototypical implementation of simulation environment based on systems theoretical concepts. Proc. of the SCS European Simulation Multiconference '90. Nuremberg, FRG: SCS Press 1990, pp.75-80
4. F. Pichler, H. Schwärtzel (eds.): CAST Methods in Modelling. Berlin: Springer Verlag (1992)
5. B.P. Zeigler: DEVS Representation of Dinamical Systems: Event-Based Intelligent Control. Proc. of the IEEE 77 (1), Jan 1989, pp. 72-80
6. B.P. Zeigler: Object-Oriented Simulation with Hierarchical, Modular Models. London: Academic Press (1990)
7. B.P. Zeigler: Multifacetted Modelling and Discrete Event Simulation. London: Academic Press (1984)

Systems Concept for Visual Texture Change Detection Strategy

Muñoz J., Garcia C., Alayon F., and Candela S.

Department of Informática y Sistemas.
Campus Universitario de Tafira
University of Las Palmas de Gran Canaria.
Las Palmas, Canary Islands, 35017 Spain

Abstract. The goals of this paper is to develop effective algorithms to achieve acceptable performance in what we call Visually Detectable Defects (V.D.D); and also to define from systems concepts a test strategy. The main problem in V.D.D. has been finally identified as that of the detection and labelling of subtle changes in the texture of an image. In consequence, none of the standard procedures for texture discrimination gave good results, so that increasing the complexity of the process was decided upon, to provide a new decision level.

1. Introduction.

As it has been pointed out in previous works, Automatic Visual Fault Detection up to 100% by a Camera-Computer Chain is not possible according the visual computer analysis of results, which showed the need of multisensory information, and also the existence of not strictly formalizable rules.

There are, however, a class of non-trivial (i.e. not holes or tears) defects which are amenable to visual automatic treatment, which, obviously, are those which are computer-visually detectable.

2. Strategy for fault detection.

A strategy for test of the algorithms for the extraction of descriptors within windows, for automatic window definition and for implement decision algorithms is illustrate in figure 1. We start with pictures of selected samples which were assessed as V.D.D. after their observation in the computer monitor.

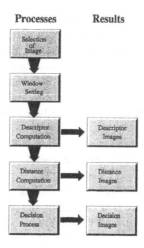

Fig. 1. Illustration of the strategy for tests

2.1 Window setting.

Next, for a set of tests, a program for the determination of the size of a window is applied, appropriate to the background texture, which is almost periodic. The program proceeds as follows:

  a) Determine intenseting distributions along a line, for several lines.
  b) Smooth the distributions (low pass filter).
  c) Determine distances, d_i between adjacent maxima.
  d) Adjust the frequency distribution of d_i to a gaussian; compute σ, μ and the confidence interval to neglect disperse distances; compute the mean distance, d_m.

The size of the windows is then set-up to kd_m, where typical value of k is 2.

2.2 Image partition and descriptors computation.

This set of programs perform the image partition and the computation of descriptors within each window by walking the computed window over the picture, with or without overlapping, performing a reduced -or not- generalized convolution.

   Now, the generation of descriptors images is performed, by computing various parameters to characterize each window. Descriptors used in the tests up to the moment have correspond to programs which compute statistical parameters and Fourier transforms as follows: Histogram, Average, Variance, Energy, Entropy, Skew, Median, and Armonic Content. There are also descriptors images which are constructed by computing a weighted function of several of the above parameters, such as the statistical euclidean distance between each point in the parameter space and some systematically selected other point in its periphery.

Descriptors images are typically of much lower resolution than the original image, which means that the process goes up on what in Artificial Vision is known as the "Visual Pyramid". Descriptor images are themselves considered as new "pictures" where a classification process is to be performed, in two steps, that can be recursive.

2.3 Distance computation and decision processes.

Decision criteria are based in computing distances between points in each descriptor image. For each point, a periphery is defined by the points in the vertical, horizontal and diagonal directions, as illustrated in figure 2.

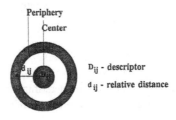

Fig. 2. Distance computation

Next, a"distance", dij, is computed between center and periphery such that if dij is over a certain percentage of the value of the descriptor at the center (or a selected one in the periphery), the coordinates of the center are labeled "good" or "bad". This new type of generalized convolution is thereafter walked on the Descriptor image to generate a Primary Decision Image.

An aspect which is pending to formalize is the way to choose the percentage of change to take the decision. Heuristically, it appears that a value around 40% is appropriate for the different descriptors, though it might change for different types of skins. The procedure is implemented as follows:

a) Let D_{ij} be the descriptors of Center and D_{p1}, Dp2..Dpk the descriptors of the periphery.
b) Compute distances as absolute values $|Dij-Dp1|$, $|D_{ij}-D_{p2}|$, ... $|D_{ij}-D_{pk}|$ and select the maximum distance, d_{ij}. Let D_{pm} the periphereral descriptor which corresponds to that distance.
c) Proceed according the following rule:

```
IF  d_ij < C * min(D_ij,D_pm) THEN
     label ij "0"
ELSE
     label ij "1"
END IF
```

min is the minimal of the two D_{ij}, D_{pm} and C is the established percentage or threshold.

A primary Decision image is then obtained, having the same resolution than the original Descriptor Image, except for the border effects, having values of only 0 and 1.

Multidescriptor Decision Function. When applying the above decision criteria, it has been observed that there are points for which d_{ij} do not overcame the threshold, but that they are very close to it for different descriptor images, so that there is a high probability that the labeled is wrong. The problem is then to design a decision criterium that would take into account that proximity for several types of descriptors, so that when it happens, the probability of labeling the point as 1 increases. The implementation goes as follows:

a) Compute

$$P_{ij} = \sum_{l=1}^{l=L} K * \frac{|D_{ij}^l - D_{pm}^l|}{\min(D_{ij}^l, D_{pm}^l)}$$

where L is the number of different descriptors images and D_{ij}^l, D_{pm}^l are as above for descriptor type l.

b) When P_{ij} is over a threshold, point ij is labeled "1" (defect), "0" otherwise.

Recursive processing. The set up for recursive process by mean of which distance pictures are to be inputs to new distance-pictures generation process is available, that will provide for the equivalent of higher variable directional derivations on which to act by the decision processes. This means that in the computer set up, descriptor images are substituted by first order distance pictures. Results on this are pending on tests.

Decision Images Logical OR. This is, in fact, an additional step in the decision process, which corresponds to the idea of layered fault detection, where each layer tries to "catch" a fault, so that the logical "OR" of the layers will provide for a safer final decision.

The last step in the decision process consists in performing a logical OR of the Primary Decision Images obtained for each type of descriptor. The idea comes for considering the overall System as a layered fault detector, such as it was illustrated in figure 3.

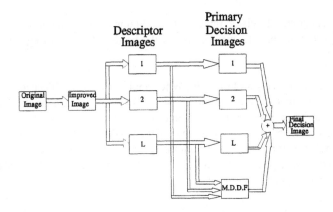

Fig. 3. Illustration of overall process

3. Tests Description Results and illustrations.

Statistical techniques. These series of tests is based on first order statistical parameters, which, in general, are sensitive to tonal properties. They correspond to moments of order q with respect to the origin or to the average, or even a function of them. Its general expression is:

Selection of Imag. Results presented here correspond to images with the following computer characteristics: size (resolution) 1024*1024 (in the picture 512x768, CCIR) level of gray: 256

Window Setting. Window size is 32*32 pixels.

Descriptor Computation. Parameters to generate descriptors are:

Average of the gray levels within each window:

$$\overline{I} = \frac{1}{N}\sum_{i=0}^{r-1} i*h(i)$$

Variance respect to the average:

$$\sigma^2 = \frac{1}{N}\sum_{i=0}^{r-1} (i-\overline{I})^2 h(i)$$

Illustrations.

Fig. 4. Original image (cow leather)

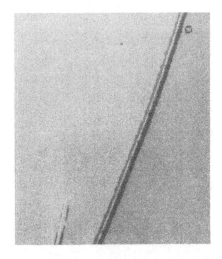

Fig. 5. Detection result

Fig. 6. Original image (cow leather)

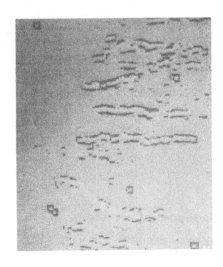

Fig. 7. Detection result

Fig. 8. Original image (pig leather)

Fig. 9. Detection result

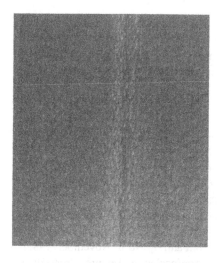

Fig. 10. Original image (pig leather)

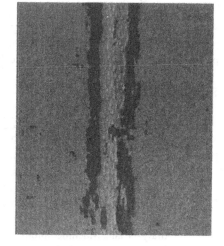

Fig. 11. Detection result

4. General conclusions.

Results show that the above are potentially useful procedures for V.D.D., though more research should be immediate, specifically on: a) Detections by increasing the overlap. b) Tests whit preprocessed filtered pictures. c) Test new descriptors. After the careful analysis of the catalog provided for defects in leather skins and the diagnosis of the human experts, and the research efforts in Computer Vision to cope with the automation of said processes, three main conclusions are reached:

1. there was a kind of optimistic appreciation by computer vision experts that all details of human expertise will be formally provided or accessible, and that all of them should be strictly visual. Typical answers by experts to the question "but is this really a defect and how could it be detectable? were like this: same times is should be considered as a defect and many times it escapes to our consideration".

2. Even though the initial fuzzy problem of Automatic Visual Detection of Leather Faults is not going to be solved 100%, this work is to provide important contributions to both, formalize and strictly determine what is a V.D.D. and what is not, and to open wide new area of applications of artificial systems to detect faults where the diagnosis rests on the detection of subtle changes of texture of an image (i.e. wood, fabrics, metallurgical products and similar).

3. In any case, the process of automatic computation of V.D.D. by camera-computer chains have a high computational cost.

Acknowledgement.

Autors would like to acknowledge the guidance of Roberto Moreno Díaz during the preparation this work. The work presented in this paper has been supported under proyec Brite 2195, and the catalog used belong to GUCCI (leather company).

Bibliography.

[1]   Candela, S. Transformaciones de Campo Receptivo Variable en Proceso de Imagenes y Visión Artificial. Tesis Doctoral. Nov 1987.

[2]   Candela, S., Muñoz, J., Alayon, F., Garcia C. Un Sistema de Visión para la Detección de Defectos.Actas de Panel'92 p 240-247. Las Palmas de G. Canaria.1992. Universidad de Las Palmas de Gran Canaria.

[3]   Conners, R. Identifying and Locating Surface Defects in Wood: Part of an Automated Lumber Processing System. IEEE Transactions on Pattern Analysis and Machine Intelligence, Vol.Pami-5, No 6, November 1983.

[4]   Haralick, R. Statistical and Structural Approaches to Texture. Proccedings of the IEEE, vol 67, no 5, May 1979.

[5]   Muñoz Blanco, J. A."Jerarquizaci ón de estructuras de nivel bajo y medio para reconocimiento visual. Aplicaciones a texturas y formas. Tesis Doctoral. 1987.

[6]   Rao, K., Ahmed, N. Orthogonal Transforms for Digital Signal Processing. IEEE International Conference on Acustics, Speech and Signal Processing P 136-40, 1976.

[7]   Santana, O., Candela, S., Moreno-Diaz, R. Computer non-linear and Algorithmic Simulation of Static Retinal Processes. 6th International Congress of Cybernetics and Systems. Paris. 1984.

[8]   Unser, M., Coulon, F. Detection of Defects by Texture Monitoring in Automatic Visual Inspection. Proceedings of The 2md International Conference on Robot Vision and Sensory Controls November 1982. Stuttgart, Germany.

[9]   Wang, Li.; D. C. He. "A new statistical approach for texture analysis". Photogrammetric Engineering and Remote Sensing, vol. 56, Nº 1, pp 61-66. 1990.

[10]   Wang, L.; and D. He. "Texture Classification Using Texture Spectrum". Pattern Recognition". 1990.

[11]   Weszka, J. S.; Dyer and A. Rosenfeld."A Comparative Study of Texture Measures for Terrain Classification. IEEE Transaction On Systems, Man, and Cybernetics. 1976.

Qualitative Computation with Neural Nets: Differential Equations Like Examples

A. Delgado*, R. Moreno-Díaz** and J. Mira*

*Dpto. Informática y Automática. Facultad de Ciencias. UNED.
28040 Madrid. SPAIN
T: 34-1-3987155, Fax: 34-1-3986697

**Dpto. Informática y Sistemas. Facultad de Informática. ULP
35017 Las Palmas de G.C. SPAIN
T:34-28-458751, Fax: 34-28-351756

Abstract: Linear and non-linear local computation with self-programming facilities is the more used model of biological neural nets. The diversity, specificity and complexity of anatomo-physiological contacts (dendrite-dendrite, axon-axon, axon-dendrite,...) and the variety of local processes carried out by those contacts make one think of authentic subcellular microcomputation.

To illustrate the enormous computational capacity asociated to a neuron we present the masks necessary to solve the more usual equations of classical physics (Newton, Diffusion,...) and compare with the dendritic field of a Purkinje cell.

Size, form and symmetries in the anatomy of receptive fields are interpreted as responsable of specific spatio-temporal filtering, orientation and movement detection. Furthermore, there is experimental evidence of other algorithmic local functions which do not have an analytical counterpart and consequently are not considered in this paper.

1 Introduction.

By neural computation with universal artificial neurons (*UAN*) we mean calculations carried out by tridimensional architectures consisting of a great number of elements with a high degree of connectivity which locally carry out analog and algorithmic functions (linear sum followed by non-linear comparison and/or conditional branching on the output of relational operators followed by a FIFO delay). Here programming is replaced with learning (self-programming) through training [6, 7, 10, 12, 9].

In this paper we will provide a way of using this sort of granular and distributed computation in the formulation and solution of some problems in qualitative physics. To illustrate the methodology behind this proposal, the more usual differential equations in classical physics are formulated and solved using multilayer networks of localy connected *UAN* units. Then, the values of weights and local operators can be relaxed and a learning algorithm can be used to adjust the parameters of the net.

The basic idea in this paper is that every *UAN* realizes a local computation which is a formulation of the anatomy of its dendritic and axonic fields, of the spatio-temporal distribution of the value of the excitatory and inhibitory connections, and of the threshold function. Each local anatomy (convergent, divergent, lateral inhibition, columns, barrels, loops, etc...) has an associated family of possible computations.

From this hypothesis we deduce the fact that every differential equation (*DE*) admits a local formulation computable by a multilayer net of *UAN's*. To design it, all we need is to calculate the *masks* represented by the coefficients of the corresponding equation in *finite differences*. What is really significant in the local calculation with *UAN's* is that by producing small variations in the parameters of the net we obtain new computations which do not have an associated *DE* but which can be considered as "inspired" by it and giving rise to a family of *qualitative processes* [3, 4, 14] that clearly display the deep physical reasoning behind the initial formulation in terms of qualitative *DE's*. In other words, the local formulations with *UAN's* generalizes the physical "laws" by allowing a greater diversity in the volumes, R and R^*, of the input and output spaces on which it carries out the sampling of data, as well as in the weights associated to the different samples. A new degree of generalization is achieved when we expand the set of local operators.

The local computation structure for a neuron of the ith layer is defined by the sampling masks R_i and $R_i{}^*$ in the FIFO memories of its input and output spaces according to the diagram in figure 1.

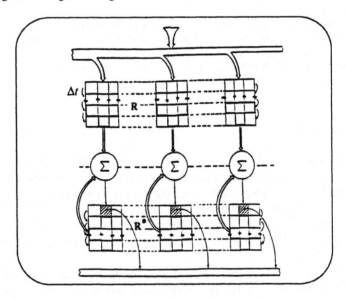

Fig. 1. Computing structure for a layer of recurrent UAN's with FIFO local memories.

The process of obtaining the masks R and R^* stems from the DE, writes the corresponding expressions in finite differences, and passes the values of the coefficients to the corresponding coordinates of the two masks, remembering that the structure of the input and output spaces are FIFO-type. Finally, the problems of initial and/or boundary conditions are specified. With this we obtain an initial prototype of layered net on (x,t) for the treatment of signals, or on $(x,y;t)$ for the treatment of stacks of images. The effect of granularity $(\Delta x, \Delta y; \Delta t)$ is evaluated by means of this prototype. The next stage depends on the domain of applications (for example, medical images) and includes the mechanisms of *adaptation* with the adjustment of parameters *(learning)*. Let us consider the bidimensional case (x,t) for reasons of simplicity in the drawing.

2   Newton's "Neural" Machine

Of all the possible DE we might have selected as an example of a solution by means of neural nets of local computation, we have chosen first Newton's equation. We can approach the Poisson or Laplace equations in the same manner. All of these cases are particularly interesting due to the fact that they are representative of the local formulation of physical processes in terms of simple analytical operations between "neighbours". This neural interpretation of Physics is thought-provoking and was initiated in our surroundings by R. Moreno, A. Delgado and J. Mira around 1982. Shortly after, R. Moreno and F. Martin Rubio published an example of Newton's machine [11] and, more recently emphasis has once again been placed on the neural representation of physical laws and on the integration of the specific circuits which resolve them [2].

Newton's *"neural machine"* must resolve the following equation in a local manner:

$$F = m\frac{d^2x}{dt^2} = m\left[v(t) - v(t-1)\right], \quad \Delta t = 1$$
$$v(t) = x(t) - x(t-1)$$

therefore:

$$F = m\left[x(t) - 2x(t-1) + x(t-2)\right]$$

The corresponding input and outputs masks of figure 1 are show in the box of figure 2.

In other words, to carry out the local calculation of its value at a specific point and instant, Newton's neural machine uses its value at this point in the previous instant $x(t-1)$, and two instant before, $x(t-2)$. These two numbers *(2-1)* along with the input coefficient *(1/m)* define the computation.

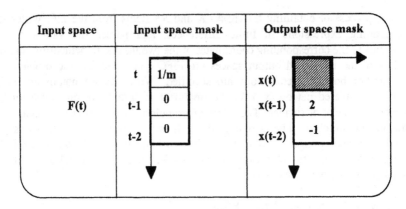

Input space	Input space mask		Output space mask	
	t	1/m	x(t)	
F(t)	t-1	0	x(t-1)	2
	t-2	0	x(t-2)	-1

Fig. 2. Mask in the input and output spaces.

There is a variety of possible preliminary conclusions which can be derived from this simple example:

2.1. Note that only if the numbers are 2 and -1 do we obtain the DE. However, at the level of local calculation, we could consider a multitude of analogous cases. For example, keeping the structure (2r,-r) and varying the value of r. We would then have the calculations (1.8,-0.9), (-2.2,-1.1), etc... If this result is interpreted in anatomical terms with over 60,000 synaptic contacts with variable weights, it is easy to glimpse the quality and magnitude of neural computation, and this just at the analog level.

2.2. This form of neural computation is parallel computation in the strict sense (SIMD type). There is no calculation which lasts longer than the minimum local calculation. A computation is strictly parallel when the global calculation lasts as long as the local calculation.

2.3. The inconvenience of the previous item is the cost in number of neurons. The granularity of the sampling in order to pass to finite differences specifies the necessary neural density and the density of the synaptic contacts. Each point of calculation, $(x=x1, x2, ... xn)$, requires a processor.

2.4. The solution of the DE by means of a net of UAN's needs n numbers which can be:

a) n values at one point (problem of initial conditions)

b) n values at various points (problem of boundary conditions)

Obviously, these n values -independent and arbitrary- are the order of the DE which corresponds to the initial filling of a strip of the input and output spaces.

2.5. Shapes of R and R^*. Note that the Newton net is a vertical net in which there are one or two layers of neurons (depending on the type of implementation) but in both cases each neuron only "speaks" with its memory; R and R^* take on values strictly above and below the unit of calculation. There is neither dialogue nor cooperativity since the responses of the neighbours at previous instants are not considered.

Let us now examine the implementation. In the general model of our theoretical framework (figure 1) which does not limit the number of delays in the output FIFO, the implementation is as shown in figure 3.a. If we accept the most usual limitations of the analog neuron model (which only considers one delay per layer), we need two layers. In that case, we get the circuit in figure 3.b where the first layer calculates the "speed" and the second the "position". Figure 3.c illustrates the result of the digital simulation of this analog net for $F=10$ and $m=1$ with initial conditions $x(0)=x'(0)=0$, 50 neurons operating for 30 intervals of time have been used.

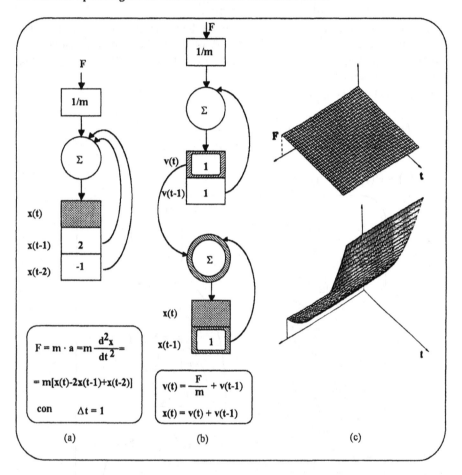

Fig. 3. Implementation of the Newton neural net. a) With FIFO local memory in the output space. b) With two layers of only one delay per layer. c) Results of the simulation for F=10 and m=1.

The general solution of second order nets is shown in figure 4 with K_0, K_1 and K_2 as coefficients and $f(t)$ the "forcing" function. As in the previous case implementation can be made using our general frame with FIFO local memory of two delays (fig. 4.a).

$$K_0 \frac{d^2 x(t)}{dt^2} + K_1 \frac{dx}{dt} + K_2 x(t) = f(t)$$

$$x(t) = \frac{f(t)}{\sum\limits_{i=0}^{2} K_i} + \frac{2K_0 + K_1}{\sum\limits_{i=0}^{2} K_i} x(t-1) + \frac{K_0}{\sum\limits_{i=0}^{2} K_i} x(t-2)$$

To implement it in two layers we re-group the terms, thus obtaining the net in figure 4.b. Note that if $K_0 = m$, $K_1 = 0$ and $F = 0$, we obtain the harmonic oscillator net "without external input", which oscillates from the initial conditions and according to the values of K_2 and m.

$$K_0 \left[x'(t) - x'(t-1) \right] + K_1 x'(t) + K_2 x(t) = f(t)$$

$$x'(t) = \frac{f(t)}{\sum\limits_{i=0}^{1} K_i} + \frac{K_0}{\sum\limits_{i=0}^{1} K_i} x'(t-1) - \frac{K_2}{\sum\limits_{i=0}^{1} K_i} x'(t)$$

$$x(t) = x'(t) + x(t-1)$$

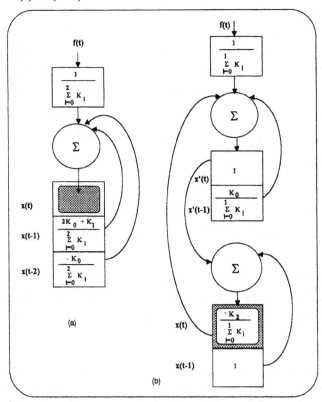

Fig. 4. General solution for second order nets. a) One layer solution. b) Two layer implementation.

Figure 5 shows the results for $K_0=2$, $K_1=0$ and $K_2=-10$. If we write the *nth* degree *DE* in terms of *n DE* of the first order we always can synthesize them with *n* layers of single delay *UAN's* in the usual way.

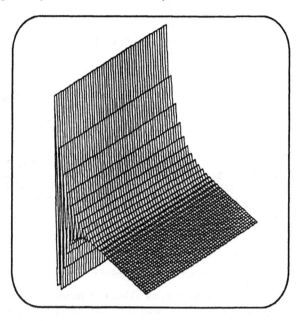

Fig. 5. Results of simulation for the armonic oscilator case with $K_0=2$, $K_1=0$ and $K_2=-10$.

3 Continuity, Waves and Laplace "Neural" Machines

The continuity equation (parabolic), the wave equation (hyperbolic) and the Laplace equation (elliptical) are three examples of local calculation which is realizable by neural nets in which dialogue between neighbours and cooperativity are clearly manifest. This is due to the fact that there are now partial derivatives or - similarly - there are formulations in partial derivatives because they are the differential manifestation of a cooperative phenomenon of local calculation.

Starting with the continuity equation we can obtain its expression in differences and from here, the mask of the output space, as shown in figure 6.a.

$$\frac{\partial u}{\partial t} = K \frac{\partial^2 u}{\partial x^2} \quad \Delta x = 1, \ \Delta t = 1$$

$$u(x,t) = \frac{1}{1-K} u(x,t-1) - \frac{2K}{1-K} u(x-1,t) + \frac{K}{1-K} u(x-2,t)$$

In the same way for the diffusion equation we have,

$$\frac{\partial u(x,t)}{\partial t} = K_1 \frac{\partial^2 u(x,t)}{\partial x^2} + K_2 u(x,t)$$

$$u(x,t) = \frac{1}{1-K_1-K_2} u(x,t-1) - \frac{2K_1}{1-K_1-K_2} u(x-1,t) + \frac{K_1}{1-K_1-K_2} u(x-2,t)$$

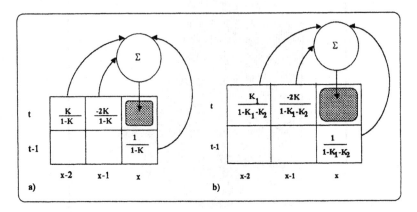

Fig. 6. Continuity and Diffusion equations. a) Mask of the continuity equations. b) Diffusion equation obtained changing (1-K) for (1-K$_1$-K$_2$) and K for K$_1$.

The introduction of a term in $u(x,t)$ (as is the case in the diffusion equation for minority carriers in a semiconductor) only modifies the value of the coefficients of the masks, but not its shape. Thus, if $K_1=1/Dp$ (Dp being the diffusion coefficient for holes/hollows) and $K_2=1/(Dp.\tau p)$ (τp being the average life span) we would obtain the diffusion equation (figure 6.b). This expression agrees with the previous one, substituting $(1-K)$ for $(1-K_1-K_2)$ in the denominators and K for K_1 in the numerators (fig. 6.b).

Figure 7 shows the results of simulation for the initial conditions $x(t-\Delta t)=1$. Note that when there is a connection between the response of a module and that of its neighbours at the same instant, it is necessary to distinguish the delay of the net ΔT and the local computation time Δt with $\Delta T=n\cdot\Delta t$ where n is the number of modules (granularity). The *stationary* in a computation of the net is not reached until n iterations. For the external observer who perceives the net with a sampling period of ΔT, the net computes in parallel. For the observer in the proper domain (Δt), the net computes serially. The difference lies in the "dialogue" time which transmits environmental and initial conditions. This result is the consequence of the existence of partial derivatives in time and can be used to illustrate the need to define families of computations before synthesizing the neural net which supports them.

Let us return to figure 7 which illustrates three situations in the output space at the first, eighth, and twentieth instant, along with a summary of the diffusion

374

process obtained by juxtaposing the partial results. We could repeat this process in an analogous manner for the Wave and Laplace equations.

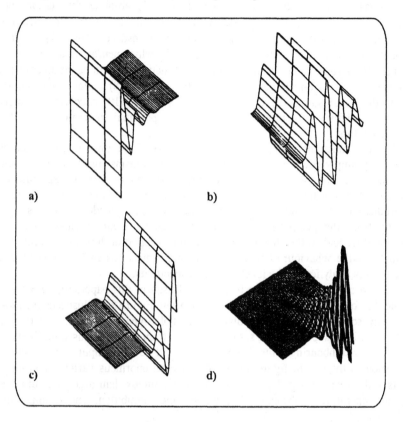

Fig. 7. Results of simulation. a) State in t=1. b) State in t=8. c) State in t=20. d) Summary of the process.

4 Some Neurophysiological Implications

Adult neurons are cells spacialized in the acquisition, integration, transformation, generation, transport and distribution of states of excitation which manifest themselves as slow potentials or action potentials in the form of pulses which recur with a frequency proportional to an excitation level exceeding a certain threshold value. However, the diversity of contacts (axon-soma, axon-dendrite, axon-axon, dendrite-dendrite and soma-soma), along with the specificity and complexity of the variety of local computation carried out by those contacts, make one think of authentic subcellular microcomputation. This is to say that even before arriving at the level of the neuron and considering only processes existing within its dendrite field, there is already sufficient experimental evidence to formulate a non-linear model with the following functions [1, 5, 13, 9] : (1) Convergent processes from

field, there is already sufficient experimental evidence to formulate a non-linear model with the following functions [1, 5, 13, 9] : (1) Convergent processes from specific areas of the dendritic and axonic fields, (2) absolute facilitation and inhibition (*if ... then*). (3) Algebraic sum (excitation, inhibition). (4) Rectification. (5) Analog multiplexed product. (6) Logarithmic transducers and (7) specific functionality (depending on "where", "when" and "for which neurotransmitter").

These "data route" and "local computation" functions has to be complemented with memory and "wave-form generation" functions such as (8) analog delay on/over axon, dendrite and soma, (9) digital delay on synapse, (10) absolute refractory period, (11) relative refractory period, (12) slow potential generators and (13) "spike" generators.

All these functions operate repetitively in space and time within a single neuron. They amply cover all the computational mechanisms used by analog and digital electronics and by computer architecture. Furthermore, there is experimental evidence of other functions which do not have an analytical counterpart and require the introduction of a conditional to complete the neural model. This results - surprising from the perspective of analog and digital computation- is easy to understand if we look at the chemical and electronic support of the electric level used to characterize the behaviour of neurons. That which appears as an "*if ... then*" at the electric level is analytical at the level of protein specificity.

Through comparison with the linear model used in this paper, figure 8 try to illustrate the computational capacity of a real neuron. Figure 8.A illustrates the local connectivity scheme necessary to solve a hyperbolic-type differential equation in partial derivatives. We only need sampling values from two previous time intervals of the closest neighbouring units. What, then, would be the computational capacity of nets such as the one in figure 8.B which have an enormous variety of synaptic contacts and, at each point, local functions more complex than algebraic sum (i.e: nonlinear operations, absolute facilitation and inhibition, neurotransmitter specificities...)?.

Figure 8.B shows synaptic contacts with dendrites from Purkinje cells in the cerebellar cortex. The axon of a basket cell (c) sends descending collaterals to form baskets (a1) around the soma of a Purkinje cell (a) and ascending collaterals (indicated with arrows) to synapse in the molecular layer. In turn, a ganglion cell (b) sends dendrites to the molecular layer [1]. The complexity and degree of connectivity (a granular cell makes contact with more than 50 neighbouring Purkinje cells and one Purkinje receives more than 90.000 contacts from granular axons) along with the geometric precision of these contacts and its density and configuration-change with development and experience would point to the search for much more complex models of neural computation than those currently in use, including the analogic one used in this paper.

The computational interest of this analog formulation of the neural function lies in that we can use the anatomy of dendritic and axonic fields to predict - at least partially - the function of the net. The inconvenience is the low semantics of the operators it handles. This recommends its use in peripheral functions of the sensory pathways, where the function of the net is a pre-process with extraction of spatio-

always a spatio-temporal kernel that will satisfy them within this family of computations.

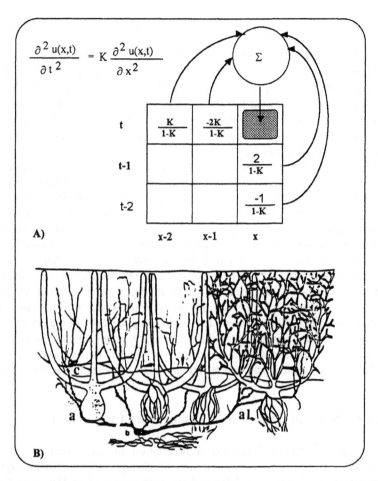

Fig. 8 .Comparative illustration of the computational capacity of Purkinje cells with more than 90,000 contacts in relation with the limited needs for hyperbolic differential equations. See description in the text.

Thus, qualitatively, dendritic fields of the same sign (excitatory or inhibitory) always define low-pass filters in space, in time, or in space-time. The larger the extension of the kernel, the more low-pass the function, as we are reminded in the similarity theorem for Fourier transforms (compression in the space-time domain is equivalent to expansion in the transformed domain and vice-versa). We have mentioned extension but not shape or symmetry. Extension, shape and symmetry in space-time are the three basic computational properties of dendritic and axonic fields.

The shape begins to be computationally richer when excitatory regions alternate with inhibitory regions. In its most basic form, a kernel [-1, 1] derives in space and $\begin{bmatrix} -1 \\ 1 \end{bmatrix}$ derives in time. When they are made symmetrical [-1, 2, -1] or $\begin{bmatrix} -1 \\ 2 \\ -1 \end{bmatrix}$, the centro-peripheric structure -responsible for spatio-temporal derivations and consequently for the high-pass and band-pass filter functions (ON, OFF, and ON-OFF cells)- arises. It is usual to initially think of the sampling areas in the input space (R), but it is important to remember that the theoretical framework includes the same structure for the output space (R^*) which represents the connectivity in the feedback pathway (recursive nets).

What is relevant from the neurophysiological perspective is not the analytic expression of the kernel, but the need for a certain type of excitatory-inhibitory configuration that is able to realize a function. This way, broad areas (be they excitatory or inhibitory) always obtain average values, alternating areas generate differential calculus and, in extreme cases, harmonic analysis in spatial and/or temporal frequencies. Neurons with large receptive fields will be tuned to low frequencies, and those with small receptive fields, to high frequencies. The symmetries in the anatomy of the receptive fields, R and R', are associated with properties of orientation, of directional movement detection, and of measurement of speed. Finally, it must be understood that all these properties are calculated on the pre-process provided by the previous layer. Thus, bipolar cells in the retina operate on the illumination logarithm after the transducer function in the photoreceptors. Ganglion cells operate on the pre-process of the bipolar, horizontal and amacrine cells. The primary cortical cells operate on the results of the ganglion process, and so on.

5   Some Systems Theory and Engineering Implications.

Every modular theory is based on a description of the system in terms of fundamental modules and composition laws which represent the different types of connectivity between these modules and the nets generated in this way. Usually analysis, synthesis and stability theorems are examined. Given a global function, which net of modules and connections will realize it?. Inversely, given a net and "knowing" the functions of its fundamental modules, along with existing interconnections, what is the emergent global function?.

In the nearly "trivial" case of ordinary differential equations we have both theorems (direct and inverse).

In the case of neural computation the problem becomes more complicated since a large part of the function is unknown and, the rest is printed in the connections of the net and these are plastic. In order words, structure and function coincide and there is no external programming. In terms of a modular theory of systems the question is, -what type of systems possess the computational (structural and functional) requirements for the synthesis of which it would be possible and advisable to use neural networks?. Taking into account this question, qualitative computation with neural nets is suitable in the fields of signals and images processing and control situations where the environment is not well known. Note that not all functions are segmentables maintaining causation and interdependences between local computations.

Again, the most serious problem from the systems theory point of view is the lack of *analysis* and *synthesis* theorems for different families of global computations.

Acknowledgment

We acknowledge the economical support of the Spanish CICYT under project TIC-92/0136.

References

1. T.H. Bullock, R. Orkand and A. Grinnell. Introduction to Nervous Systems. W.H. Freeman and Company. San Francisco. (1977).
2. L.O. Chua, L. Yang and K.R. Krieg: Signal Processing Using Cellular Neural Network. Journal of VLSI Signal Processing 3, pp. 25-51. Kluwer Academic Publishers. Boston. (1991).
3. D.F. Forbus: Qualitative Process Theory. Artificial Intelligence, 24. pp. 85-168. (1984)
4. B. Kuiper: Qualitative Simulation. Artificial Intelligence, 29. pp. 289-388. (1986)
5. J.Y. Lettvin et al.: What the Frog's Eye Tells the Frog's Brain. Proceedings of the IRE. Vol. 47, No. 11, pp. 1940-1959. (1959).
6. W.S. McCulloch and Pitts: A logical calculus of the ideas immanent in nervous activity. Bulletin of Mathematical Biophysics, Vol. 5, pp. 115-133. Chicago Univ. Press. (1943).
7. W.S. McCulloch: Embodiments of mind". The MIT Press. Mass. (1965).
8. J. Mira and A.E. Delgado: Linear and Algorithmic Formulation of Co-operative Computation in Neural Nets. In F. Pichler, R. Moreno Díaz (eds): Computer Aided Systems Theory. Lecture Notes in Computer Science, 585. Berlin : Springer 1992, pp. 2-20.
9. J. Mira et al: Towards More Realistic Self-contained Models of Neurons: High-order, Recurrence and Local Learning. In J. Mira et al (eds): New Trend in Neural Computation. Lecture Notes in Computer Science, 686. Berlin: Springer 1993, pp. 55-62.
10. R. Moreno Díaz: Deterministic and probabilistic neural nets with loops. Mathematical Biosciences 11, pp. 129-136. (1971).
11. R. Moreno Díaz y F. Martín Rubio: Incidencia de la Informática en la Educación. Radio y Educación de Adultos. n° 6. pp. 5-8. Las Palmas de G.C. España. (1987)

12. D.E. Rumelhart, G.E. Hinton and R.J. Williams: Learning Internal Representations by Error Propagation. In D.E. Rumelhart and McClelland and the PDP Research Group (eds): Parallel Distributed Processing: Explorations in the Microstructure of Cognition, Vol. 1: Foundations. The MIT Press. Mass, 1986.
13. F.O. Schmitt and F.G. Worden: The Neuroscience Fourth Study Program. The MIT Press. Mass. (1979).
14. D.S. Weld and J. de Kleer (eds): Reading in Qualitative Reasoning and Physical Systems. Morgan Kaufmann Pub. San Mateo. Californa. USA (1990)..

Interpretation-Driven Low-Level Parameter Adaptation in Scene Analysis

M. Kilger, T. Dietl

Siemens AG, Corporate Research and Development
Munich, Germany

Abstract. A hierarchically structured real-time image sequence processing architecture is presented in which low-level parameters are adapted according to the results of the high-level scene interpretation. This adaptive parameter control ensures an optimised performance in varying operating conditions. These general concepts are applied to traffic monitoring, where long outdoor sequences have to be analysed. The algorithms have been extensively evaluated on our real-time image processing system.

1 Introduction

Image processing routines use a set of low-level parameters that significantly influence the obtained scene interpretation. For analysis of long outdoor sequences with varying conditions (e.g. illumination conditions), different parameters produce different results. Obviously, the parameters have to be adjusted to appropriate values depending on the current conditions.

Two methods have been suggested in the past:
1. Analysis of the input sequence w.r.t. the noise [18].
2. Resegmentation of an image region depending on the hypotheses generated from a first segmentation. This method was introduced in the system GOLDIE [14].

The first method is a very low-level signal processing task and therefore is computationally expensive. It is not feasible on a real-time system if costs have to be minimized. The second method looks more attractive, but it was only applied to a single image. We extend the approach to image sequences and apply it to traffic monitoring, where an analysis of very long outdoor sequences in real-time is necessary.

Of course several other authors have worked on traffic monitoring. But most authors [1]-[4], [8], [15], [16] have used only short sequences due to the high computing power needed. In this case no parameter adaptation is necessary, because it is simple to optimise the parameters manually w.r.t. that short sequence. The group of authors [10], [17] who made extensive field tests mention that parameter adaptation is necessary, but they did not give any details.

In section 2 the algorithms of the system are briefly presented. More details are given in [6] and in [11]. In section 3 the adaptation techniques are presented, section 4 contains the implementation and in section 5 the results are shown. Finally in section 6 some conclusions are drawn and future work is outlined.

2 Overview of Algorithms

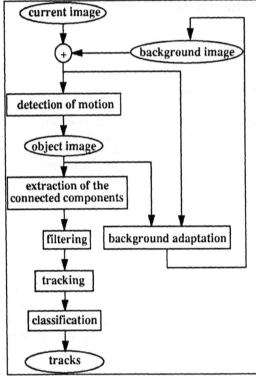

Fig.1: Flow chart

In Fig. 1 a flow chart of the algorithms in the system is given. For detection of the vehicles first the regions containing motion are detected. A binary object image is calculated. The connected components (segments) are extracted from this object image. These segments, which represent regions of motion, are filtered according to their size and width in real world coordinates using inverse perspective mapping. The bounding box is used as a model for the assumed vehicle. The tracking of these bounding boxes is based on a Kalman filter with a constant velocity state model. As a tracking feature the center of the front edge of the vehicle is used. Finally the vehicles are classified into three different classes: trucks, cars and motorcycles. This classification is based upon the width of the bounding boxes.

An example, which illustrates our adaptation concept, is the control of the binarisation threshold and the background update coefficients in the algorithm for the detection of motion. Therefore, the detection of motion and the background adaption routines are presented in more detail below.

2.1 Detection of Motion

The detection of motion is based upon a background image that is continuously updated.

A binary object image (see Fig. 4) is generated by subtracting the current image (see Fig. 2) from the background image (see Fig. 3) and making a threshold decision.

More formally,

$$\left| B_k(p) - I_k(p) \right| > thr \implies M_k(p) = 1$$
$$\left| B_k(p) - I_k(p) \right| \leq thr \implies M_k(p) = 0 \tag{1}$$

where $I_k(p)$ is the current image at time k and pixel p, B is the background image, M is the binary object image and thr is the threshold.

2.2 Background Adaptation

The background image is continuously updated using recursive filtering where regions of motion and regions of no motion are treated separately. The equation for the background adaptation is given by

$$B_{k+1} = B_k + \left[\alpha_1 \cdot \left(1 - M_k \right) + \alpha_2 \cdot M_k \right] \cdot \left(I_k - B_k \right) \tag{2}$$

where α_1 and α_2 are the update coefficients. An update within regions of motion is necessary because regions in the background where vehicles had parked and then moved away or vice versa (parked vehicles have to be treated as background in our sense) would otherwise be erroneously considered to represent regions of motion. These coefficients significantly influence the quality of the generated background.

3 Adaptation Algorithms

The strategy used to obtain the scene interpretation is straightforward. The low-level image processing produces results that are used to interpret the observed scene.

In our approach (more details are given in [5]) results from high-level scene interpretation (e.g. concerning traffic flow, etc.) are fed back to a module called the

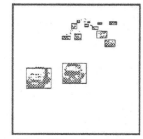

Fig.2: Current image   **Fig.3:** Background image   **Fig.4:** Object image with corresponding segments

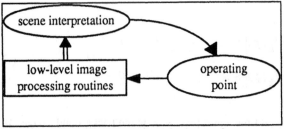

Fig. 5: Interpretation-based control

operating point. This module controls all parameters used by the low-level algorithms. This interpretation-based control (see Fig. 5) is used to adjust the parameters to appropriate values depending on the scene interpretation. This strategy is demonstrated below by applying it to different parameters used in the image processing routines.

3.1 Adaptation of the Binarisation Threshold

The quality of the binarisation threshold *thr*, which determines how much a pixel in the current image must differ from the background to be considered as being in motion, can be judged using the results of the segmentation algorithm. The numbers of segments before and after the filtering are determined and combined into a ratio R:

$$R = \frac{all\ detected\ segments}{all\ valid\ segments} \tag{3}$$

Fig.6: The treshold is too low

R is significantly influenced by the binarisation threshold: A threshold that is too low can be recognised easily because too many irrelevant segments occur (Fig. 6). In a slow control loop we lower the threshold to a value that is still high enough to prevent this from happening.

If the threshold is too high (see Fig. 7) the segments are often fragmented, only parts of the vehicles are detected, or the vehicles are not detected at all. The control tries to decrease the threshold until many small segments occur.

3.2 Adaptation of the Background Update Coefficients

The change of the update coefficient α_2 for regions containing motion depends on the threshold *thr*. The value for α_2 has to be very small to avoid disturbing the background more than absolutely necessary. On the other hand, a minimal update (by at least one greyvalue) has to be made for the described reasons.

Fig.7: The threshold is too high

The update of α_2 is restricted by a formula, which represents a compromise between these conflicting requirements:

$$\alpha_2 * \underbrace{\left| I_k - B_k \right|}_{thr} \geq 0.5 \tag{4}$$

Note that due to the rounding of greyvalues, an update will be performed if the greyvalue is bigger than 0.5.

In the worst case the difference between the current and the background image is equal to the threshold. α_2 is therefore set to

$$\alpha_2 = \frac{1}{2 * thr} \tag{5}$$

The update coefficient α_1 (no motion detected) is linearly adapted according to the value of *thr*. If the threshold is high, no-motion regions are erroneously declared as background, then the coefficient α_1 has to be low. On the other hand if the threshold is low, no-motion regions are safe to be background, then α_1 has to be high. [7] and [9] have empirically found that the coefficient α_1 should be in the range of 0.05 to 0.2 to get good results. Choosing a range from 10 to 60 for the binarisation threshold *thr* (again empirically determined) in an image with 256 greyvalues, the following update coefficient α_1 is chosen:

$$\alpha_1 = 0.23 - 0.003 * thr \tag{6}$$

3.3 Adaptation of the Update Rate; Background Storage and Reuse

A further example where this method has been applied is the adaptation of the background update w.r.t. the current traffic flow. In heavy traffic a lower rate is used because large areas in the background are covered by vehicles. To determine the traffic flow the segmentation results are used.

Additionally, after a certain time with no vehicles detected the background is supposed to have reached a high quality and is stored for further use. When heavy traffic is detected it can be used instead of the current background. If the stored background does not meet the requirements of the current situation any more it is discarded.

4 Implementation

The algorithms were implemented on a low-cost system. A PC with an Intel i486 processor and two plug-in boards, a frame-grabber, and a signal processor board (Motorola DSP96002) is used. The low-level image processing algorithms such as detection of motion and background adaptation are done on the signal processor, while the high-level algorithms run on the i486. The algorithms are written in C. With an image resolution of 256x256 a frame rate of 3 - 5 Hz can be achieved.

5 Results

The algorithms were tested in various traffic and illumination conditions for different video sequences lasting several hours with the following results:

Constant threshold:	Average associations of tracks:
15	6,0
20	6,8
25	7,8
30	7,6
40	7,6
50	6.9
60	6,7
Controlled threshold	8,3

Fig. 8: Average associations

- The tracking of the segments that is done in a subsequent processing level was markedly improved. The number of successful associations is a criterion of the quality of the algorithms. A high value points to a long retention of the tracks during their way through the observed scene. It was increased by 10% by controlling the threshold. Fig. 8 shows the mean associations where different constant thresholds and a controlled threshold are applied. It can be seen that the highest value is obtained by controlling the threshold.

- The quality of the background was significantly enhanced. The average deviation of the background in heavy traffic from a background with the same camera position but no traffic was lowered from 10 to 4 grey values per pixel. (see Fig. 9: Background before adaptation and Fig. 10: Background after adaptation)

- Especially in heavy traffic, up to 30% of the processing time was saved due to
  1) the adaptation of the update rate and
  2) the background storage.

- In Fig. 11 to Fig. 13 the behaviour of the threshold over the time is shown. It is stable over a period of time (see Fig. 11) and converges to different values for different scenes after a few steps (Fig. 12 and Fig. 13).

Fig. 9: Background before adaptation

Fig. 10: Background after adaptation

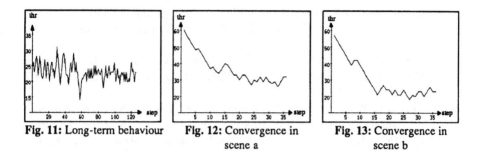

Fig. 11: Long-term behaviour   **Fig. 12:** Convergence in scene a   **Fig. 13:** Convergence in scene b

6 Conclusions and Future Work

Methods for low-level parameter adaptation based on high-level scene interpretation have been presented. In our special application in traffic scene analysis we were able to show that these concepts significantly improve the overall performance.

Evidently our approach, where interpretation-based information is fed back to control subordinate processing levels, has a general character and hence a broad range of application.

The next step that has to be made is the adaptation of the algorithms used to the observed scene, i.e. the best suited algorithms have to be selected depending on the scene interpretation (see Fig. 14).

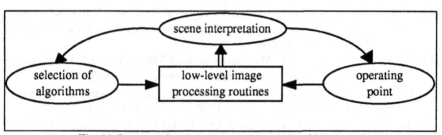

Fig. 14: Interpretation-based parameter and algorithm control

References

[1] K.D. Baker, G.D. Sullivan, Performance Assessment of Model-Based Tracking, *IEEE Workshop on Applications of Computer Vision,* Palm Springs, CA, USA, Nov. 30 - Dec. 2, 1992.

[2] A. Bielik, T. Abramczuk, Real-time wide-traffic monitoring: information reduction and model-based approach, *Proc. 6th Scandinavian Conference on Image Analysis, pp. 1223-1230,* Oulu, Finland, 1989.

[3] J.M. Blosseville, C. Krafft, F. Lenoir, TITAN: a traffic measurement system using image processing techniques, *Second International Conference on Road Traffic Monitoring,* London, UK 7-9 Feb. 1989.

[4] K.W. Dickinson, C.L. Wan, Road traffic monitoring using the TRIP II system, *Second International Conference on Road Traffic Monitoring*, London, UK 7-9 Feb. 1989.

[5] T. Dietl, Adaption von Parametern der Low-Level Bildverarbeitung unter Verwendung der Ergebnisse höherer Verarbeitungsebenen, *Diplomarbeit, TU München*, Dez. 1992.

[6] W. Feiten, B. Bentz, G. Lawitzky, I. Leuthausser, How to achieve perception of real-world scenes: a case study, *Computer Aided Systems Theory - EUROCAST '91. Second International Workshop. Proceedings*, Krems, Austria, 15-19 April 1991.

[7] R. Gerl, Detektion bewegter Objekte in digitalen Bildfolgen natürlicher Szenen mit unbewegtem Hintergrund (In German), *Diplomarbeit TU München*, 1990.

[8] A.D. Houghton, G.S. Hobson, N.L. Seed, R.C. Tozer, Automatic vehicle recognition, *Second International Conference on Road Traffic Monitoring*, London, UK 7-9 Feb. 1989.

[9] K.P. Karmann, A.v. Brandt, Moving object segmentation based on adaptive reference images, *Proc. of EUSIPCO 1990*, Barcelona, 1990.

[10] D.M. Kelly, Results of a field trial of the IMPACTS image processing system for traffic monitoring, *Proc. of 6th International Conference on Road Traffic Monitoring and Control*, 28-30 Apr., 1992.

[11] M. Kilger, Video-based traffic monitoring, *Proc. of 4th International Conference on Image Processing and Its Applications*, pp 89-92, Maastricht/The Netherlands, Apr. 1992.

[12] M.Kilger, A Shadow Handler in a Video-based Real-time Traffic Monitoring System, *IEEE Workshop on Applications of Computer Vision*, Palm Springs / CA (USA), Nov-Dec 1992.

[13] M. Kilger, Heavy Traffic Monitoring in Real-Time, *Proc. 8th Scandinavian Conference on Image Analysis*, Tromsoe, Norway, May 1993.

[14] C.A. Kohl, A.R. Hanson, E.M. Riseman, Goal Directed Control of Low-Level Processes for Image Interpretation, *IU Proc. DARPA, Vol.2*, 1987.

[15] D. Koller, Detektion, Verfolgung und Klassifikation bewegter Objekte in monokularen Bildfolgen am Beispiel von Straßenverkehrsszenen (In German), *Dissertation, Universität Karlsruhe*, 1992.

[16] W. Leutzbach, H.P. Bähr, U. Becker, T. Vögtle: Entwicklung eines Systems zur Erfassung von Verkehrsdaten mittels photogrammetrischer Aufnahmeverfahren und Verfahren zur automatischen Bildauswertung (In German), *Technischer Schlußbericht*, Universität Karlsruhe, 1987.

[17] P.G. Michalopoulos, Vehicle detection video through image processing: the Autoscope system, *IEEE Trans. Veh. Technol. (USA) vol. 40, no. 1*. Feb 1991.

[18] D. Park, D.R. Hush, Statistical analysis of a change detector based on image modeling of difference picture, *ICASSP 90. 1990 International Conference on Acoustics, Speech and Signal Processing*, Albuquerque, NM, USA 3-6 April 1990.

General Systems Theory as a Framework for Model-Based Diagnosis [*]

Zdeněk Zdráhal

Human Cognition Research Laboratory
The Open University, Walton Hall,
Milton Keynes. MK7 6AA.

Abstract. The aim of the paper is to provide an overview of various alternatives to consistency-based diagnostic reasoning and to pinpoint their relationship with General Systems Theory (GST). The inconsistencies between the observations of the system to be diagnosed and predictions provided by its model are used to calculate all possible culprits, i.e. to diagnose the system. It is demonstrated that different variations of the basic diagnostic problem are defined by the choice of the epistemological level of the system and the model. Proposing the most informative next measurement is an important part of diagnosis. The paper also indicates how the techniques of model-based diagnosis can be applied to program debugging.

1 Introduction

Various versions of the model-based diagnostic task have recently been specified and various solutions have been suggested. However, all these tasks are derived from the basic problem description introduced in [13, 3, 5]. In this paper we are interested in a class of techniques for diagnostic reasoning known as the *consistency-based* approach. The simplest version of the diagnostic problem can be described as follows:

> A system and its model are given. The model is described in terms of its structure and a behaviour of components, the system (usually) by means of its variables and their values. The observed data of the system differ from predictions provided by the model. It is assumed that the difference is due to a malfunction of some system components. The goal is to find a minimal set of model components whose abnormal function makes the predictions *consistent* with the observations. This set is called a diagnosis for the system.

---

[*] This research is a part of the Vital project which is partially funded by the ESPRIT Program of the Commission of the European Communities, as project 5365. The partners in the Vital project are: Syseca Temps Reel (F), Bull Cediag (F), Onera (F), The Open University (UK), University of Nottingham (UK), Royal PTT Netherlands (NL), Nokia (SF), University of Helsinki (SF), and Andersen Consulting (E).

We will show that in accordance with the various definitions of model and system introduced in [10], it is possible to distinguish several alternatives of this basic problem.

Diagnostic reasoning consists of two steps repeated in a loop: (1) Calculating diagnoses from current information, and (2) gathering new information to refine the current diagnoses. The loop terminates either if there is no additional information to be obtained, or if we are for any reason satisfied with the current diagnoses (e.g. because the diagnosis is a singleton). In step (1) all potential diagnoses for which the predictions are consistent with the observations are computed. The problem description makes use of the concepts *system* and *model* without explaining them, relying on their common sense meaning. Our objective is to show that by means of the common diagnostic paradigm and specific definitions of system and model we can specialise to other, already known diagnostic techniques. Diagnostic reasoning is explained in section 2. In step (2) the most promising variable to be measured next is proposed, taking into account the diagnoses calculated so far. Possible approaches will be briefly discussed in section 3. In section 4 we will demonstrate how the model-based diagnostic approach can be used for program debugging.

The ultimate aim of the first part of this paper is to show that model-based diagnosis (which is a current fashionable in artificial intelligence) has its roots firmly in the General Systems Theory. The idea of using the system theoretic approach to analyse model-based diagnosis has already been presented by Zeigler in [16] who is not, however, focused on exploiting epistemological levels of system definitions.

In the second part of this paper we want to demonstrate the possibilities of expanding the described diagnostic techniques beyond traditional areas of hardware diagnosis. We have applied this approach to reasoning about errors in computer programs, i.e. to program debugging. This application benefits from the earlier program visualisation and debugging work described in [7, 6].

2   Computing Diagnoses

The General Systems Theory (GST) provides a uniform framework for our analysis. Let us briefly recall how the GST defines systems, models and related concepts. Our explanation is considerably simplified; for a detailed description see Klir [10].

The GST organises systems into a hierarchy of *epistemological levels*. At the lowest level the system is derived from an investigated *object*. Systems at a higher epistemological level entail all knowledge of the corresponding systems at lower levels and, in addition, possess some knowledge which is not contained at the lower levels. The object is a distinguishable part of the world subject to our investigation. At the lowest epistemological level a system is defined on the object by selecting relevant *attributes*. The attributes are mapped by means of an *observation channel* into *state variables*. Individual observations of the same variable are distinguished by means of *support variables*. They are time, space, population or their combinations. For a given object the *object system* is defined as a set of state variables and a set of support variables. Object systems belong to *level 0* at the epistemological hierarchy of systems. Systems at epistemological level 0 describe only possible values, i.e. the ranges of variables. They are however dataless. Real values of each variable are defined in terms of a function from the support variables to the corresponding set of state variables. This function assigns unique values to each support. The *data system* is defined as an object system and a function providing real data. Data systems are at *epistemological level 1*. Given a data system we can describe some support-invariant constraints by means of which values of variables can be generated. The class of possible constraints depends on properties of the support variables. If the set is ordered, constraints can also include neighbours (predecessors, successors) and consequently capture changes (dynamics) over the support set. There are different methods for expressing support invariant constraints. However the basic idea remains the same: Individual data are expressed in terms of a formula for their calculation. A *generative system* is defined as an object system plus some kind of generative constraints. Generative systems entail data systems and therefore are at *epistemological level 2*. A *structure system* is a set of systems at epistemological levels 0, 1 and 2 which share the same support variables. The systems that form a structure system are denoted as its elements. They may be interconnected by means of some common variables. Given a structure system we define an overall structure system by means of all variables of all elements. The elements of a structure system can be viewed as subsystems of the overall structure system, the overall structure system is a supersystem with respect to its elements. Structure systems have been introduced in order to facilitate (i) an integration of large systems from their elements and conversely, (ii) decomposition of large systems into manageable subsystems. Structure systems are at *epistemological level 3*. Similarly systems at epistemological level 4, 5 ... - metasystems - are defined in terms of systems at lower levels. A model is a system which is in a modelling relation with the original. The modelling relation is very often an isomorphism.

Diagnostic problems are usually represented in terms of first order predicate calculus (FOPC) with equality [13, 5]. The model is expressed in terms of first order sentences of the predicate language. The constants of the language correspond to components of the model.

First, let us assume that only the correct behaviour of components is modelled. However, some components may be faulty. Their normal or abnormal behaviour is expressed in terms of the $Ab(c)$ predicate which states that the component c manifests an abnormal behaviour. Only those components that have included the Ab predicate can be diagnosed as faulty. Typically the sentence with the Ab predicate is written as

$$A(c) \land \neg Ab(c) \rightarrow C(c)$$

which means that if the component has property A and is not abnormal then it has also property C. If A is the class name of the component then the formula says that, if c is a normal A, its behaviour is constrained by C. In the GST inter-pretation the formula states that the normal component c which belongs to class A is modelled by a system C. We usually consider C to be a generative system. For example the multiplier M which calculates the product of two numbers is described as follows:

$$mult(M) \land \neg Ab(M) \rightarrow output(M) = input_1(M) * input_2(M).$$

This notation means: if M is a multiplier which is not abnormal, then the output of M equals to the product of the $input_1$ of M and the $input_2$ of M. In fact this is just an instance of a general formula

$$(\forall x)(mult(x) \land \neg Ab(x) \rightarrow output(x) = input_1(x) * input_2(x))$$

which describes multipliers. The implication guarantees that if the left-hand side holds so does the right-hand side. The Ab predicate serves as a kind of switch. If the x is abnormal, i.e. it does not manifest the correct behaviour, the formula is satisfied by the semantics of the implication. This means that the constraint described by the right-hand side does not apply and the formula is effectively cut off. Formulae without the Ab predicate describe constant part of the model, they must be satisfied unconditionally since they cannot be eliminated in any way. Thus the model description is a set of the formulae with or without the Ab predicate and the system is represented by values of variables measured so far. The conflict is a logical inconsistency of the model and the system. Diagnoses are calculated from conflicts by eliminating some formulae of the model until consistency is restored. This is done by selecting some components as abnormal, i.e. by assigning value true to their Ab predicate. By means of this mechanism the corresponding formulae are cut off. If the system and the model are consistent no conflict is found and the diagnosis is empty.

Diagnostic problems which model only the correct behaviour can be repre-sented by a set of Horn clauses (a subset of the FOPC language). It can be shown [8], that Horn clause theories are computationally easier to solve. The algorithm for finding the first diagnosis is polynomial while for non-Horn theories the sim-ilar algorithm is exponential. However algorithms for finding all diagnoses are exponential for both types of theories.

Defining system and model at various epistemological levels [10] leads to different ways of approaching the problem. However, they can be expressed by means of a common modelling paradigm with the system and its model operating in parallel. The block diagram is shown in Figure 1.

There are several conditions which are usually only implicitly assumed al-though they are crucial for the solution to be justified.

When solving diagnostic tasks we assume that both the system to be diag-nosed and the model are structure systems (epistemological level 3) and that a structural isomorphism between the system and the model exists. However in the case of the system to be diagnosed only the data system is observable; i.e. only values of some variables are available. By adjusting the model we are trying to find a minimal set of model components whose abnormal behaviour

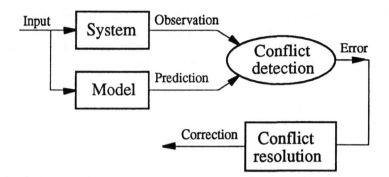

Fig. 1.: Architecture with a parallel model

explains the error. This set is called the diagnosis. Now, provided that the presupposition of structural isomorphism holds, the diagnosis can be extended to the faulty components of the system. The isomorphism guarantees that the reasoning about (inaccessible) components of the system derived from calculating with (accessible) components of the model is sound. In the case of the basic diagnostic problem, the modelling relation, i.e. the relation between the system and its model, is monomorphism from the model to the system. The components which are not assigned are diagnosed as faulty.

The diagnostic methods described so far have assumed an error-free observation channel. The data provided by the system are supposed to be measured without any error and differences between them and predictions calculated from the model are ascribed to faulty components. However it is quite possible to commit an observation error. The system may function properly and the inconsistency between the data and the predictions may be caused by corrupted measurements. In such a case the task is to validate data and to eliminate inconsistent values. The problem can be solved within the same framework with the correction in Fig. 1 being now related to the system. Let us introduce new constants of the language, say $m_1, m_2, ..., m_N$ for N suspicious measurements. Measurement m_i is represented in terms of the formula

$$\neg Ab(m_i) \rightarrow Var_i = Val_j,$$

which means: Unless the measurement m_i is abnormal (i.e. false) the variable Var_i has value Val_j . This notation makes it possible to combine both erroneous measurements and faulty components since it does not make any distinction in the representation of potentially erroneous parts and treats them symmetrically. The potentially faulty measurements are also described by Horn clauses and thus they do not increase computational complexity.

Diagnostic reasoning which takes into account only models of correct behaviour proved to be inadequate in many situations [14]. The reason is that these models do not necessarily capture all important domain-specific knowledge and therefore logically sound inferences do not guarantee the correctness of the diagnosis when it is interpreted in terms of the real world problems. It has been demonstrated in [14] that additional information about possible faults helps to eliminate physically impossible diagnoses. Each component is described by the model of correct behaviour and by a set of fault models. The fault models are incorporated into the FOPC language as follows:

$$A(c) \wedge \neg Ab(c) \rightarrow C(c)$$
$$A(c) \wedge Ab(c) \rightarrow C_{F1}(c) \vee C_{F2}(c)$$
$$A(c) \wedge Ab_1(c) \rightarrow C_{F1}(c)$$
$$A(c) \wedge Ab_2(c) \rightarrow C_{F2}(c)$$
$$Ab(c) \rightarrow Ab_1 \vee Ab_2$$

where $C(c)$ is the model of correct behaviour of component c, $C_{Fi}(c)$ is the ith fault model of component c and predicate $Ab_i(c)$ expresses that c is abnormal by ith way. The formulae therefore mean: If the component is not abnormal, then its behaviour is constrained by model C. If the component is abnormal, then apply either model C_{F1} or C_{F2}. If the component is abnormal in the first way then apply C_{F1}, similarly for the second abnormal way. The final formula means that if the component is abnormal then it is abnormal either by the first or by the second way.

The structure model of the whole system is composed of generative models for each of the components. There are several models for each component (modelling its correct as well as fault behaviour) which can be combined together to produce a structure model and therefore there is a class of structure models applicable to the task. The number of models increases with the number of fault models for each component. The diagnoses are calculated as isomorphisms between the class of models and the system.

The use of fault models has its pros and cons. The main advantage is the elimination of unrealistic diagnoses. On the other hand the search space of all possible diagnoses increases combinatorically and yet one cannot be sure of including all fault models. As it has been shown the description of fault models leads to non-Horn clauses. The problem has, however, a natural interpretation in terms of the GST concepts of generative and structure models.

We have seen that fault models are mainly constructed at the level of generative systems. The requirement to describe components with all possible faulty models is rather difficult to fulfil. It is not sufficient to complete the component by a catch-all mode which would come into operation if neither the model of correct behaviour nor any fault model satisfy. The problem also exists at a conceptual level. We can hardly know all the possible ways in which a component can become faulty but we should not be required to do so. We just know that some situations are impossible.

Having in mind the GST concepts let us speculate about possible ways of improving the computational efficiency. The piece of common sense knowledge about impossible situations can be represented as a *structure fault model* - an interconnection of generative model of components that must not exist in the model simultaneously. From the diagnostic reasoning viewpoint a structure fault model differs from fault models of individual components. One intuitively feels that structure models are a more powerful tool to prune the combinatorial explosion of diagnoses. We have found that a similar approach has already been used in [8] where the term *physical impossibility* refers to structure fault models. The structure fault model is described as

$$\neg(C_{i_1}(c_1) \wedge C_{i_2}(c_2) \wedge \cdots \wedge C_{i_n}(c_n))$$

where $C_{i_j}(c_j)$ is the i_jth model of component c_j. Structure fault models are described in terms of Horn clauses and, as proved in [8] they retain the expressive power of component fault models and improve computational efficiency.

3 Proposing measurement

Diagnostic reasoning is a sequential process. In order to refine partial diagnoses a new measurement of some system variable is needed. Generally, given current knowledge about the case, we want the next measurement to provide the maximum amount of information [3]. The information is measured in terms of entropy. It is necessary to carry out one step look ahead considering all possible outcomes of all available variables and calculate the expected entropy.

In diagnostic reasoning conflicts occur because one part of the model predicts for variable Var_i value Val_j and another part predicts a different value Val_k. The measurement of Var_i will verify which of values Val_j or Val_k is correct or whether Var_i takes some new, up to now unexpected, value. The expected entropy is defined as

$$H(Var_i) = \sum P(Var_i = Val_j) \, H(Var_i = Val_j)$$

the sum being calculated over all possible values of Var_i. The maximum amount of information is achieved by measuring the variable which minimises the expected entropy.

Each potential diagnosis has an associated probability, which is updated as additional information is processed. Representing possible candidate diagnoses in terms of the diagnostic lattice is convenient for entropy calculation. For a given variable Var_i the candidate diagnoses are split into several subsets: (a) the candidates already eliminated; they are not interesting any more and their probability equals 0, (b) supporting set S_{ik} which predicts $Var_i = Val_k$, the probability for each Val_k being calculated by means of the Bayes formula, and (c) uncommitted set U_i which does not predict any value for Var_i; the probability is evenly distributed over all possible values and updated by means of the Bayes formula.

It is possible to demonstrate that U_i consists of sets of those components which, if abnormal, completely decompose the model so that the variable Var_i remains in an isolated subsystem, while S_{ik} preserves connections needed to derive $Var_i = Val_k$. Since the calculation of entropy is computationally expensive and moreover multiple component diagnoses are of a low probability, various simplifying assumptions have been introduced [9, 4]. Alternatively, the properties of possible structural decompositions of the model have been used in [15] to optimise the calculation of the original entropy-based criterion.

4 Diagnostic Reasoning About Programs - Debugging

In this section we will show how the model-based diagnostic technique may be used for program debugging. We will follow the original idea of comparing the system and its model. The system corresponds to the intended program specification and the model is the code itself. If the code does not meet the specification the program is faulty. The discrepancy initiates a diagnostic process - debugging. There are various ways of describing the correct behaviour of the program. For the purpose of debugging we assume that test examples with known results are available so that they can be compared with results calculated by the program. The goal of debugging is to localise the bug, i.e. to find the faulty component of the program. The diagnostic engine for bug localisation is schematically shown in Figure 2. The program to be debugged is executed by an interpreter of the

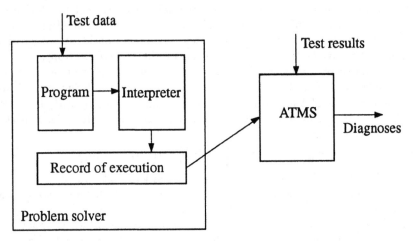

Fig. 2.: Diagnostic engine for dependency-directed debugging

programming language and each program step is intercepted and stored as an execution record. Thus the complete execution history is available for reasoning. In accordance with the traditional problem solving architecture the arrangement shown in Figure 2 can be understood as a problem solver which carries out the specialised computation and communicates results to a truth maintenance system (ATMS in this case) which is responsible for book-keeping. For convenience the described debugging tool has been implemented in Common Lisp which already provides hooks for implementing user-supplied debugging tools. The ATMS calculates how values achieved during the program execution depend on individual programming steps. The steps are user-defined. At the finest level the steps may be individual instances of form evaluation, but usually we prefer a more abstract level. We provide a selection of functions which are taken as primitive debugging steps. The information about dependencies the ATMS receives has the usual form of premises, assumptions, justifications and nogoods (contradictions). The *premise* is a piece of data which we are not prepared to consider as a bug. For example the input data to the program are usually taken as correct and if the results of the execution are erroneous the program is blamed. The *assumptions* are the program steps which are either correct or buggy. They are instances of the selected functions. If the result calculated by the program differs from our expectations (i.e. from known results of a test example) the goal is to find the instance of the function call which can be responsible for the bug. In other words the assumptions are the components we want to use to model the computation. A Lisp function applied to data returns a value. This value is *justified* by the instance of the function call (assumption) and by the arguments (usually data). The assumptions used for calculations of arguments are propagated to the result. If some test datum differs from the corresponding value calculated by the program the two values contradict one another and cannot co-exist. The contradiction is communicated to the ATMS as a *nogood*. We will demonstrate the process of debugging on a simple example.

Example

The program to be debugged is supposed to calculate the roots of the quadratic equation $ax^2 + bx + c = 0$. We know the results for $a = 4$, $b = 5$ and $c = 1$, they are $x_1 = -1.0$ and $x_2 = -0.25$. The roots are calculated by a function roots-of-quadratic-equation which is expressed in terms of functions calc-x1, calc-x2, denom and discr. These functions have been selected as primitive debugging steps. They are defined as follows:

```
(defun discr (a b c)
   (- (* b b) (* 4 a c)))
(defun denom (a)
   (* 2 a))
(defun calc-x1 (b ds dn)
   (/ (+ (- b) ds) dn))
(defun calc-x2 (b ds dn)
   (/ (- (- b) ds) dn))
(defun roots-of-quadratic-equation (a b c)
   (let ((dn (denom a))
         (ds (sqrt (discr a b c))))
      (values (calc-x1 b ds dn) (calc-x2 b ds dn))))
```

After evaluating (roots-of-quadratic-equation 4.0 5.0 1.0) we get the values of two roots x_1 and x_2 together with all dependencies, i.e. x_1 depends on (discr denom calc-x1) and x_2 depends on (discr denom calc-x2). Since each selected function is used in the calculation only once, we do not need to distinguish instances of the function. However, if we extended the choice for multiplication (*) we would have to distinguish between two instances of multiplications in discr and the multiplication in denom.

Assume that some of the functions differ from the definitions above, i.e. there is a bug. Let us follow possible bug location scenarios. There are several possibilities:

- x_1 calculated by the program is not correct, e.g. $x_1 = 37.0$. The ATMS receives the information about the contradiction (nogood) between program calculated value $x_1 = 37.0$ and user supplied value $x_1 = -1.0$ and calculates from the dependency of x_1 the following 3 single-fault bugs: { (discr) (denom) (calc-x1) }. The bug in any of these function instances may produce the observed discrepancy. The diagnostic engine proposes x_2 to be investigated next, since its value is expected to provide the best discrimination.

- x_2 calculated by the program is not correct, e.g. $x_2 = 54.0$. Analogically, the ATMS is informed about another inconsistency and calculates new diagnoses which combine the effect of both conflicts discovered so far. New diagnoses will be: { (discr) (denom) (calc-x1 calc-x2) }. There are 2 single bug diagnoses and one double bug one.

- x_2 calculated by the program is correct, e.g. $x_2 = -0.25$. The ATMS gets the information confirming the result and updates the diagnoses again. With this diagnostic information the diagnoses are: { (calc-x1) (discr calc-x2) (denom calc-x2) }.

The diagnostic reasoning continues and the diagnoses are further refined. This example demonstrates how model-based diagnostic techniques can be applied to program debugging.

The construction of dependency relations has been studied in [11, 12] however our approach seems to provide a more uniform and complete treatment of the problem. For small programs this techniques works. We can reason about iteration as well as recursion. Reasoning from the execution history does not represent any constraint - the memory requirements are acceptable. Scaling up towards big programs has not been yet tested. The complexity of all model-based techniques mentioned in this paper is however exponential, since we still calculate some morphism from one set to another. There are various possible ways of coping with large diagnostic problems. We believe that the hierarchical organisation of program debugging will be a natural solution. GST will certainly be a very useful methodical guide for hierarchical decomposition.

5 Conclusions

In this paper we showed close relations between model-based diagnosis and the GST. We have mentioned, however, only the most typical diagnostic tasks. There are others for which the GST will also be a useful tool. Model-based reconfiguration (see [1]) is an example of a problem, which can also be derived from the basic paradigm shown in Fig. 1.

The experiment with program debugging is an application of the model-based techniques to a new field. The important advantages of our approach are: (1)

the user does not build any abstract model of the program since the debugging makes use of the source code, (2) the diagnostic algorithm reasons about single as well as multiple faults, (3) due to the results achieved in the area of model-based diagnosis the approach is systematic with the attempt to optimise the diagnostic process and (4) the method is easily extendible to other programming languages. The possibility of scaling up this technique so that it can be used for debugging big programs remains to be proved.

References

1. Crow J. and Rushby J.: Model-based Reconfiguration: Towards an Integration with Diagnosis, Proceedings of the AAAI 91, Vol. 2, pp. 836-841
2. de Kleer J.: An Assumption-based TMS, Artificial Intelligence 28, 1986, pp. 127-162
3. de Kleer J. and Williams B.C.: Diagnosing Multiple Faults, Artificial Intelligence 32 (1987), pp. 97-130
4. de Kleer J.: Using Crude Probability Estimates to Guide Diagnosis, Artificial Intelligence 45 (1990), pp. 381-391
5. de Kleer J., Mackworth A.K. and Reiter R.: Characterizing diagnoses and systems, Artificial Intelligence 56 (1992), pp. 197-222
6. Domingue J., Motta E. and Watt S.: The Emerging Vital Workbench, to appear in Proceedings of the European Knowledge Acquisition Workshop (EKAW'93).
7. Eisenstadt M., Price B.A., and Domingue J.: Software Visualization as a Pedagogical Tool, to appear in Instructional Science, also Technical Report 93, HCRL, The Open University, November 1992
8. Friedrich G., Gottlob G. and Nejdl W.: Physical Impossibility Instead of Fault Models, AAAI 90, Boston, MA, 1990
9. Freiteag H.: A Generic Measurement Proposer, in Expert Systems in Engineering - Principles and Applications, International Workshop, Vienna, Austria, September 24- 26, 1990, Lecture Notes in AI 462, Springer-Verlag, 1990, pp. 79-89
10. Klir G.J.: Architecture of Systems Problem Solving, Plenum Press, New York, 1985
11. Korel B.: PELAS - Program Error-Locating Assistant System, IEEE Trans. on Software Engineering, Vol 14, No. 9, September 1988
12. Kuper R.I.: Dependency-Directed Localization of Software Bugs, TR 1053, MIT AI Laboratory, May 1989
13. Reiter R.: A Theory of Diagnosis from First Principles, Artificial Intelligence 32, 1987, pp. 57-95
14. Struss P. and Dressler O.: Physical Negation - Integrating Fault Models into General Diagnostic Engine, in Proceedings IJCAI-89, Detroit, Vol.2, 1989, pp.1318-1323
15. Zdrahal Z.: Using Candidate Space Structure to Propose the Next Measurement in Model Based Diagnosis, to be published in Proceedings of DEXA 93 conference, Prague, 6-8 September 1993, Springer-Verlag
16. Zeigler B.P.: System Formulation of a Theory of Diagnosis from First Principles, IEEE Trans. on Reliability, Vol 41, No. 1, 1992

Technical Applications of Knowledge-Based Systems

P. Kopacek

Institute for Handling Devices and Robotics
Technical University of Vienna
Vienna, AUSTRIA

Abstract. The paper deals with the industrial application of knowledge based systems as a consequent result of computer aided methods of system theory. Such systems are well known since many years, but concrete industrial applications are more or less missing until now. On two particular realized applications some problems in developing and using of such systems are discussed. First the minimization of the tool changing and adjustment time of a continuous rolling line producing hollow steel sections is carried out by a knowledge based system called BERASYS and second an expert system for diagnosis in a robotized assembly cell is discussed.

1 Introduction

Knowledge based systems offer an efficient tool to reduce the dependence of the product quality from the knowledge of some experts. Processes can be operated by normally skilled persons. They can get faster knowledge about a system by consulting a knowledge based system. Moreover a knowledge based system might be able to find an appropriate solution faster than an expert. Solutions can be suggested in case of a very complex process, which would be too complicated to find, even for an expert. The last decision for running a process, still depends on the machine operator or expert.

The application of knowledge based systems in automation can be divided in the following fields:

⊗ Diagnosis and prognosis (e.g. early recognition of errors and maintainance).

⊗ Control tasks (e.g. realtime systems control, knowledge based control design, decision support systems)

⊗ Planning (e.g. automatic supply of workplanes)

⊗ CAD (e.g. interpretation of CAD data, data exchange of CAD systems).

2 Application Examples

The two examples discussed in the follow are from the field of diagnosis and from the field of control and especially decision support systems.

2.1 Automation of a Tool Changing Process:

Cold formed structural steel hollow sections are a tool for load bearing structures; e.g. in steel construction, automotive engineering, bridge building, crane construction, etc. The cross sections can be circular, square or rectangular. They are produced on rolling lines consisting of 18 to 24 standing device and a cutting device (Fig. 1). Usually a stand of rolls consists of a frame with 4 rolls which are adjustable either manually or electrically by motors. The rolls are driven by electrical motors. If a new profil has to be produced the rolls are to be changed or adjusted or only to be adjusted. Currently the adjustment is done by machine operators who need a lot of experience to move exactly the rolls on stands which are responsible for faults.

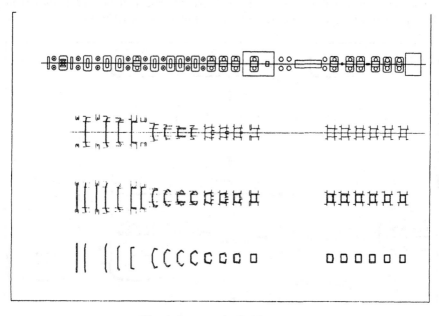

Fig. 1. Layout of a Rolling Line

The main idea for a low cost automation concept of this tooling process is to develop a kind of knowledge based system. This system called BERASYS makes suggestions during the adjustment process. The knowledge based system BERASYS consists of a knowledge base, divided in a qualitative art and a quantitative art.

Usually the tooling process is carried out manually. The rolls are adjusted manually according to the last historical data. After producing 10 - 20 m the process is stopped

- the section is measured - and in most cases a readjustment of some stand of rolls is necessary.

There were two main reasons for the development of the knowledge based system "BERASYS":

⊗ The determination of a mathematical model for the process either in a theoretical or experimental way is impossible because of the complexity of the process.

⊗ Reasons for inaccuracies of the section are available only in verbal form as facts and rules.

Therefore the main idea for a "low cost" automation of this tooling process is to develop a kind of knowledge based system. Not only the last data should be stored from a lot of data from the past (historical data). From these data the optimal adjustment parameters for each stand of rolls are calculated and shown at a monitor, respectively could be used for automatic adjustment (if there are electrical drives for the rolls adjustment available). The results of this adjustment - measured data from the product - are stored in the computer and serve for calculating the next adjustment data. The hardware consists of PCs (386 compatible), connected by a LAN.

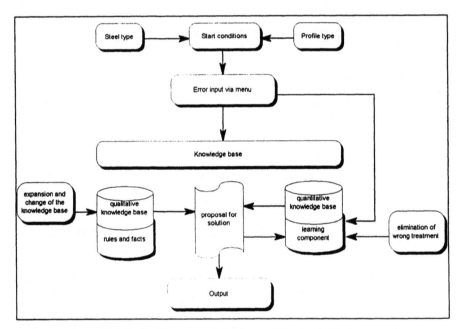

Fig. 2. Architecture of the Knowledge Base System

The basic structure of "BERASYS" is shown in Fig. 2. Main part is the knowledge base consisting of a qualitative part (facts and rules) and a quantitative part (classical knowledge). The last is generated automatically by the learning component. Only if a distinct decision is checked by both parts of the knowledge base it will be indicated.

The expert knowledge was accumulted by using different methods of knowledge acquisition and was transformed into logical rules constituing the so called "qualitative" part of the knowledge base. which is responsible for locating the reason of faults.

To get the information about the amount of the adjustments, a learning component creates a deviation factor for each material, stand, error-value and type of hollow section. The qualitative output of the system has to be multiplied with this factor. Getting enough factors for the different amounts, the learning component regresses a function, which describes the relations between the quantitative output of the expert system and the necessary deviation factor for the actual hollow section, material and stand. After a period of gathering practice, the system offers suggestions which exactly take care of the unknown elasticity and temper, depending on different materials. The splitting of rule-based knowledge and the deviation factors makes it easy to adapt the knowledge based system to changes of the process.

The rule-based part of "BERASYS" first was developed in a logical language (FORLOG), because rule based problems are best to be treated with logical programming. The learning component was realised in a procedural language (C).

First tests in practice were showing a good conformity of the proposals of the knowledge based system and the experts treatment.

2.2 Fault Diagnosis in a Robotized Assembly Cell

As a second example serves an expert system for diagnosis in a robotized assembly cell. The assembly cell taken into consideration consists of two industrial robots including the necessary tool and parts storages as well as transportation systems. This assembly cell is supervised by one person who is responsible for handling errors, feeding all necessary units including loading and unloading of head sinks. The installed diagnosis system including a small expert system is implemented on a PC in addition to the supervising system. The system allows less skilled operators to choose a propriate action in order to recover errors. In addition this knowledge based system serves for education of operators by simulation of different conditions of the assembly cell.

Primary parts of welding transformers were assembled by hand in 10 working stations and 3 test stations. Unfortunately from the technical point of view it is impossible to carry out all assembling operations automatically. Some of the parts are of a large size, some of them are "flexible"; e.g. cables. Therefore only a part of the necessary assembling operations can be carried out automatically from the technical as well as the commercial point of view.

The primary part consists of a heat sink whose dimensions are determined by the attached electronic parts as well as two printed circuits.

The necessary operations in the automatic assembling station are:

⊗ applying of a heat conductivity paste for the electronic parts onto distinct areas of the heat sink

⊗ attaching of two transistors. 4 diodes. 2 resistors and 1 thermocouple

⊗ screwing on of these electronic parts

⊗ attaching of the printed circuit 1 - the 8 pins of the diodes have to be inserted in the holes of the printed circuit 1

⊗ screwing of the printed circuit 1 onto the diodes

⊗ screwing of 3 power cables to the printed circuit 1

⊗ soldering of some cables and connections onto printed circuit 1

⊗ attaching and soldering of a capacitor onto print 1

⊗ attaching of a pressboard plate and the printed circuit 2

⊗ screwing on of these two parts.

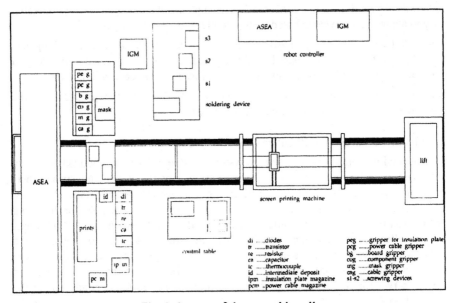

Fig. 3. Layout of the assembly cell

The layout of the assembling device for these tasks is shown in Fig. 3. It consists of two robots, the storage units for the parts to be assembled, the screen printing machine, the necessary transportation devices, and the input-output station.

After a working period of approximately half a year first experiences show that it might be very difficult for the operators to find out quickly the reasons for a disturbance in the process. For example the best skilled operators had also huge problems. Therefore it was decided to develop a low cost diagnosis system for these tasks. Fig. 4 shows the implementation of the diagnosis programme in the control strucutre of the assembly cell.

The diagnosis module need information about the state of the process from the controller. These informations are available mainly in the status erase of the controller and are moved in the module by means of a temporary file. The diagnosis module is activated from the control programme automatically. Because of uncontrollable movements of the robots the start of the diagnosis module is only possible if the process is in a save state (e.g. break mode). The user interface of the system is very simple because the computer is situated on the shop floor level. This kind of knowledge base is called rule based. The user has the possibility to add some new rules and closes in the programme in a very simply way. The system is able to classify the errors in the following categories:

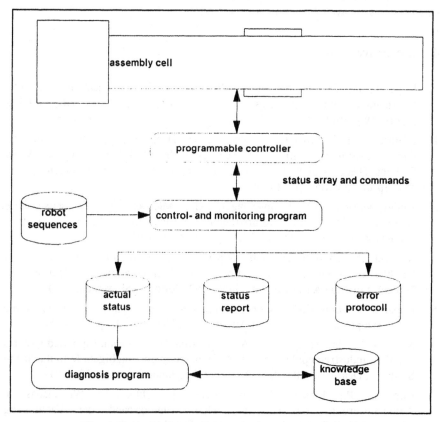

Fig. 4. flow of information: control - diagnosis module

Error of the screwing device, error in storage units, error of the soldering device, robot programme error, gripper error, pallette error, etc. The actual error is localized by the knowledge based in an iterative way. The next development step was decided to add learning component in the system.

3 Summary

The two examples discussed shortly in this contribution are only a first step in the direction of technical applications of methods of artificial intelligence in form of expert system or knowledge based systems in industrial applications. At that time hardware and software facilites are available the direction of low cost concepts especially suitable for small and medium sized companies. Meanwhile the number of such applications growing up dramatically especially in the field of electrical and mechanical engineering. Might be possible that in the nearest feature real knowledge based systems are available for a reasonable price.

4 Literature

1. Kopacek, P.; Girsule, N.; Probst, R.: PCs in CAM Education. Preprints: Information Control Problems in Manufacturing Technology, (INCOM'89), Madrid 1989, Vol. 1, p. 269-272

2. Halmschlager, G.; Mletzko, F.; Trippler, D.: Decision Supported Systems and their Industrial Applications. Proceedings of the Workshop "Computer Aided Systems Theory - EUROCAST'91", Lecture Notes in Computer Science, Vol. 585, Springer Verlag, p. 585-591

3. Ambrojewitsch, V.; Halmschlager, G.; Mletzko, F.: Wissensbasierte Automatisierung von Rohrstraßen. Berichtband des 7. Österr. Automatisierungstages, ÖPWZ Vienna, 1991, p. 95-101

4. Kopacek, P.; Probst. R.: Assembly of primary parts of welding transformers in a flexible. robotized cell. Proceedings des 1. Internationalen Meetings on "Robotics in Alpe-Adria-Region". June 1992, Portoroz, Slowenien, p. 148-153

5. Kopacek. P.: Low Cost Factory Automation. Plenary paper: Preprints of the IFAC Symposium on "Low Cost Automation". Vienna 1992, p. 273 - 280

6. Kopacek, P.; Halmschlager. G.; Ambrojewitsch, V.: Decision supported systems and their industrial applications. IFAC Workshop on "Intelligent Manufacturing Systems", Dearborn, MI. USA, Oct. 1992, Pergamon Press, p. 177 - 181

7. Kopacek, P.; Probst. R.; Wernstedt, J.; Otto, P.: Wissensbasierte Automatisierung von Rohrwalzstraßen. Tagungsband der Fachtagung AUTOMATISIERUNG, Feb. 20-21, 1992, Dresden, pp. 44

8. Kopacek, P.; Probst, R.: Robotized Assembly cells in "Low cost" CIM Concepts. Proceedings of the 23. Intern. Symposium on "Industrial Robots", Oct. 1992, Barcelona, Spain. p. 191 - 195

CASE – Computer-Aided Systems Engineering, a New Approach for Developing IM-Systems with Special Consideration of CIM Systems

Martin Zauner

Department of Systems Engineering and Automation
Scientific Academy of Lower Austria
Krems, Austria

Abstract. The paper deals with a new approach for developing Information Management (IM) Systems with special consideration of CIM systems. For example. CIM systems are well known since many years. but there is no recommondation how to develop them well. Concrete industrial applications are still rarely seen and the existing ones are mostly realized as special systems. So. two significant examples show, how to design integrated informations management systems for different applications. The idea is, to connect the strategy of open systems with object orientated systems design. It is show on one example that deals with hospital information systems (HIS) and another one that shows a computer integrated manufacturing (CIM) system.

Keywords: RDBMS. active database management systems. client/server concepts. CIM concept. IM concepts. prototyping. right-sizing. open systems.

1 Introduction

Unfortunately until nowadays a lot of CIM workers do not accept the high complexity of integrated manufacturing processes. CIM is not only a technical problem. Rather the organisational and economical influences as well as the education level of the CIM-working group members mainly control the progress of CIM-development process. These high level based, company orientated goals are the kernel for each well developed CIM system. CIM-system developers have to build up a logical shell to cover and rule over the described difficulties. Under consideration that CIM-systems are usually designed for special company structures it is obviously that CIM-systems are widely unique. The bad acceptance of CIM systems is attributed mostly to non realizable offers of system-houses as well as non realistic requirements of the companies. State nowadays is that hardware infra structures within the companies support local area networks and high performance server-machines but without data, information or application integration between the departments. Considering, that the discussion now was only about integrated manufacturing processes, the question appears: how to design Integrated Information Management Systems (IIMS)? What is missing that this problem could not be solved until now? This paper dicusses an approach for the structured development of IIMS with special consideration of CIM-systems. Therefore one main criteria in the software-engineering process, the reusability of objects, was taken under special account. Due to the fact, that handling as well as developing CIM systems and their maintainance during their life-cycles are the problem of the companies and also the problem of CIM systems developers a model is shown how to handle all necessary activities.

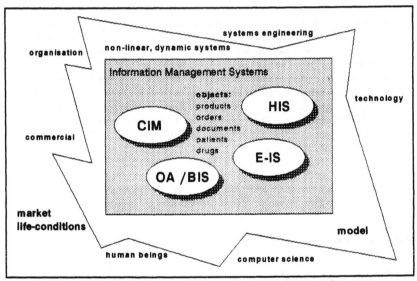

Fig. 1. Views to Integrated Information Management Systems

2 Problem Classification and Objectives

Meanwhile most of the technical problems have been solved, but how to handle difficulties in organisational functions interfaces? So, as you see it´s necessary to combine technical and organisational structures. The reasons, why CIM projects are not economical respectively efficient are usually lost informations. To solve this problem first of all an information flow structure is defined. An information flow structure attributed to organisation structures is defined as Information Flow Network Structure (IFNS) as follows:

S_{IIMS} = IFNS

IFNS = IFNS ifns_node(IS, to_ifns_node, from_ifns_node,
up_ifns_node, down_ifns_node)
| ifns_node(IS, to_ifns_node, from_ifns_node, up_ifns_node,
down_ifns_node)

IS = IFNS
| system (A, B, C, d, e)

where

S_{IIMS} ... System (*Integrated Information Management System*)

IS ... Information Structure,

to_ifns_node ... interface vector to the IFNS node,

from_ifns_node ... interface vector from the IFNS node,

up_ifns_node, down_ifns_node ... connection between the levels,

system (A, B, C, d ,e) ... input-, output-, dynamic matrix, initial- and state
processing vector.

On this level all possible informations, processes, data interfaces and organisation structures are described. It is obviously that the IFNS contains more than one hierachical level (vertical approach). As one advantage the management of versions is mentioned. The depth within one ifns_node depends directly on the organisational structure. Processes as well as input and output interfaces which are ordinary defined parts of the organisation structure as tasks, operations, events or triggers are described as ifns_node parameters. The design of the information system is shown because each ifns_node is able to activate all attributed ifns_nodes independently (horicontal approach).

Based on the horicontal approach the information flow can be optimized. In order to find an optimal S_{IIMS} each ifns_node has to be optimal (information structure). The assumption to define an optimal ifns_node is to optimize input and output interfaces, the internal processes and the stability of each node.

To point out finding an optimal CIM solution the data interfaces (*d*) and data structures (ds) for all departments (*dept_i*) are fixed. The common data pool (*dp*) is

$$(1) \qquad dp := (d_{dept(1)} \times d_{dept(2)} \times ... \times d_{dept(n)}),$$
$$n = \text{number of departments.}$$

One requirement is to connect to the common data pool (*dp*) via a logical central data access interface. For data fields that are only used inside departments without any external connection (*df*) it is not necessary to be stored within the central data pool. Decentral, transaction supported concepts are often better in this case. The organisation wide data pool (c_dp) is shown as follows.

$$(2) \quad c\_dp := (dp + df_{dept(1)} + df_{dept(2)} + ... + df_{dept(n)}),$$
$$n = \text{number of departments.}$$

The division between the horicontal and vertical view splits up the data model and the organisational model. Based on the independent data model and organisational model it is possible to design information management systems that are independent of data models. As the last step, the amount of data as well as conditions proving and keeping the consistence in data management systems determine the requirements for operating systems and hardware capacities.

The main objectives are the possibilities to offer
- a method for the design of information management systems,
- useability for different organisational structures (i.e. CIM Systems, Hospital Information Management Systems),
- horicontal optimizations can be done before the connection to the data level is done,
- modularity within this design offers high flexibility for systems engineering processes,
- protocol based data exchange improves full advantages of object-orientated structures,
- open interfaces between modules,
- migration from existing and "old" data structures to new data model structures,
- widest independence of hardware components - operating system compatibility,
- support of development standards (ISO/OSI, ANSI, Software Engineering Methods, ...),
- systems requirments: minimize network traffic, improve performance, rules for complex validations, referential integrity.

From the commercial point of view this model offers additional advantages.
- Hardware components are bought "just in time",
- introduction of the new system is done step by step,
- decreasing prices in hardware technologies are fully supported,
- employees are not overtaxed,

- investigations can be activated in financial calculations,
- existing software and hardware components can be further used,
- integration of standard software packages is offered.

3 The Design of Integrated Information Management Systems

The way, how to desgin Integrated Information Management Systems (IIMS) follows the idea to connect top-down and bottom-up development activities. First of all, a so called *meta-model* is necessary. It allows to structure the main information flows and to define high-level interfaces. The top-down method supports to specify detail tasks inside the already exisiting information flow processes. On the other hand there are data structures resulting from software applications or from forms. In addition to those data structures several practical experiences belonging to machine handling as well as life-conditions and many more influence the design of systems. The experiences of these, mostly company specific restrictions, shape the systems bottom up design.

The know-how to design and implement the IIMS is the well designed process communication. To explain it in detail, process communication is split up into two domains the *process* and the *communication*.

The connection between processes should map the information flow of the target system. Further the IIMS contents system-processes, too. Those are responsible to control all process activities. A monitor ensures the right reacting of the system-processes especially in cases, when inconsistent system-states occur. It is responsible for the rollback activities and the systems recovering back to a deterministic, consistent state. The communication is recommended to be realized as potocoll function. So, the communication between tasks is dynamic and can react on time when updates in system-structures are done.

3.1 Client/Server Design and the Possibilities of Active Data Bases

Modern system architectures follow the goals to use ressources optimal (*rightsizing*). Parameters therefore are the costs of performance (hardware) and maintainance, system stability (decentral concepts), modularity and compatibility (open-systems). Client/server systems designs support the use of decentral processes, running on external computers, that are connected to central host machines. These so-called host machines are more and more down-sized from classical mainframes to high-performance workstations. This step is possible, because user-processes are running on small, intelligent front-end computers, mostly PCs. One advantage is, that the host-computer is unemployeed from the user-tasks and is only busy to run the data-access interactions and intelligent system processes that are responsible for monitoring and information flow sending. The application oriented task of these new generation of host-machines is that they are used as data-base and information-flow servers.

Client/server designs offer a very modular systems architecture design. The great advantages are not only the use of low-cost respectively right-sized hardware-units but the possibility to change or upgrade all units inside the system. The central idea of data base concepts within client/server applications is now extented to active data base concepts. System stability and event triggers are handled as a central data base function. The consistence of the system, modelling conditions are now additional to transaction handling possibilities offered by data base designs. In cases, when active data base systems are not available now - for examples in industrial applications - a shell covers these tasks in well designed solutions, in other cases - mostly grown up structures - these tasks are hidden in the applications itselfes.

3.2 Information Flow Design

The concept supports full decentral task-to-task communication. Finally, the user tasks are running on intelligent front end computers, mainly PCs.

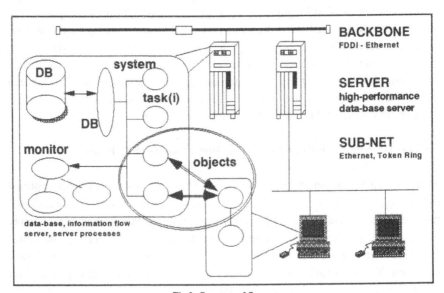

Fig.2. Concept of S_{IIMS}

Because there are many tasks necessary to implement a information management system, the concept, of course, supports the use of more than one data-base and information flow server machine.

In this case the information flow processes itselves are reacting as client and server tasks. Each process is directly or indirectly attributed to an object. The communication between all of the objects is done via system wide defined protocols. Indeed, the communication is send from a front end object via serveral system objects to the target object (mostly the data base). The result is afterwards sent back to the client. An additional, but high sensitive responsibility is located in the system processes that are responible to keep the systems consistance and integrity.

4 Hospital Information Management System - A Case Study

Based on the approach to design Information Management Systems, an example for the information flow model of a Hospital Information System is shown. A hospital information structure connects the domains of administration, medicin and nursing activities.

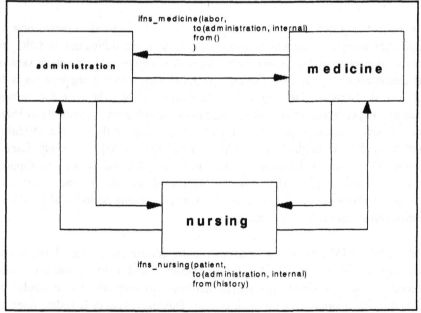

Fig. 3. Object orientated information flow, considering history states

Because of the high complexity of interactions between these domains and inside of each domain, it is very difficult to develop an integrated HIS. The idea is, to model each patient as single object, that is allowed to use several service processes (e.g. stations, doctors). The informations flow of the object *patient* is directly (*dynamically*) influenced by each used service process.

Further it is possible to compute a list of periods of time, each object spent with several service processes and a list of ressources (e.g. operation material, drugs) each object used. With these information, an automatic calculation (and bill computation) depending on performance figures and classified from international standards (e.g. ICD-9, SNOMED) can be done. Knowledge based systems, implemented as decision support systems (DSS) or expert systems for diagnosis help the doctors to find right decisions and optimize the patients treatment as well as possible. Further the social system saves a lot of money, because costs of doctors, administration, nursing, drugs and many more are concentrated on the right illnesses and treatments. The correct treatment of patients also minimizes cost intensive side-effects which are often mentioned lots of years after leaving the hospital.

5 CIM Systems

Computer Integrated Manufacturing (CIM) is another typical exampel, how to implement complex information systems. Nowadays, the difficulties in CIM are not the connection of data respectively information seen from the points of view of the customers or the products but the integrated, synchronized organisation wide data access. Because, data are more and more available, the base for making decisions is prepared. Within those integrated information systems, knowledge based modules should be placed to support human beings in their works. Modules, implemented with methods of Artificial Intelligence (AI), especially Expert Systems (XPS) for applications in automation, are still mainly used to support decisions. For example, they can be integrated usefully in the short-time production planning process, or processes, that take a long period of time, where human beings start to lose information.

The modular CIM system was developed under certain restrictions. First, it was necessary to include standard software-packages and existing software-tools. Second, it was necessary to use special hardware components. These conditions pushed the development of an open system. This open system is widely open to software-packages, because of the unique and standard data access interface

(SQL). The interfaces to hardware components are only the operating systems, used in standard access modes. So for example, all self implemented modules are strictly programmed in ANSI standard. If it is not possible to update existing programms or to support a data base SQL-interface, a data communication management (DCM) interface is offered to garantee a connection to SQL data base systems. The operation systems used mostly are UNIX for server machines, because of the multi-tasking features and DOS for front end PCs. This combination allows very cost efficient solutions.

To describe this CIM concept well from the systems-technical point of view it basically should be devided into the emphasis of the logical and physical concept. The logical concept is based on a company-wide data model that is implemented using a relational data base management system (RDBMS). Due to the well defined data access interface (ANSI, product independent SQL) the concept was implemented with *Informix* as well as *Oracle* RDBMS. It supports a highly integrated and modular linked software-system that offers server CIM modules that are necessary for real integrated information and data flow.

The CIM concept allows to connect all standard software-packages. Usually for example CAD packages are purchased, because it makes no sense to develop well proved and well suitable software once more. The common data pool, respectively the RDBMS is connected via a standard-SQL interface. Here should be mentioned, that it is necessary to administrate those data in the common data pool, that are also used from other organisational units.

Further CIM specific modules for data flow integration are developed, as follows:
- production control system (PCS) for short time planning activities, management of high-priority orders, detail informations according to drawing, updates, assignment of operations to group control systems,
- group control system (GCS) for the connection to the NC-machines, tools management, NC-working place,
- operating characteristics akquisition (OCA) (personell-, orders-, qualitiy-data),
- structured drawing and nc-programm simulation,
- alteration administration.

Special consideration should be placed on CAQ-modules. This concepts, also developed to support ISO 9001 activities and furthermore to be certified, includes modules for quality assurance. These are for example goods recieve, test means, test orders - generation as well as management anc including the PPS, maintainance, FMEA, repair and many more.

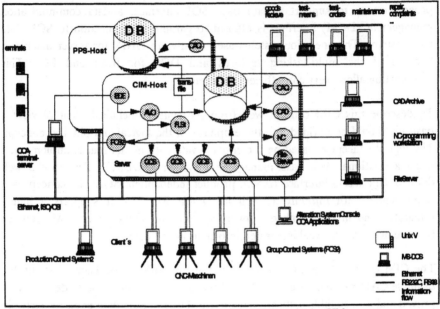

Fig. 4. Integrated information and data flow within a modular CIM concept

The calculated dimensions for memory, storage, CPU-performance including a security factor depend mainly on the amount of data. Depending on the decentral and dedicated processes the transaction concept and dedication of the data bases are done.

The system is running in the company with use of one CIM/CAQ Unix data base server and five PPS Unix data base and application server machines. The PPS systems is running with about 70 terminals. The client applications are all running under DOS using PCs. The CIM system need three 286 PC for each PCS, 8 386 PCs for GCS working places, two OCA concentrators, serveral terminals for OCA, about 20 CAD workstations, about 15 CAQ workstations and one 386 PC for the nc-programming workstation. Because the concept is hardware independent the system is portable to different hardware units as shown in figure 5. While the first CIM system is finished successfully since one and a half years with a very high degree of acceptance the implementation of the system based on the hardware units as shown in figure 5 is running now.

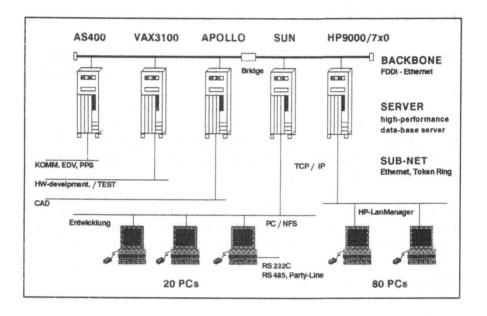

Fig. 5.Hardware infra-structure of a modular CIM structure

6 Summary

A successful development of Information Management Systems and a real integration of all belonging units is possible nowadays. The design of those systems considers existing systems, especially standard ones. One factor of the quality scale is as a matter of fact the possibility to migrate existing data from the old to the new systems. Furthermore a approach for developing integrated, complex systems supports their design and implementation. Development tools, as well as version management, LAN based repositories and a very high degree of reusability of software offers a quick but well defined software solution. The right approach to develop systems allows to model a high level systems design. This one is the base to make a systems-design, including the possibility to optimize information flows. This well structured process is done first, without the detail connection to data fields. Step by step the top-down and bottom up-design is done and it is a systems specific decision, when it is connected. This approach is used until now in the design of modular and high integrated processes in the fields of production, manufacturing and assembly as well as quality assurance. The high level of abstraction allows to model different target systems, too. As two examples, the models of computer integrated manufacturing (CIM) systems and hospital information systems (HIS) are discussed.

7 Literature

Zauner, M., Kopacek P. (1992). Data Integration in a CIM Concept for Small and Medium Sized Companies (in German), Proceedings, *CAT´92 Computer Aided Technologies*, pp. 34-42, Stuttgart.

Kopacek, P., Zauner, M. (1992). CIM for Small and Medium Sized Companies. Plenary Paper. Preprints of the IFAC Workshop on "A Cost Effective Use of Computer Aided Technologies and Integration Methods in Small and Medium Sized Companies", pp. 1-6, Vienna.

Zauner, M., Kopacek, P. (1993). Client/Server Architecture - Requirement of Intelligent CIM Solutions for Small and Medium Sized Companies. International Congress on Information and Communication Technology for the New Europe. (organized by ADV, working group for data processing). Volume II, pp. 871-882, Vienna.

Modelling and Analysis of Complex Stochastic Systems: from a Specific Example to a General Environment

G. J. Marshall

GEC Hirst Research Centre, Wembley, Middlesex, U.K.

A. Behrooz

Department of Management Systems, Bournemouth University, Bournemouth, U.K.

F. M. Clayton

GEC Hirst Research Centre, Wembley, Middlesex, U.K.

Abstract. A model of a standard cordless telephone system is presented and analysed. Some results are presented describing the behaviour of the system both qualitatively and quantitatively. By considering the problems involved in treating this complex stochastic system in some generality, we proceed to derive a general environment capable of supporting the modelling and performance analysis of a range of such systems.

1   Introduction.

The aim of this paper is to describe a rather general environment to support models of complex stochastic systems and to facilitate their analysis. We begin by describing the modelling and analysis needed to solve a specific, but typical problem. By recording the entire process of solving the problem, and not just the particular method by which it was eventually solved, we can expose and explore many of the issues relevant to the solution of this and similar problems.

From this general approach to one problem, we proceed to derive a general environment that could support:

•   models of complex stochastic systems,

•   methods of reducing the complexity of these models to a point at which they can be analysed in a reasonable time, and

•   a means of analysing the performance of the models.

In addition, we consider the possibility of including some 'intelligent' means of selecting the most appropriate reduction method once a model has been described to the environment.

2   The Specific Problem.

The problem to be examined concerns the modelling of the standard cordless telephone system known as CT2 [1]. This system provides a central point capable of connecting to the public telephone network up to 40 simultaneous telephone calls made from portable telephones operated within a distance of approximately 100 metres of it. The aim of the modelling is to examine the behaviour of the system, and to provide answers to questions concerning the occupancy of the system, the probability of not being able to make a call because the system is busy, the likelihood that a new call will 'grab' a channel already in use for another call, and so on.

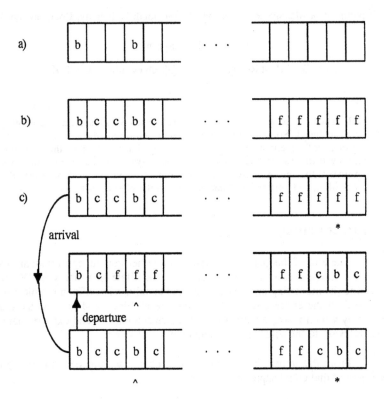

Fig. 1   a) The 40 channels with some busy. b) Busy (b), contaminated (c) and free (f) channels. c) Transitions caused by arriving and departing calls.

The formalism we adopt for the modelling is the *state transition network* [2]. This type of representation has been used in many domains, not least in Telecommunications, from which our particular example is drawn. CT2 operation may be modelled in a 'brute force' way by assigning a state to each possible configuration of the 40 channels in terms of which are busy and which are not, as illustrated in Figure 1a. In fact, we characterise each channel as being either busy, contaminated or free, because when a particular CT2 channel is in use it contaminates its neighbouring channels by 'spilling' a little of its signal into them (again, see Figure 1b). CT2 automatically gives the 'best' (least contaminated) channel to any new call, so that its behaviour is affected by the pattern of contaminated channels as well as the pattern of busy channels in the current configuration. A transition between states is caused by one of the events corresponding to the arrival of a new call or the termination of an existing one, as shown in Figure 1c.

Despite the fact that the resulting state transition network is far too large to be used directly as a model, it can be described accurately. It has 2^{40} states, one for each pattern of 40 busy or unused channels. The number of transitions into and out of a state can be described compactly by denoting the number of busy channels by B, the number of contaminated channels by C, and the number of free channels by F. (Note that $B + C + F = 40$.) For a given state, the number of successors resulting from call departures is B, and the number due to arrivals is given by:

if $F \neq 0$ then F else C.

Thus, the total number of successors is:

if $F \neq 0$ then $B + F$ else $B + C$,

or

if $F \neq 0$ then $40 - C$ else 40.

A similar argument shows that the number of predecessors is the same as the number of successors.

Thus, at this level of abstraction, we have a model which can be described precisely with a state description, a number of states, and an interconnection pattern. Further, each transition can be labelled with a probability of occurrence obtained from the probability of the event causing the transition. Unfortunately, the model is too large and complex to store in a memory of any reasonable size, let alone to analyse. Before any further progress can be made, it is necessary to reduce the state transition network in some way.

2.1 Reduction.

Natural methods for reducing the complexity of a state transition network include partitioning the network, pruning states and combining states. If a network can be partitioned, the partitions may provide sensible 'super-states' for a reduced network. Rarely occurring states, for example, can be pruned to reduce the size of the network. States with common properties or symmetries can be combined, although they must also have the same numbers of transitions if their combination is not to distort the fabric of the network.

When examining the cordless telephone system, the prime focus of interest is not the pattern of occupancy of the 40 channels, but the *number* of busy channels.

The formulae obtained above for the numbers of predecessor and successor states show that configurations having the same number of busy channels do not necessarily have the same number of predecessors and successors, and so cannot be combined. But configurations in which both the number of busy channels and the number of contaminated channels are the same do share common numbers of predecessors and successors. Consequently, the state transition network for CT2 may be reduced in a valid and meaningful way by combining states which have the same number of busy channels and the same number of contaminated channels. This gives (B, C) as the the descriptor for the states of the reduced system. The reduced state space is illustrated in Figure 2. It contains 535 states, a number which although large is manageable and represents a dramatic reduction.

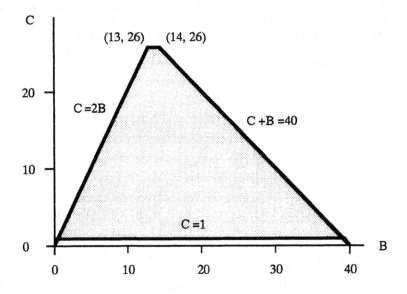

Fig. 2 The reduced state space. The state descriptor is (B, C).

However, the reduced transition network is unlabelled. The transitions of the original network can be labelled in a way related to the events which trigger them, but the labels do not carry over in any coherent manner to the reduced network.

2.2   Analysis of the Reduced Network.

The reduced state transition network immediately allows us to describe qualitatively some broad aspects of the behaviour of CT2. A segment of the state space may be identified within which the trajectories of the *initial behaviour* of the system are confined: it is shown in Figure 3. When CT2 is first started, with all its channels free, it can obviously assign an uncontaminated channel to all new calls for some initial period, the duration of which depends on the pattern of call arrivals and terminations. These trajectories all begin at state (0, 0) and end at or between (14, 26) and (20, 20) on the line B+C=40, having been confined to the segment bounded by C = B and C = 2B. The reason the initial trajectories are confined in this way is that, while uncontaminated channels are available, there must be at least one and at most two channels contaminated by each busy channel.

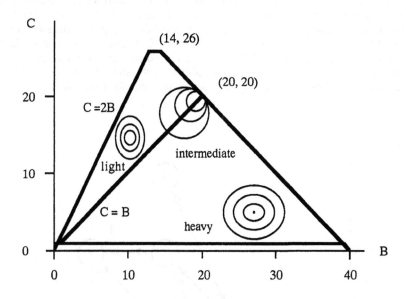

Fig. 3 Reduced state space, showing the sector containing the intial trajectories, and trajectories for continual light, intermediate and heavy traffic loadings.

The trajectories in the state space resulting from continual light, intermediate and heavy loadings may be distinguished, and are also shown in Figure 3. When lightly loaded, the system will follow trajectories confined to the initial segment and centred on the mean number of calls in progress. For heavy loadings, the trajectories will be centred on the mean level and confined to the bottom right corner of the state space. With intermediate loadings, the trajectories will cross the border between the two segments, usually leaving the initial segment through the state (20, 20).

Now, qualitative analysis is all very well but quantitative analysis is often necessary. To make possible a quantitative analysis, it is necessary to have a meaningful labelling of the transitions in the network. As we have seen, the labelling of the transitions of the original network cannot be used to derive a labelling for the reduced network in this case. Experience suggests that it not possible in many similar situations. There is, however, a general method of deriving a labelling for the transitions of the reduced network. It depends on being able to count the numbers of states which have been aggregated into the super-states of the reduced network. These numbers, along with natural laws of conservation, allow each labelling to be determined as the quantification of the 'state flow' along the transition: these state flows may readily be converted to transition probabilities. The upshot of this is a general method for labelling the transitions of a reduced state transition network which can be used as long as the number of states from the original network combined to form each super-state is known. Once the transitions are labelled, the state probabilities may be determined by well-known methods involving the derivation and solution of a set of simultaneous equations [2]. In fact, the states of the reduced CT2 network (and of the reduced networks obtained in other similar problems) are aligned in distinct layers. This enables the equations for each layer to be formulated and solved in turn for the probabilities of the states in the layer, which is considerably easier than solving one large set simultaneously. (See also, for example, [3].)

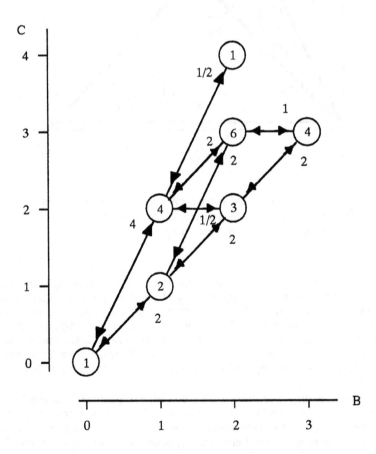

Fig. 4 Part of a state transition network, showing the number of configurations combined in each state and the 'state flows' between states.

By way of an example, a part of the state transition network for a cordless telephone system having only six channels is shown in Figure 4. The transitions are labelled with their 'state flow' numbers. With a given average loading, E, the probability of occurrence of the state (0, 0), representing the probability that no call is in progress in this system, is:

$$p[(0, 0)] = \frac{68E^5 + 344E^4 + 711E^3 + 806E^2 + 516E + 144}{12E^7 + 136E^6 + 556E^5 + 1219E^4 + 1625E^3 + 135E^2 + 660E + 144}$$

3   Towards an Environment.

An environment should be able to support models appropriate to a range of complex stochastic systems and, in the state transition network, we have a widely used formalism. But initial models are often too large to store explicitly. However, as shown above, a state transition network can be represented implicitly by giving the state descriptor and some knowledge about the operation of the system which should include, at least, descriptions of the events that trigger transitions from state to state.

There are various general approaches to reducing the complexity of a state transition network. Redundant states may be omitted; states that, in a given context, are identical may be merged; and states that are similar to some degree may be combined with a corresponding controlled loss of information. Semantics are preserved with all these approaches, so that the states of any reduced network would have known meanings. In turn, the meaning of results obtained by analysing the network would be known.

The operational techniques available for reducing the complexity of a state transition network include those mentioned above, which are:

- clumping together states within each partition of a partitioned network,

- pruning rarely occurring states, and

- combining states on the basis of their properties or their symmetry.

When a technique is applied mechanically, the meanings of the states of the reduced network will not be known unless the technique corresponds directly to an approach which preserves semantics. Additionally, a specific reduction technique is appropriate only if there is some underlying constraint on the affected states. Partitions can only be established if equivalence classes exist; pruning can only take place if there are rarely exercised states; and states cannot be aggregated into super-states unless they have the same numbers of input and output transitions or the structure of the reduced network will be a distortion of that of the original . Thus, there are attributes of states and constraints on states which may, for a given system, suggest a mapping between an approach that preserves semantics and a standard reduction technique. The existence of states with common properties, symmetries, and patterns of input and output transitions, for example, will suggest a style of aggregation. The failure of a proposed grouping to ensure that all the states to be aggregated share a common pattern of transitions demonstrates that it is inappropriate. These ideas provide a basis for constructing an environment that can, for a given state transition network, detect whether a proposed reduction method is inappropriate and, at best, automatically determine an appropriate reduction method.

After an appropriate reduction method has been either determined automatically or selected by the user, the environment should proceed to generate the reduced model. The next step is to carry out an automatic analysis of the model and, as shown above, there are general procedures that may be followed, first, to label the reduced network and, second, to analyse the network. When states are combined, for example, the labelling procedure depends on counting the numbers of states from the original network that are combined in each state of the reduced network. These numbers can be determined automatically from the knowledge given to the environment in describing the operation of the system. The procedure for analysis is well established.

4   Conclusions.

The way in which the CT2 problem was approached suggests that a rather general environment for modelling and analysing complex stochastic systems may be created from the following ingredients:

•   An implicit description of the initial model of a complex stochastic system,

•   a single formalism for modelling in the form of the state transition network,

•   a variety of methods for state space reduction each with associated constraints on their use,

•   an 'intelligent' means of selecting a reduction method appropriate to a given model, and

•   a single, standard method for the analysis of the reduced network.

The environment could provide the following forms of support for its users:

•   Suggestions for an appropriate reduction method.

•   Automatic testing of the appropriateness of a reduction method selected by the user.

•   A guarantee of meaningful results.

•   A choice of ways in which results are to be presented.

References.

1. MPT 1375: Common Air Interface specification. Department of Trade and Industry, (May 1989)

2. J G Kemeny and J L Snell: Finite Markov Chains. Van Nostrand (1969)

3. I F Akyildiz: Product form approximations for queueing networks with multiple servers and blocking. IEEE Trans. Computers 38, 99-113 (1989)

System Theoretic Approach to Migration of Project Models

Karl Kitzmüller

Institut für Systemtechnik und Automation,
Johannes Kepler Universität Linz,
A-4040 Linz, Austria

Abstract. An integral component of a Software Engineering Environment(SEE) is a project model which is a description of a class of intended processes. Changing the SEE must not necessarily mean to loose the already established project model. Via migration it is possible to transfer the project model from one SEE to another. In order to manage the migration process a meta model to describe project models, based on the theory of many sorted algebra, is introduced. As an example, the theory is applied to describe parts of the meta models of Softlab´s MAESTRO*II* and IBM´s ADPS.

1 Introduction

A Project Model describes the development process its dependencies and the rules. It is a very important company knowledge base which should be described in a computer independent way. Due to the rapid evolution in computer science, especially concerning hardware, it is then possible to stick to the same project model by simply changing the underlying Software Engineering Environment (SEE).

Software development project models have been around for many years, describing the activities, the results and their relationship of developing. The complexity of the development process, the multitude of intermediate and final results and the number of team members and resources involved require computer support. Computer support not only helps the user to follow the established process, it also provides the user with various support functions. At the present time, several commercial computer based SEEs, which explicitly allow the definition of a customer specific project model (so called *General Purpose Environments*), are available [11,1]. Examples for such SEEs are IBMs ADPS (Application Development Project Support), Softlabs MAESTRO*II* or Delphis SEE.

In Linz we investigate in migration and portability of project models from one SEE to another. To understand and solve the problem we carefully have to distinguish between several system levels.

- *Level 2*: A project is an instance of a project model
- *Level 1*: A project model (i.e. SEtec, Merise, V-Model) is an instance of a meta model
- *Level 0*: A meta model is the base of the above levels and specifies the functionality of the SEE.

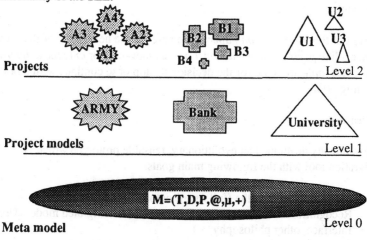

Fig. 1. Different levels of models

2 General Procedure of Migration

The project "*migration of project models*" is not only a theoretical discussion, we are working on a concrete migration at the industry side. Concerned with the migration project are three phases which have to be executed in the given sequence. Especially the first phase, the system theoretical modelling, is considered in this paper.

- Theoretical Investigation
- Prototype for the Migration
- Migration as Procedure

2.1 Theoretical Investigation

Currently there is no generally accepted method how to describe development processes and process models. Migration, considered as a similar process to the selection of a process model interpreter, has to start with an analysis of the meta-model of the involved environments. We use formal methods especially the theory of *many sorted algebra* [5]. We have chosen a *n* tupel like the MASP (Model for Assisted Software Processes) introduced by [6]:

$$S=(T,D,U,M,REL)$$

where:

T := set of task classes
D := set of deliverable classes
U := set of user (actor) classes

M   := set of finite state machines
REL := relationships between the object classes

After having defined the meta models of the involved SEE (i.e. ADPS and MAESTRO*II*), we have to investigate if there exists any transformation between these meta models.

The result of this *migration on paper* is the base for the next phase of the migration. Statements like *parts of the meta model are not moveable* or *only the deliverable model is moveable* are input for the decision to stop or to continue with a prototype for the migration.

2.2 Prototype

We have already developed an evolutionary, reusable prototype which is the basis for the migration tool with the following main goals:

- Check the limits of the (full / part) automatic migration
- Recognition and recovery of information in the migrated model (User-Interface, other philosophy, ...)
- Investigation of different system architecture of the automatic migration tool

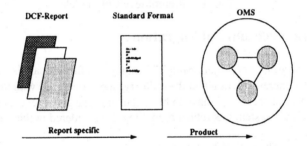

Fig. 2. System-Architecture of the Migration tool

After the first discussion at the customer side and after the first technical problems of the automatic transformation we detected the following points:

- Problem of volume: 140.000 records and more
- Migration in two parts: internal format : product adaptation
- Different rights concept, based on the underlying system (TSO, UNIX)
- Learning effects of ADPS and MAESTRO*II*
- Different model reports on customer side → Customising
- Different kinds of navigation through the networks

2.3 Migration as procedure

The first step of the real migration at the customer side, is the presentation of the already developed prototype. We want to talk about the results and the user requirements based on the existing migrated project process model. In the future we want to be able to migrate ADPS based project models to other SEEs like MAESTRO*II*, Delphis SEE, EPOS[11,1] based on the experience of our prototype.

3 Definition of a Project Process Model

In this paper we always talk about project process models, which are similar to the well known and already published Software Process Model [7]. The only difference is the point of view. Process models concentrate their consideration to the process i.e. which tasks have to be performed in which order. Project process models are a more special view and focus on the project idea, i.e. the connection architecture between users, tasks, deliverables and their life-cycles.

3.1 A Meta Meta model to describe Project Process Models

Meta-Meta-Model: The conceptual model, the meta-meta model for Project Process Models (the model for level 0 in fig. 1), is on the level where we are talking about entity-sets, relationship, attributes and so on. In literature there are many different forms of description. We choose the terminology of database description, defined by [15] who talks about an Entity-Relationship-Database. Another also very interesting form, because of the fact that MAESTRO*II*, one of the involved environments supports an OMS (Object Management System), is the definition of an OMS introduced by [12,16].

OMS Definition. The conceptual base of the OMS consists of 3 components: object classes, attributes and relationships which are considered at two levels,
- Model level (Plan for a database - Conceptual scheme)
- Instance level (Actual data present in the database - Instances)

which is similar to the scheme and instance terminology by Ullmann. It is similar to the on-line LAN repository introduced by IBM which stores and handles all project information.

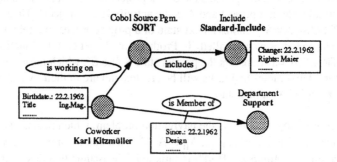

Fig. 3. Typical part of an OMS database

In fact of the different terminology used by the authors, the following table relates the terminology used for MAESTRO*II* to the standard definition of the Entity-Relationship-Model of Ullmann.

Ullmann[15]	Softlab[12,13]
entity	object (instance)
entity-set	object-class
relationship	relation
functionality of relationship	type of relation
scheme	class (type)

3.2 A Meta model for Project Process Models

We describe the Project Process Model as an incompletely specified automaton (machine)

$$P=(STA,ACT,TRA)$$

where

STA: is a set of global states of a project model P. The global state of a project process model is defined by the current contents of the project database. Examples for states of a project model are the initial state "Project initialised" and the state "Project finished", which in most cases is the main goal of a project.

ACT: is a set of activities which changes the state of the project process [8]. [4] defines activity as *"Unmanaged, but perhaps interactive and controlled process step useful in carrying out tasks"*. Typical activities are: Execute and change planning, assign and withdraw rights, create and delete relationships, modify attribute values. In general every activity can be described by its precondition and post condition. Preconditions are some kind of logical conditions which should be satisfied in order that an activity can be performed. Post conditions describe the result of an activity. Precondition and post condition correspond to the input- and output parameters of an activity in the sense of [8].

TRA: is a one step transition function defined as

$$TRA: STA \times ACT \rightarrow STA$$

which changes the global state S of a project model. Let us assume S_0 as an initial state of a project and A is the activity "start project". Then $TRA(S_0,A)=S'$ denotes a next state of a project process model which can be interpreted as a state called "In Production". In general the function $TRA(S,A)=S'$ transfers the project state S into S´, which means all mandatory deliverables and tasks will be created and are in their own state for example "Awaiting start".

One aspect of a project model is the navigation problem, which means one should find a sequence of activities $A_1, ..., A_n \in ACT^*$ such that a number of goals (i.e. "Minimize time", "Minimize effort") is satisfied. We assume that this kind of problem can be solved by applying the methods of knowledge based systems. During the execution (interpretation) of a project there are many critical states i.e. users leave the company, resources are damaged, planned and estimated dates differ, the project model interpreter (Software Engineering Environment) is inconsistent, subsidiaries cannot deliver. In such situations a knowledge based navigation, which is based on the project model combined with the whole experience of already finished or executed projects should give help to the responsible management. The help may be a simulation of which activities can be executed and what their effects to the project are.

Now we continue with the description of the global state of a project model $S \in STA$ in form of a many sorted algebra

$$S = (T, D, U, M, REL)$$

In order to describe the elements of S, we introduce a more general definition of the conceptual level used to describe the individual elements of S. A relational scheme RS is defined as:

$$RS = RS(A_1, ..., A_n)$$

where A_i is an attribute (i.e. colour, address, length..), which describes the properties of RS. To each attribute A_i a set of values is assigned, which is called a domain of an attribute and denoted as $Dom(A_i)$. Usually, the domain for an attribute will be a set of integers, character strings, real numbers, but we do not rule out other types of values.

An instance IRS of a relational scheme is represented by:

$$IRS \subseteq DOM(A_1) \times .... \times DOM(A_n)$$

and at the data store level the instances of IRS are given by

$$irs = (a_1, a_2, .., a_n)$$

Based on this concept we introduce the individual elements **T, D, U** of the state S of a project model as follows.

$$TASK = TASK(N_T, A_1, ... A_n, S_T)$$
$$T \subseteq Dom(N_t) \times Dom(A_1) \times ... \times Dom(A_n) \times Dom(S_T)$$
$$t = (n_t, a_1, .. , a_n, s_t)$$

Two attributes $N_{(T,D,U)}$ and $S_{(T,D,U)}$ are of special interest. N_T is the name of the task class. It is used to identify the task class and has to be unique in a project process model. S_T is interpreted as the current state of a task for example "in production", "interrupted" or "awaiting start".

T: is an instance of TASK and is called task-class. A task is *"a managed process step. Tasks are carried out by performing activities, some of which may lead to the definition and assignment of other tasks"* [4]. Typical attributes of a task class are: estimated effort, planned start-date, last accounting-date.

$$\text{DELIV} = \text{DELIV}\ (N_D, B_1, \dots B_m, S_D)$$
$$D \subseteq \text{Dom}(N_d) \times \text{Dom}(B_1) \times \dots \times \text{Dom}(B_m) \times \text{Dom}(S_D)$$
$$d = (n_d, b_1, \dots, b_m, s_d)$$

D: is an instance of DELIV and is called deliverable-class. A deliverable class is *"a product created or modified during a process either as a required result or to facilitate the process"* [9]. SD is a current state of a deliverable class, a typical assigned state automaton is shown in fig.5.

$$\text{USER} = \text{USER}\ (N_U, C_1, \dots C_k, S_U)$$
$$U \subseteq \text{Dom}(N_u) \times \text{Dom}(C_1) \times \dots \times \text{Dom}(C_k) \times \text{Dom}(S_U)$$
$$u = (n_u, u_1, \dots, u_k, s_u)$$

U: is an instance of USER and is called user-class. A user class is an entity set, which performs process steps. Typical User classes(also known as roles) are Manager, Developer, Observer, Administrator and so on.

REL: is a set of relationships which defines the connections between the involved parts of the project model like tasks and deliverables. We suggest at least

$$\text{REL} = R_{tt} \cup R_{dd} \cup R_{td} \cup R_{dt} \cup R_{uu} \cup R_{ut} \cup R_{ud} \cup R_{ua} \cup R_{mt} \cup R_{md}$$

Before we continue with detailed description of the set REL, we recall some basic properties of relationships. The properties of relationships are very important and serve as a type of framework with the meta model and they can be interpreted as a set of axioms which have to be satisfied in every global state of the project process model. In other words it is assumed that the properties of relationships are not changed during execution of project process.

Let $R \subseteq A \times B$ be an arbitrary relationship (relation) and R is said to be:

- **FUN**: a *function* or N:1 relation if the following condition is satisfied:
  $(a,b) \in R$ and $(a',b') \in R$ and $a=a'$ then $b=b'$
- **BFUN**: a *bijective function* or 1:1 if R and its inverse R^{-1} is a function
- **HIR**: R is a *hierarchical* relation if A=B and R forms a tree
- **REF**: R is *reflexive* if A=B and $(a,a) \in R$, for every $a \in A$

R_{tt}: $R \in R_{tt} \mid R \subseteq T \times T$

CON_t	Consists_of	HIR	describes the hierarchical decomposition of tasks
ISP_t:	Is_part_of		$ISP_t = (CON_t)^{-1}$
SUC	Successor		describes the sequential order of tasks
PRE	Predecessor		$PRE = (SUC)^{-1}$

R_{dd}: $R \in R_{dd} \mid R \subseteq D \times D$

CON_d	Consists_of	HIR	describes the hierarchical structure of deliverables
ISP_d	Is_part_of		$ISP_d = (CON_d)^{-1}$

$R_{td}: R \in R_{td} \mid R \subseteq T \times D, R_{dt}: R \in R_{dt} \mid R \subseteq D \times T$

PRO	Produces		describes deliverables which are created by tasks
ISPRO	Is_prod_by		$ISPRO = (PRO)^{-1}$
REQ	Requires		describes deliverables which must be available for tasks before work can be done on the selected task
ISREQ	Is_req_by		$ISREQ = (REQ)^{-1}$

$R_{uu}: R \in R_{uu} \mid R \subseteq U \times U$

DOM	Dominates	HIR	describes the hierarchical structure of users
ISDOM	Is_dom_by		$ISDOM = (DOM)^{-1}$

Remark: For example: a administrator dominates a managers. A manager dominates developers. Developers dominates observers.

$R_{ut}: R \in R_{ut} \mid R \subseteq U \times T, R_{tu}: R \in R_{tu} \mid R \subseteq T \times U$

COM$_t$	Is_responsible		describes "which user is responsible for a task"
ISCOM$_t$			$ISCOM_t = (COM_t)^{-1}$

$R_{ud}: R \in R_{ud} \mid R \subseteq U \times D, R_{du}: R \in R_{du} \mid R \subseteq D \times U$

COM$_d$	Is_responsible		defines the rights of a user class to the assigned deliverable class
ISCOM$_d$			$ISCOM_d = (COM_d)^{-1}$

$R_{ua}: R \in R_{ua} \mid R \subseteq U \times ACT, R_{au}: R \in R_{au} \mid R \subseteq ACT \times U$

PER	User_performs		describes the users activities
ISPER	Is_performed_ by		$ISPER = (PER)^{-1}$

Remark: A manager can create and delete tasks, their substructure and their relationships, a administrator can consolidate estimated and actual plan values, a developer can execute state transitions and so on. We assume that the following axiom is true which describes the connection between the DOM and PER relationships:

$$\text{If } (u,u') \in DOM \text{ then } PER(u') \subseteq PER(u)$$

The meaning of this sentence is that if one user u dominates (DOM) another user u', then the set of activities u is larger than the set of activities of user u'. For Example a manager really has to have more competence during the realisation of a project than a developer.

M: is a set of finite state machines. The finite state machine represents the life-cycle of a task or a deliverable. It consists of states, state-transition functions and output functions. Fig.4. and fig 5. give some feeling for such a life-cycle oriented finite state machine which describes dynamic aspects of a task realisation.

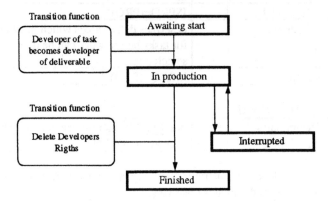

Fig.4. Typical finite state machine assigned to a task class [13]

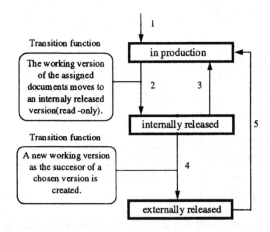

Fig.5. Typical finite state machine assigned to a deliverable class [13]

$R_{mt}: R \in R_{mt} \mid R \subseteq T \times M$

LIF$_t$	Life_cycle_of	FUN	defines the life-cycle for the assigned task

$R_{md}: R \in R_{md} \mid R \subseteq D \times M$

LIF$_d$	Life_cycle_of	FUN	defines the life cycle for the assigned deliverable

Remark: It is worth noting that a set of finite state machines which is assigned to a set of tasks and a set of deliverables has a structure like a Tree-Systolic-Architecture [10]. This structure describes a communication network between tasks and deliverables.

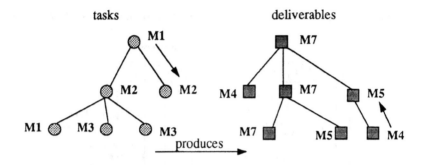

Fig. 6. Parts of the systolic connection architecture of a project model

4 Case Study: ADPS ==> MAESTRO*II*

The case study considers only one part of the project process model to give an introduction how to use the formalism described in section 3. We try to specify the deliverable classes D and their relationship R_{dd} and give some examples for the conclusions concerning the transformation (migration) from one environment to the another.

Remark: APDS is used in 35 companies in Germany, most of these companies try to migrate to another SEE. In Linz we investigate in the migration from IBM´s ADPS to Softlab´s MAESTRO*II*. Both companies, IBM and Softlab, are working together with us. In the next future we want to be able to migrate customer at Germany. Our Prototype of the migration tool was developed and tested in co-operation with the "Bayrische Vereinsbank" in munich.

4.1 Source Environment: ADPS

A deliverable class in ADPS represents a document at the lowest level of the tree. A document is a physical file, i.e. COBOL-Source or Flow-Chart. Deliverables at a higher level are only groups which are used to break down the complexity of the product. The consequence of the connection between documents and deliverables is, that the same kind of document, i.e. notes, diary, introduction, Pascal source, is used under different names on different nodes of the tree.

$$DELIV=DELIV(N, A_1, ..., A_n, S_D)$$

N: key attribute to identify the deliverable class which consists of 8 CHAR

$A_1 .. A_n$: freely defined attributes like date-stamp, effort.
Remark: One attribute called "description of a deliverable class" is very important for the migration. It is a string of 50 characters which has not to be unique.
SD: state as a special attribute

The relationship $R_{dd} = D \times D$ is realised with Gorn´s formalism [14] shown in fig. 7.

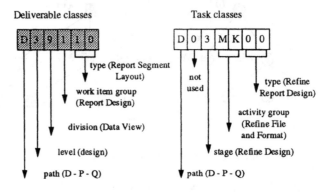

Fig. 7. Realisation of hierarchical relations in ADPS

4.2 Destination Environment: MAESTRO*II*

PCMS, a MAESTRO*II* component, is used for the management of tasks and deliverables for projects.

$$DELIV=DELIV(N, A_1, ..., A_n, D_1, ..., D_n, S_D)$$

N: is the key attribute to identify the deliverable class and consists of 32 characters
$A_1 .. A_n$: freely defined attributes of the underlying OMS
$D_1 .. D_n$: a special kind of attributes called document-type
Remark: Examples for document types are C-Source, Diary, Flow-Chart. That means a document-type is identified by his name and can be assigned as attribute on every deliverable class independent from the hierarchical order.

$$R_{dd} = D \times D \text{ is realised via an OMS.}$$

4.3 Migration (Transformation)

The prototype for the migration procedure is already implemented in PROLAN2, a MAESTRO*II* specific program language. At the current time we are able to migrate the following parts of the model:
$$DELIV, TASK, R_{tt}, R_{td}, R_{dd}$$

including attributes like document templates, help information and others.

Specific Problems during the migration of the delieverables are:

- Name mapping: ADPS-name=(8 char, 50 character description)
  MAESTRO*II*-name=(32 characters)
  not unique: i.e. a deliverable class with the description notes
- Connections between document-types and deliverable classes (see fig. 8.)
  ADPS deliverable class → MAESTRO*II* deliverable class and an assigned
  document-type with the corresponding name
- Hierarchical structure: R_{dd}
  Transformation from Gorn´s formalism into OMS relations.

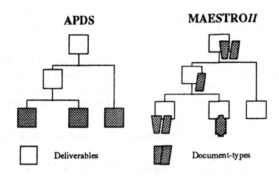

Fig. 8. Document types and deliverables

Conclusion: With respect to the deliverables a principle automatic migration from
ADPS to MAESTRO*II* is possible. Due to the fact of the different concepts of deliverables and document types, a redesign of the migrated model is not necessary but
recommended.

5   Summary

In the paper we introduce a first version of a project process model with results in a
systolic connection architecture. The current model has to be extended with more
components like actors, tools and resources. Another very interesting part which will
be discussed in the near future is the knowledge based navigation through such a
complex model.

Acknowledgement

I would like to acknowledge many helpful discussions with Ireneusz Sierocki and
Witold Jacak.

References

1. Balzert H.: CASE Systeme und Werkzeuge.- BI Wissenschaftsverlag, 1992
2. Chroust G.: Modelle der Software-Entwicklung Aufbau und Interpretation von Vorgehensmodellen.- Oldenbourg Verlag, 1992
3. Chroust G.: Application Development Project Support (ADPS) - An Environment for Industrial Application Development.- ACM Software Engineering Notes, vol. 14 (1989) no 5, pp. 83-104
4. Downson M., Nejmeh & Riddle W.: Fundamental Software Process Concepts in Proceedings of the first European Workshop on Software Process Modelling, CEFRIL, Mailand, Italy. Mai 1991, pp 15-38, AICA (Italian Computer Association)
5. Gallier J.H.: Logic for Computer Science, Harper & Row Publishers, 1986
6. Griffith P., Oldfield D., Legait A., Menes M., Oquedo F.: ALF: Its Process Model and its Implementation on PCTE.-Bennett K.H. (ed.): Software Engineering Environments - Research and Practice.- Ellis Horwood Books in Formation Technology, 1989, pp. 313-331
7. Humphrey W.S.: Managing the Software Process.Addison-Wesley Reading Mass. 1989
8. Leymann F.: A Meta Model to support Modelling and Execution of Processes.- Trappl R. (ed.): Cybernetics and Systems Research 92, World Scientific Publishing Singapore 1992, pp 287-294.
9. Lonchamp J.: A Structured Conceptual and Terminological Framework for Software Process Engineering.- Proceedings of the Second International Conference on the Software Processes, Februar 1993, Berlin, pp. 28-40. IEEE Computer Society Press, 1993.
10. Pichler F., Schwärtzel H.: CAST Methods in Modelling, Springer Verlag, 1992
11. Schulz A.: CASE Report ´92, GES mbH, 1992
12. Softlab GmbH.: Manual: The OMS Modellers Guide, Softlab GmbH, München 1993
13. Softlab GmbH.: Manual: The PCMS Modellers Guide, Softlab GmbH, München 1993
14. Sierocki I.: On Determinig the k-Nerode Equivalence For Tree Automata Inference. - in Lecture Notes of Computer Science, vol. 410, Springer Verlag, Berlin, New York, 1990
15. Ullman J.D.: Principles of Database and Knowledge-Base Systems Vol I., Computer Science Press, Rockville, Maryland 1988
16. Zdonik B.S.: An Object Management System for Office Application.- Shi-Kuo Chang (ed.): Languages for Automation, Plenum Publishing Corporation, 1985

Computer-Aided Analysis and Design of Sequential Control in Industrial Processes

Kai Zenger
Helsinki University of Technology
Control Engineering Laboratory
Otakaari 5 A, FIN-02150 Espoo, Finland

Abstract

Methods of program design for the automation of practical industrial processes are discussed. The Grafcet standard is used as a means to reduce the complexity of large programming tasks and to formalise the state-machine structure often used in the control of discrete-event dynamic systems. Knowledge-based techniques by means of the programming language Prolog are introduced to enable logic reasoning as a tool to be used in program construction and verification.

1   Introduction

Traditional control theory has mainly been based on the analysis of continuous models of unit processes, which have then been used in the design of feedback controllers to meet some performance criterion. However, in industrial process plants only a part of the operation can be described as a combination of such control loops. Usually, the total operation of the plant is based on sequential steps of unit operations, so that the automation has to deal with discrete-event type logic control.

Despite the extensive amount of research concerning discrete-time mathematics and discrete-time control systems the design methods for the automation of industrial processes are still quite undeveloped. The main concept in the use of most digital automation systems is the 'configuration of control loops', which is the traditional philosophy of feedback design of continuous processes. Programmable logic controllers (PLC's) on the other hand are basic tools for the solution of combinational logic problems, and the possibility to use them in large programming tasks like in the automation of the whole plant has become possible only recently.

From the technological point of view the extensive use of PLC's today is slightly bizarre. This is because although the developments in microprocessor technology have been fully utilized in the hardware design of PLC's, the programming methodology has not actually made any significant progress since the early 70's [1]. In the automation of large process plants the programs constructed by using relay diagrams (ladder diagrams) are difficult to understand, modify, troubleshoot, and document.

The awkwardness with the automation of larger process entities by using PLC's becomes clearly apparent in the design of programs for the control of sequential processes [2]. For example, in a process plant it is totally unacceptable that an occasional failure in instrumentation leads to a situation, where the program gets stuck because of a missing feedback signal from an instrument. Instead, the program should react to the error situation properly and safely, and give an error message to the operator. Also, it must be possible to continue the program from the error situation or just stop that particular program depending on the choice of the operator. Unfortunately, the traditional programming technique of PLC's does not support the construction of programs performing in this manner.

The design problem has been made easier by the introduction of the Grafcet standard, which gives conceptual tools to the description of the sequential problem by means of a state machine structure. Several PLC families nowadays allow the construction of programs directly in this fashion, which is a significant improvement, but has also made it possible to demand a deeper theoretical analysis of practical design techniques of sequential programs.

Although straightforward in principle, the design task may grow very rapidly, because many different situations including abnormal operation of the plant must be taken into account. Side-effects of control actions e.g. remain easily undetected in the planning stage of the program, which together with incomplete exception handling may cause difficulties later even with plants, which have been in automatic operation for a long period of time. Knowledge-based reasoning is thus a necessity for detecting these 'pitfalls' during the design phase.

The aim in this paper is to discuss different approaches to be used in planning, verification, monitoring, and simulation of boolean-like programs of PLC's. Specifically, the Grafcet standard is described as a means to formalize the design of sequential programs.

2 Control of Continuous vs. Discrete Systems

In control theory the main concepts of design are *models*, *cost functions*, and *algorithms* to be used in the determination of *control signals*. A model which approximates the input-output behaviour of the plant is obtained by using some identification method. The desired operation of the plant is formulated by means of a cost function. The actual control problem is to determine a control signal which minimizes the cost function.

In discrete-event control systems the methodology should be analogous to the continuous time control. A plant model would then be used in order to calculate control signals for the plant so that a specified cost function would be minimized. However, in discrete-event systems the problem is much more difficult, because concepts such as models and cost functions cannot be easily expressed in such mathematical form that the conventional methods of control design could be utilized. In fact, there are several kinds of models for discrete-event systems such as Petri nets, finite state machine models, communicating sequential process models, and queueing network models, but these are used for different kinds of purposes, and a general theory to be used in analysis and controller design is lacking.

In general, dynamic models can be divided into two classes: continuous variable dynamic systems (CVDS) and discrete-event dynamic systems (DEDS). The mathematical state of a CVDS model is replaced by an event-driven physical state in a DEDS model. The structural properties, i.e. controllability, observability, and stability are well-defined in continuous time models, but for discrete-event models there may not even be similar definitions, or these depend on the type of the model. Hence, it is easy to understand, why general methods for identification, estimation, and controller design are much more difficult to develop for discrete-event systems.

The problems of theory are apparent in the practical design and automation of industrial processes. In general, large process entities consist of unit operations, which can hardly ever be modelled by either a CVDS or a DEDS model. The total operation is a combination of both of these basic models. Automatic control is needed for opening and closing valves and starting and stopping pumps, interlocking, safety logic, switching on and off stabilizing controllers etc. It is not easy to deduce, how the programs of process computers (digital automation systems, PLC's etc.) should be constructed in order to minimize errors and the time used for programming.

At present, the automation of 'discrete' parts of the processes in industry plants is made separately from that of processes suitable for 'continuous' control. The main emphasis in the design is on the discrete set of states and transitions between them. The idea of looking at the process in this way as a state machine may be sound, but for large processes there arise problems like incomplete exception handling, unforeseen side-effects in the operation, and the general complexity in the design.

Tools for design, programming, and program verification would be a challenge and a good target for computer aided design. It is believed that elements from the theories of Artificial Intelligence, Operations Research, and System Theory are needed to achieve the above goal.

3   Grafcet

Logic controllers can be described by Grafcet, which is a standard for the specification for discrete automata, see e.g. [3]. When introduced in 1977 it was meant to be a method to describe the logical structure of program flow in PLC controllers. Numerous PLC families today allow the direct programming of sequential control problems by using graphical languages which resemble Grafcet. The problem is that most of these languages use Grafcet only to enable the programmer to use a simple state machine structure. However, the description power of Grafcet would be much more wider.

A diagram of a logic controller (a PLC for example) is presented in Fig. 1. The operation in the control section is described by Grafcet. The logical flow of control is dependent of both internal and external events, and the controller

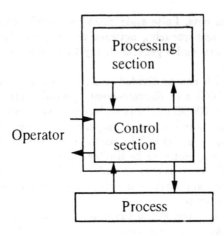

Figure 1. Logic controller

outputs are determined according to the Grafcet program. The processing section can be regarded as the controller device itself; it takes care of calculations, timings, etc. which are needed by the application program. The processing section is responsible for the operation of the controller according to the program specified by Grafcet.

A Grafcet is a diagram having two types of *nodes* viz. *steps* and *transitions*. Transitions are connected to steps and steps to transitions by *directed links*, see Fig. 2. A step can be either *active* or *inactive*. The state (or marking) of the Grafcet is given by the active steps. The marking changes when a transition is *fired*. Grafcet is in many aspects similar to and under certain assumtions even equivalent to interpreted Petri nets [3].

A transition is called *fireable*, if all the steps preceding the transition are active and the boolean variable associated to the transition (*receptivity*) is true. When a transition is fired, all the preceding steps connected to the transition are inactivated and all the steps following it are activated. The firing is assumed to have zero duration and obey the following rules [3]:

- All fireable transitions are immediately fired.

- Several simultaneously fireable transitions are simultaneously fired.

- When a step must be simultaneously activated and inactivated, it remains active.

The inputs to a Grafcet are *conditions* or *events*. Conditions are boolean functions of external or internal variables. Events are rising or falling edges of

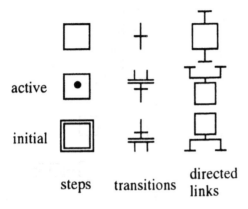

active

initial

steps   transitions   directed links

Figure 2. Basic elements of Grafcet

external variables. The outputs of a Grafcet are *actions*, which can either be level actions or impulse actions. Level actions are boolean states, which always have a finite duration. For example, the action "pump on" attached to a step means that the pump is running only when the step is active. The corresponding impulse action, which is a boolean variable, would be "start the pump". The pump would then be running also after the step is inactivated.

The receptivities attached to transitions can be internal logic conditions, external events, or a combination of both. An example of a Grafcet is presented in Fig. 3.

With Grafcet it is easy to specify program structures like sequentiality, parallelism, and multiple choice. The use of it as a specification of discrete-event control offers several advantages and possibilities when compared with traditional programming of PLC's. It becomes possible to use knowledge-based methods in the program design, program verification, and monitoring. The first step in this direction would be achieved, if the correctness of a program could be proved.

The above question can be examined by considering the stability of a Grafcet diagram. A state is called stable, when no transition can be fired until the next external event occurs. In order the Grafcet to work properly, a stable situation must always be reached in a finite time. In fact, there are two interpretations of how to decide, when actions of the Grafcet diagram must be executed [4]: According to the first interpretation actions must be executed immediately, but in the second case not until a stable situation has been reached. The problem with the first and more straightforward interpretation is that the program will fail, if it never reaches a stable situation.

One way to overcome this problem is to define Grafcet as a synchronous process. The outputs are synchronous with inputs, and all actions and communications are instantaneous. The idea is that programs can be expressed in

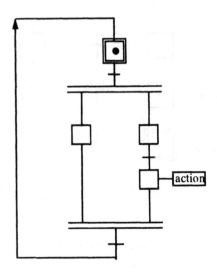

Figure 3. Example of a Grafcet diagram

a mathematical form that enables logical reasoning. The same holds in the case of the two above interpretations of program execution.

There are synchronous languages that can be used in the analysis of sequential programs, which is a step towards program verification and even automatic program generation. One of these languages is SIGNAL [4], which uses simple operators to define the structure of a program. For example, it is easy to express the following definition of one step and one transition by using SIGNAL:

```
X(i,t)=(X(i,t-1) and not T(i,t)) or
(X(i-1,t-1) and T(i-1,t))
```

In the equation X(i,t) is the state of step i at time t, and T(i,t) is the state of the transition following step i.

Similar equations are written for each step of the Grafcet program. SIGNAL then provides tools for simulation of the program and further provides a graph of conditional dependencies, which can be used to check possible deadlocks or other logical errors in the program. However, it is clear that—at least at the present stage—it does not provide an actual proof of program correctness.

From the structural point of view the correctness of a specification given by Grafcet can be proved, when certain additional assumptions and rules in the construction of the diagram are taken into account [3]. The power of Grafcet then lies in the fact that it becomes possible to reduce the complexity of the application program by using *hierarchical* structures. These are introduced in a natural way by *macroactions*.

Macroactions like force, forcing, freezing, and masking [3] utilize the possibility of several programs having global influences on each other. If all programs are

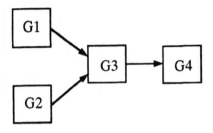

Figure 4. Interdependent Grafcets

specified with (different) Grafcets, the interdependency between them is easy to define by using macroactions, which brings the programming of large entities at a much higher level. Fig. 4 shows an example of Grafcets having effects on each other. It is important to understand that macroactions do not actually increase the specification power of Grafcet, because all programs with macroactions can be reduced to ordinary Grafcets in a straightforward manner. Hence, the large programs with multiple interconnections can be 'proved' to be correct in the same way as ordinary Grafcets.

The use of the above concepts in PLC's would make it possible to reach a new conceptual level in software development for automation. It is possible to define new *program environments*, in which many 'unnecessary' details are hidden from the application programmer. For example, typical (anticipated) error situations, supervision of actuators, alarms to operators etc., can be automatically included in the application program. In case of a malfunction of a device in a plant, the program can be interrupted and continued when appropriate. The application programmer can thus concentrate on the logical sequence of operations to control the plant.

4 Computer-Aided Design

The old-fashioned programming method of PLC's should be replaced with more advanced languages, which would make the programming effort more process-oriented and in the same time more interesting to the end-user. The new programming paradigm described in the previous section would be one alternative, but there are others as well. In general, knowledge-based methods could be used to describe the controller of a discrete-event system in a way that allows logic reasoning. This would give another possibility to bring CASE tools into program construction and verification.

Generally, the information needed for the construction of the plant model can be summarized as follows:

- Structural information about the plant.

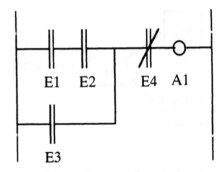

Figure 5. Ladder diagram

- Operational characteristics of the components.

- Information on the laws of physics (mass and heat balances etc.).

The design procedure would then consist of the following steps: modelling, analysis, controller design, simulation and verification. It should be emphasized that the goal of bringing computer-aided tools into the planning phase as well as into the verification of the logical correctness of the programs opens a much wider perspective than described earlier.

One method of 'logical' design with PLC's would be to use the declarative formalism of PROLOG, which has a predicate structure so that it is easy to define the logical connections between input and output variables. The combinational ladder diagram shown in Fig. 5 could be described as follows

```
A1:-E1,E2,not(E4).
A1:-E3,not(E4).
```

The modelling of the whole process requires the possibility to use numerical information in the description also. A formalism according to Fig. 6([5],[6]) could be

```
connected(valve(V1),pump(P3)).
connected(pump(P3),node(1)).
connected(valve(V3),node(1)).
connected(node(1),tank(T4)).
```

The corresponding program in PROLOG, which would make logic reasoning possible, is

```
sys ([V1,V3,P3,T4]):-
    valve(V1_in,V1_out,V1),
    pump(V1_out,P3_out,P3),
```

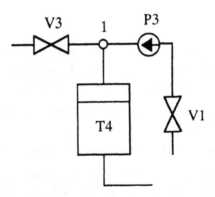

Figure 6. An example process

```
valve(V3_in,V3_out,V3),
tank(T4_in,T4_out,T4),
sum([P3_out,V3_out,T4_in],0).
```

The idea in the above example is to generate the PROLOG code automatically from the process specification. Logical reasoning then becomes possible using that formalism. However, since part of the plant can be described only with a continuous time model, the problem of using knowledge-based methods in analysis is more difficult. There have been attemts to use *qualitative simulation* to overcome the difficulty, see e.g. [5], but it is questionable, whether these methods are mature enough to be applicable in real (large) process plants.

A further goal for the use of the model would be to use logic reasoning in the design of control of the plant. The control specification could be created e.g. in the form of Grafcet.

5 Discussion

The paper has shown the need for improved design methods for the automation of industrial processes. A unified approach, which could be used in design with programmable logic controllers or with other kinds of process computers, would be needed in order to combine process analysis, process oriented programming, program verification, and monitoring. Knowledge-based techniques based on Grafcet and on Prolog were discussed as possible candidates for further development. The research in this topic is still quite immature, and the results are far from process engineering practice.

References

[1] Boullart, L.(1992). *Using A.I.-Formalisms in Programmable Logic Controllers.* In: Boullart, L., Krijgsman, A., and Vingerhoeds, R. A. (Eds.) *Application of Artificial Intelligence in Process Control.* Pergamon Press.

[2] Warnock, I. G.(1988). *Programmable Controllers, Operation and Application.* Prentice Hall.

[3] David, R., Alla, H.(1992). *Petri Nets and Grafcet.* Prentice Hall.

[4] Marce, L., Le Parc, P.(1993). Defining the Semantics of Languages for Programmable Controllers with Synchronous Processes. *Control Eng.Practice, Vol. 1. No. 1. pp. 79-84.*

[5] Välisuo, H.(1991). Discrete-event Control and Monitoring of Industrial Process Plants. *Proceedings of the First IFAC Symposium of Design Methods od Control Systems,* Zurich, Switzerland.

[6] Välisuo, H.(1992). Computer-Aided Design of Plant Automatics. *The Joint Finnish-Russian Symposium on Computer-Aided Control Engineering,* Espoo, Finland.

Index of Authors

Printing: Weihert-Druck GmbH, Darmstadt
Binding: Buchbinderei Schäffer, Grünstadt

Lecture Notes in Computer Science

For information about Vols. 1–690
please contact your bookseller or Springer-Verlag

Vol. 726: T. Lengauer (Ed.), Algorithms – ESA '93. Proceedings, 1993. IX, 419 pages. 1993

Vol. 727: M. Filgueiras, L. Damas (Eds.), Progress in Artificial Intelligence. Proceedings, 1993. X, 362 pages. 1993. (Subseries LNAI).

Vol. 728: P. Torasso (Ed.), Advances in Artificial Intelligence. Proceedings, 1993. XI, 336 pages. 1993. (Subseries LNAI).

Vol. 729: L. Donatiello, R. Nelson (Eds.), Performance Evaluation of Computer and Communication Systems. Proceedings, 1993. VIII, 675 pages. 1993.

Vol. 730: D. B. Lomet (Ed.), Foundations of Data Organization and Algorithms. Proceedings, 1993. XII, 412 pages. 1993.

Vol. 731: A. Schill (Ed.), DCE – The OSF Distributed Computing Environment. Proceedings, 1993. VIII, 285 pages. 1993.

Vol. 732: A. Bode, M. Dal Cin (Eds.), Parallel Computer Architectures. IX, 311 pages. 1993.

Vol. 733: Th. Grechenig, M. Tscheligi (Eds.), Human Computer Interaction. Proceedings, 1993. XIV, 450 pages. 1993.

Vol. 734: J. Volkert (Ed.), Parallel Computation. Proceedings, 1993. VIII, 248 pages. 1993.

Vol. 735: D. Bjørner, M. Broy, I. V. Pottosin (Eds.), Formal Methods in Programming and Their Applications. Proceedings, 1993. IX, 434 pages. 1993.

Vol. 736: R. L. Grossman, A. Nerode, A. P. Ravn, H. Rischel (Eds.), Hybrid Systems. VIII, 474 pages. 1993.

Vol. 737: J. Calmet, J. A. Campbell (Eds.), Artificial Intelligence and Symbolic Mathematical Computing. Proceedings, 1992. VIII, 305 pages. 1993.

Vol. 738: M. Weber, M. Simons, Ch. Lafontaine, The Generic Development Language Deva. XI, 246 pages. 1993.

Vol. 739: H. Imai, R. L. Rivest, T. Matsumoto (Eds.), Advances in Cryptology – ASIACRYPT '91. X, 499 pages. 1993.

Vol. 740: E. F. Brickell (Ed.), Advances in Cryptology – CRYPTO '92. Proceedings, 1992. X, 593 pages. 1993.

Vol. 741: B. Preneel, R. Govaerts, J. Vandewalle (Eds.), Computer Security and Industrial Cryptography. Proceedings, 1991. VIII, 275 pages. 1993.

Vol. 742: S. Nishio, A. Yonezawa (Eds.), Object Technologies for Advanced Software. Proceedings, 1993. X, 543 pages. 1993.

Vol. 743: S. Doshita, K. Furukawa, K. P. Jantke, T. Nishida (Eds.), Algorithmic Learning Theory. Proceedings, 1992. X, 260 pages. 1993. (Subseries LNAI)

Vol. 744: K. P. Jantke, T. Yokomori, S. Kobayashi, E. Tomita (Eds.), Algorithmic Learning Theory. Proceedings, 1993. XI, 423 pages. 1993. (Subseries LNAI)

Vol. 745: V. Roberto (Ed.), Intelligent Perceptual Systems. VIII, 378 pages. 1993. (Subseries LNAI)

Vol. 746: A. S. Tanguiane, Artificial Perception and Music Recognition. XV, 210 pages. 1993. (Subseries LNAI).

Vol. 747: M. Clarke, R. Kruse, S. Moral (Eds.), Symbolic and Quantitative Approaches to Reasoning and Uncertainty. Proceedings, 1993. X, 390 pages. 1993.

Vol. 748: R. H. Halstead Jr., T. Ito (Eds.), Parallel Symbolic Computing: Languages, Systems, and Applications. Proceedings, 1992. X, 419 pages. 1993.

Vol. 749: P. A. Fritzson (Ed.), Automated and Algorithmic Debugging. Proceedings, 1993. VIII, 369 pages. 1993.

Vol. 750: J. L. Díaz-Herrera (Ed.), Software Engineering Education. Proceedings, 1994. XII, 601 pages. 1994.

Vol. 751: B. Jähne, Spatio-Temporal Image Processing. XII, 208 pages. 1993.

Vol. 752: T. W. Finin, C. K. Nicholas, Y. Yesha (Eds.), Information and Knowledge Management. Proceedings, 1992. VII, 142 pages. 1993.

Vol. 753: L. J. Bass, J. Gornostaev, C. Unger (Eds.), Human-Computer Interaction. Proceedings, 1993. X, 388 pages. 1993.

Vol. 754: H. D. Pfeiffer, T. E. Nagle (Eds.), Conceptual Structures: Theory and Implementation. Proceedings, 1992. IX, 327 pages. 1993. (Subseries LNAI).

Vol. 755: B. Möller, H. Partsch, S. Schuman (Eds.), Formal Program Development. Proceedings. VII, 371 pages. 1993.

Vol. 756: J. Pieprzyk, B. Sadeghiyan, Design of Hashing Algorithms. XV, 194 pages. 1993.

Vol. 757: U. Banerjee, D. Gelernter, A. Nicolau, D. Padua (Eds.), Languages and Compilers for Parallel Computing. Proceedings, 1992. X, 576 pages. 1993.

Vol. 758: M. Teillaud, Towards Dynamic Randomized Algorithms in Computational Geometry. IX, 157 pages. 1993.

Vol. 759: N. R. Adam, B. K. Bhargava (Eds.), Advanced Database Systems. XV, 451 pages. 1993.

Vol. 760: S. Ceri, K. Tanaka, S. Tsur (Eds.), Deductive and Object-Oriented Databases. Proceedings, 1993. XII, 488 pages. 1993.

Vol. 761: R. K. Shyamasundar (Ed.), Foundations of Software Technology and Theoretical Computer Science. Proceedings, 1993. XIV, 456 pages. 1993.

Vol. 762: K. W. Ng, P. Raghavan, N. V. Balasubramanian, F. Y. L. Chin (Eds.), Algorithms and Computation. Proceedings, 1993. XIII, 542 pages. 1993.

Vol. 763: F. Pichler, R. Moreno Díaz (Eds.), Computer Aided Systems Theory – EUROCAST '93. Proceedings, 1993. IX, 451 pages. 1994.

Vol. 764: G. Wagner, Vivid Logic. XII, 148 pages. 1994. (Subseries LNAI).

Vol. 765: T. Helleseth (Ed.), Advances in Cryptology – EUROCRYPT '93. Proceedings, 1993. X, 467 pages. 1994.

Vol. 766: P. R. Van Loocke, The Dynamics of Concepts. XI, 340 pages. 1994. (Subseries LNAI).

Vol. 767: M. Gogolla, An Extended Entity-Relationship Model. X, 136 pages. 1994.

Vol. 768: U. Banerjee, D. Gelernter, A. Nicolau, D. Padua (Eds.), Languages and Compilers for Parallel Computing. Proceedings, 1993. XI, 655 pages. 1994.